Gravitation and Cosmology: Principles and Applications of the General Theory of Relativity

Gravitation and Cosmology: Principles and Applications of the General Theory of Relativity

Edited by August Hall

CLANRYE
INTERNATIONAL
www.clanryeinternational.com

Clanrye International,
750 Third Avenue, 9th Floor,
New York, NY 10017, USA

ISBN: 978-1-64726-677-6

Cataloging-in-Publication Data

Gravitation and cosmology : principles and applications of the general theory of relativity / edited by August Hall.
   p. cm.
Includes bibliographical references and index.
ISBN 978-1-64726-677-6
1. General relativity (Physics). 2. Gravitation. 3. Cosmology. 4. Relativity (Physics). I. Hall, August.
QC173.6 .G73 2023
530.11--dc23

For information on all Clanrye International publications
visit our website at www.clanryeinternational.com

Contents

Preface

The world is advancing at a fast pace like never before. Therefore, the need is to keep up with the latest developments. This book was an idea that came to fruition when the specialists in the area realized the need to coordinate together and document essential themes in the subject. That's when I was requested to be the editor. Editing this book has been an honour as it brings together diverse authors researching on different streams of the field. The book collates essential materials contributed by veterans in the area which can be utilized by students and researchers alike.

Cosmology is a field of astronomy that studies the origin and evolution of the universe. General theory of relativity, also called Einstein's theory of gravity, serves as the foundation for comprehending the history and large-scale structure of the universe. A number of significant predictions about the physical world such as the Big Bang origin of the universe, the existence of black holes, effect of gravity on clocks, and gravitational lensing are also based on general relativity theory. According to this theory, numerous astrophysical occurrences produce gravitational waves, which are ripples in the geometry of spacetime and these propagate at the speed of light. This theory provides a unified description of gravity as a geometric property of space and time, also known as the four-dimensional spacetime model. It also improves upon Newton's law of universal gravitation and generalizes special relativity. It serves as the foundation for the fields of relativistic astrophysics and cosmology. This book unfolds the principles and applications of the general theory of relativity. It presents researches and studies performed by experts across the globe. The book will help the readers in keeping pace with the rapid changes in this field.

Each chapter is a sole-standing publication that reflects each author's interpretation. Thus, the book displays a multi-facetted picture of our current understanding of application, resources and aspects of the field. I would like to thank the contributors of this book and my family for their endless support.

Editor

Extremal Cosmological Black Holes in Horndeski Gravity and the Anti-Evaporation Regime

Ismael Ayuso [1,†] **and Diego Sáez-Chillón Gómez** [2,*,†] ⓘ

1   Departamento de Física and Instituto de Astrofísica e Ciências do Espaço, Faculdade de Ciências,
     Universidade de Lisboa, Edifício C8, Campo Grande, 1769-016 Lisboa, Portugal; iayuso@fc.ul.pt
2   Department of Theoretical Physics, Atomic and Optics, Campus Miguel Delibes,
     University of Valladolid UVA, Paseo Belén, 7, 47011 Valladolid, Spain
*   Correspondence: diego.saez@uva.es
†   These authors contributed equally to this work.

Abstract: Extremal cosmological black holes are analysed in the framework of the most general second order scalar-tensor theory, the so-called Horndeski gravity. Such extremal black holes are a particular case of Schwarzschild-De Sitter black holes that arises when the black hole horizon and the cosmological one coincide. Such metric is induced by a particular value of the effective cosmological constant and is known as Nariai spacetime. The existence of this type of solutions is studied when considering the Horndeski Lagrangian and its stability is analysed, where the so-called anti-evaporation regime is studied. Contrary to other frameworks, the radius of the horizon remains stable for some cases of the Horndeski Lagrangian when considering perturbations at linear order.

Keywords: gravitation; scalar-tensor theories; black holes

1. Introduction

General Relativity (GR) has shown its power of prediction over more than one hundred years, and despite some important issues, it is still considered as the best description of gravity. Nevertheless, there are some fundamental questions to be answered in the future in the context of theoretical physics. From a UV completion of gravity to cosmological late-time acceleration, among other also relevant problems, the scientific community is making a great effort to afford them. In particular, black hole solutions have been widely studied in the literature, as are natural solutions of GR and have led to an important development of gravitational physics, including the famous theorems about singularities that offer a way to understand these objects and their main features better. Particularly Schwarzschild-(Anti) De Sitter spacetime arises in GR as a solution when considering a (negative) cosmological constant. The black holes described by this spacetime have been of great interest as they show a thermodynamical equilibrium when analysing Hawking radiation [1–3]. In the case of a positive cosmological constant, the Schwarzschild-de Sitter spacetime shows in general two horizons, one corresponding to the black hole event horizon and the other one to a cosmological horizon. The extreme case arises when both horizons coincide at the same hypersurface, the so-called Nariai spacetime [4], leading to an interesting structure for the spacetime and the trajectories of geodesics [5] as well as for its spectrum [6]. In addition, the stability of such extreme spacetime has been studied in [7], but when some corrections are included, an interesting phenomenon occurs, as the radius of the horizon becomes unstable and grows, what has been called black hole anti-evaporation [8]. Despite that the anti-evaporation regime was initially studied and attributed to semiclassical corrections that affect the evaporation of black holes in de Sitter spacetime when analysing the one-loop effective action [8,9], other frameworks that lead to classical instabilities that affect the radius of the horizon have been also named antievaporation, as the case of $F(R)$ gravity [10], Gauss–Bonnet gravities [11], bigravity theories [12,13] and mimetic gravity [14].

In the context of cosmology, the main issue lies on the unknown dark energy (also on dark matter), which has been widely contrasted by observational data and many theoretical models have been proposed to explain its main consequence, the late-time acceleration of the universe expansion (for a review see [15–19]). Some of such dark energy models are focused on modifications of GR, which may provide a natural solution to the problem, which might be connected to the corrections expected from some UV completions of GR, such as string theory [20]. In this sense, the simplest way of modifying GR is by introducing a scalar field, which incorporates an additional scalar mode while keeping the well known predictions by GR unbroken through screening mechanism that can be implemented by an appropriate potential—the chameleon mechanism—and by the kinetic term—the Vainshtein mechanism. In addition, scalar-tensor theories are well known and well understood, from the Brans–Dicke theory to Horndeski gravity, there is a wide range of scalar field models that have been widely analysed and used not only to provide a natural explanation for dark energy but also to get a better understanding of GR itself [21]. Generalisations of standard scalar-tensor theories have been widely studied lately, mainly in the context of cosmology, as the so-called K-essence, which presents a non-canonical kinetic term and provides a natural explanation for dark energy [22,23], or the so-called Galileons, that incorporates a Galilean-like symmetry and which can also reproduce in a simple way the late-time acceleration [24]. These types of models have in common that they may avoid the so-called Ostrogradsky instability that arises in higher order theories, which is absent in second order theories, such as the ones cited above. This class of scalar-tensor models are encompassed in the so-called Horndeski gravity [25], which represents the most general theory with second order field equations (for a review see [26,27]). Horndeski gravity is shown to be a generalisation of Galileon in its covariant form [28], which is also connected to k-essence fields [29]. Nevertheless, there have been some healthy extensions of Horndeski gravity also implying second order derivatives for the field equations [30–32]. In general, Horndeski gravity is well understood in many contexts, inflationary models have been widely analysed as well as the growth of cosmological perturbations [33–37], also consequently dark energy models can be easily implemented in Horndeski gravity [38,39], the predictions and restrictions of which are analysed [40–43]. Also in light of the era of gravitational waves [44,45], Horndeski gravity is shown to carry just an additional scalarmode [46], but the theory is well constrained by the speed of propagation of the graviton [47–49], which implies several restrictions on the full Lagrangian [50].

Also, static spherically symmetric solutions, such as black holes, have been widely studied in the literature within theories beyond GR [51–58], as they may provide a way to regularise such types of solutions [59–62], a better understanding of Birkhoff's theorem [63–65], or new direct ways for testing General Relativity [66–68]. In Horndeski gravity, there have been plenty of works where such types of solutions are studied, mainly when dealing with compact objects such as black holes [69–72], but also when assuming the constraints imposed on the full Horndeski Lagrangian by the speed of propagation of gravitational waves [73], and the stability of such types of spacetimes [71,72,74–77]. The no-hair theorem is also extended in these theories [78]. Moreover, the Cauchy problem has been analysed in Horndeski gravity by studying the hyperbolicity of the system of equations, which seems to admit a well posed initial value problem [79]. Also the stability in non perturbative cosmology has been studied in [80] as well as the gauge problem in such Lagrangians [81].

The aim of the present paper is to analyse Nariai spacetime in Horndeski gravity and the emergence of the anti-evaporation regime by studying the corresponding perturbations on the metric. Perturbations in Schwarzschild black holes and the Cauchy problem have been widely analysed in the literature within several gravitational theories [58,82–84]. Here we intend to describe how perturbations of a scalar field around a constant background value can affect the radius of the horizon of the black hole. To do so, we study the existence of Schwarzschild-de Sitter solutions in its extremal version for the Lagrangians that compose Horndeski gravity, which also show some implications on an extended version of Birkhoff's theorem for scalar-tensor theories. Finally, we analyse the stability of such solution for a shorter version of the full Horndeski Lagrangian, motivated by keeping as few free functions as possible and which coincides with the viable terms restricted by the speed of GW's.

The paper is organised as follows: In Section 2, a brief introduction to Horndeski gravity and the Nariai metric is provided. Section 3 is devoted to the analysis of the viable Lagrangians that contain the Nariai metric as a solution. In Section 4, the anti-evaporation regime is analysed. Finally, Section 5 gathers the conclusions.

2. Nariai Spacetime in Horndeski Gravity

Let us start by writing the general action that we are dealing with throughout this manuscript. This is the Hilbert–Einstein action plus the so-called Horndeski Lagrangian:

$$S_G = \int dx^4 \sqrt{-g} \left[\frac{R}{16\pi G} + \mathcal{L}_{Hr} + \mathcal{L}_m \right],$$
(1)

where $\mathcal{L}_m$ is the matter Lagrangian, which encompasses all the matter species of the system under study while the Horndeski Lagrangian $\mathcal{L}_{Hr}$ is given by:

$$\begin{aligned} \mathcal{L}_{Hr} =\ & G_2(\phi, X) - G_3(\phi, X)\Box\phi + G_4(\phi, X)R + G_{4X}(\phi, X)\left[(\Box\phi)^2 - \phi_{;\mu\nu}\phi^{;\mu\nu}\right] + G_5(\phi, X)\phi_{;\mu\nu}G^{\mu\nu} \\ & - \frac{G_{5X}(\phi, X)}{6}\left[(\Box\phi)^3 - 3\Box\phi\phi_{;\mu\nu}\phi^{;\mu\nu} + 2\phi_{;\mu\nu}\phi^{;\nu\lambda}\phi_\lambda^{;\mu}\right]. \end{aligned}$$
(2)

Here, ϕ is a scalar field, $G_{\mu\nu}$ is the Einstein tensor, $_{;\mu} = \nabla_\mu$ is the covariant derivative, $X = -\frac{1}{2}\partial_\mu\phi\partial^\mu\phi$ is the kinetic term, $G_i(\phi, X)$ are arbitrary functions of ϕ and X, and $_X$ is the derivative with respect to X. As it is well known, the Lagrangian (2) represents the most general scalar-tensor Lagrangian that leads to second order field equations despite that it depends on second derivatives of the field ϕ at the level of the action as well as on non-minimally coupling terms to the Ricci scalar. As shown in Reference [29], this is just the generalisation of the so-called covariant Galileon field, the covariant version of which loses the Galilean shift symmetry that provides its name [28]. Hence, by varying the action (1) with respect to the metric $g_{\mu\nu}$ and with respect to the scalar field ϕ, the corresponding field equations can be obtained and we can analyse how some particular spacetimes behave within this class of theories.

Throughout this paper, we are interested in studying the Nariai spacetime, which is the extremal case of the Schwarzschild-de Sitter black hole, as is shown below. The general Schwarzschild-de Sitter metric can be expressed in spherical coordinates as follows:

$$ds^2 = -A(r)dt'^2 + A(r)^{-1}dr^2 + r^2 d\Omega_2^2,$$
(3)

where $d\Omega_2^2$ is the metric of a 2D sphere, and

$$A(r) = 1 - \frac{2M}{r} - \frac{\Lambda}{3}r^2.$$
(4)

Here, $\Lambda > 0$ and $M > 0$. If $0 < M^2 < \frac{1}{9\Lambda}$, the function $A(r)$ has two positive roots r_{BH} and r_c, which correspond to the black hole event horizon and to the cosmological horizon, respectively. The global structure of this spacetime has been widely analysed in the literature [1–3]. The crucial point here is that whenever $M \to \frac{1}{3\sqrt{\Lambda}}$, the size of the black hole event horizon r_{BH} increases and approaches the cosmological horizon r_c at $r = 3M$, such that Function (4) tends to:

$$A(r) = -\frac{(r - 3M)^2(r + 6M)}{27M^2 r}.$$
(5)

This is the extremal case of the Schwarzschild-de Sitter black hole, which is known as the Nariai spacetime [4]. As shown in (5), it leads to a degenerate horizon that corresponds to the black hole one and to the cosmological one simultaneously. The causal structure of this particular case is well understood and the geodesics in such spacetime are well described in Reference [5]. Note that $A(r) \leq 0$, such that the radial coordinate becomes timelike and the time coordinate spacelike everywhere. Our aim here is to analyse the metric (3) for the extremal case in the framework of the Horndeski Lagrangian, and analyse the stability of such solution. For that purpose, let us express the metric (3) with some more appropriate coordinates, but firstly we express the extremal case as a limit in terms of a parameter $0 < \epsilon << 1$, [7]:

$$9M^2\Lambda = 1 - 3\epsilon^2 \, . \tag{6}$$

As $\epsilon \to 0$, both horizons approach each other. Then, we can choose the following coordinates [8]:

$$t' = \frac{1}{\epsilon\sqrt{\Lambda}}\psi \, , \qquad r = \frac{1}{\sqrt{\Lambda}}\left(1 - \epsilon\cos\chi - \frac{1}{6}\epsilon^2\right) \, . \tag{7}$$

In these new coordinates, and expanding at first order in ϵ, the metric (3) becomes:

$$ds^2 = -\frac{1}{\Lambda}\left(1 + \frac{2}{3}\epsilon\cos\chi\right)\sin^2\chi d\psi^2 + \frac{1}{\Lambda}\left(1 - \frac{2}{3}\epsilon\cos\chi\right)d\chi^2 + \frac{1}{\Lambda}\left(1 - 2\epsilon\cos\chi\right)d\Omega_2^2 \, . \tag{8}$$

Here the black hole horizon is given by $\chi = 0$ whereas the cosmological one corresponds to $\chi = \pi$. The spatial topology is clearly $S_1 \times S_2$. By setting $\epsilon \to 0$, the extremal case is obtained and the metric yields (8):

$$ds^2 = \frac{1}{\Lambda}\left(-\sin^2\chi d\psi^2 + d\chi^2\right) + \frac{1}{\Lambda}d\Omega_2^2 \, . \tag{9}$$

Finally, we can implement another change of coordinates that simplifies the expression (9), which is described by the following coordinates:

$$x = \text{Log}\left(\tan\frac{\chi}{2}\right) \, , \qquad t = \frac{\psi}{4} \, . \tag{10}$$

The metric (9) for the Nariai spacetime becomes:

$$ds^2 = \frac{1}{\Lambda\cosh^2 x}\left(-dt^2 + dx^2\right) + \frac{1}{\Lambda}d\Omega_2^2 \, . \tag{11}$$

The new coordinates are defined in the domain $(-\infty, \infty)$, as can be easily shown by (10).

3. Reconstructing the Gravitational Action in Horndeski Gravity

In this section, we analyse the particular Lagrangians within Horndeski gravity that reproduces the Nariai solution. To do so, we use the metric as expressed in the coordinates given in (11). As shown, Nariai spacetime can be a solution for each of the Horndeski Lagrangians as far as some constraints are assumed on the $\mathcal{L}_i$ functions.

3.1. Case with $\mathcal{L}_2$

As a first approximation to Horndeski gravity in Nariai spacetime, we will start studying the simplest case in which only $\mathcal{L}_2$ for $\mathcal{L}_{Hr}$ is considered,

$$\mathcal{L}_2 = G_2(\phi, X) \, , \tag{12}$$

which essentially is the usual term for K-essence theory. The first step will be to solve, at the background level, the equations of motion given by the Einstein tensor plus an effective energy-tensor coming from metric variations of the matter Lagrangian plus the Lagrangian defined in (12):

$$G_{\mu\nu} = R_{\mu\nu} - \frac{1}{2}g_{\mu\nu}R = 8\pi G\left[g_{\mu\nu}G_2(\phi, X) + \frac{\partial G_2(\phi, X)}{\partial X}\partial_\mu\phi\partial_\nu\phi + T_{\mu\nu}^{(m)}\right] \, , \tag{13}$$

where $T_{\mu\nu}^{(m)}$ is the energy-momentum tensor of the matter Lagrangian, and which, for the case of our interest, we are going to consider zero to focus on the vacuum, i.e., $T_{\mu\nu}^{(m)} = 0$. Therefore, this tensor equation leads to the following system of equations:

$$tt- \qquad \frac{1}{8\pi G}\frac{1}{\cosh^2 x} = -\frac{G_2(\phi, X)}{\Lambda \cosh^2 x} + \frac{\partial G_2(\phi, X)}{\partial X}\dot{\phi}^2 , \qquad (14)$$

$$xt- \qquad 0 = -\frac{\partial G_2(\phi, X)}{\partial X}\partial_t\phi\partial_x\phi , \qquad (15)$$

$$xx- \qquad \frac{1}{8\pi G}\frac{-1}{\cosh^2 x} = \frac{G_2(\phi, X)}{\Lambda \cosh^2 x} + \frac{\partial G_2(\phi, X)}{\partial X}\phi'^2 , \qquad (16)$$

$$\theta\theta- \qquad \frac{-1}{8\pi G} = \frac{G_2(\phi, X)}{\Lambda} + \frac{\partial G_2(\phi, X)}{\partial X}\partial_\theta\phi\partial_\theta\phi , \qquad (17)$$

$$\Phi\Phi- \qquad \frac{-\sin^2\theta}{8\pi G} = \sin^2\theta\frac{G_2(\phi, X)}{\Lambda} + \frac{\partial G_2(\phi, X)}{\partial X}\partial_\Phi\phi\partial_\Phi\phi , \qquad (18)$$

where dot means time derivatives and ' derivatives with respect to x. The two main issues that we intend to solve are the form of $G_2(\phi, X)$ and $\phi(t, x, \theta, \Phi)$. By combining (17) with (18), it yields:

$$\partial_\Phi\phi\partial_\Phi\phi = \sin^2\theta\partial_\theta\phi\partial_\theta\phi \qquad \rightarrow \qquad \partial_\Phi\phi = \pm\sin\theta\partial_\theta\phi , \qquad (19)$$

the solution of which is:

$$\phi = g(t, x)\left[\Phi \pm \ln\left(\cot\frac{\theta}{2}\right)\right] + f(t, x) . \qquad (20)$$

However, from (14) and (16) it is possible to deduce that $g(t, x)$ should vanish in order to keep the same dependence parameters on the left and right hand side of the equations, and therefore $\phi = \phi(t, x)$, which implies that $X = \Lambda\cosh^2(x)(\dot{\phi}^2 - \phi'^2)/2$. This is the formal way for showing that the scalar field has to be spherically symmetric as the metric is. In addition, for solving $G_2(\phi, X)$, we can use the trace equation of (13) where the scalar curvature for the Nariai metric is $R = 4\Lambda$ and therefore:

$$-\frac{\Lambda}{4\pi G} = 2G_2(\phi, X) - \frac{\partial G_2(\phi, X)}{\partial X}X , \qquad (21)$$

the solution of which is:

$$G_2(\phi, X) = -\frac{\Lambda}{8\pi G} + f(\phi)X^2 . \qquad (22)$$

However, by Equation (15), the following condition is obtained:

$$2Xf(\phi)\dot{\phi}\phi' = 0 . \qquad (23)$$

It is straightforward to show that by combining (23) with $xx-$ and $tt-$ equations, $\phi' = \dot{\phi} = 0$, such that $\phi = constant$. Hence, the solution of the background leads to the following constraint on the action:

$$G_2(\phi_0, 0) = -\frac{\Lambda}{8\pi G}. \qquad (24)$$

This solution mimics the one from General Relativity with a cosmological constant, but in this case induced by a constant scalar field ϕ. There is a special case when the coefficients for this system of equations become null and the background equation is satisfied also for non-constant and non-static scalar field solutions, which will be studied in the Appendix A. Note that despite that Birkhoff's theorem is satisfied in Brans–Dicke-like theories [63–65], where a static metric implies a static scalar field, this may not be the case for other scalar-tensor theories such as Galileons or general Horndeski scenarios [71,72].

3.2. Case $\mathcal{L}_3$

For the case $\mathcal{L}_3$, the general gravitational action is given by

$$S_G = \int dx^4 \sqrt{-g} \left[\frac{R}{16\pi G} - G_3(\phi, X) \Box\phi \right] . \tag{25}$$

By varying the action (25) with respect to the metric $g_{\mu\nu}$, the corresponding field equations are obtained:

$$
\begin{aligned}
R_{\mu\nu} - \tfrac{1}{2} g_{\mu\nu} R &= 8\pi G \left[G_{3\phi} \left(g_{\mu\nu} \nabla_\alpha \phi \nabla^\alpha \phi - 2 \nabla_\mu \phi \nabla_\nu \phi \right) \right. \\
&\quad \left. + G_{3X} \left(-\nabla_\mu \phi \nabla_\nu \phi \Box\phi - g_{\mu\nu} \nabla_\alpha \phi \nabla_\beta \phi \nabla^\alpha \phi \nabla^\beta \phi + 2 \nabla^\alpha \phi \nabla_{(\mu} \phi \nabla_{\nu)} \nabla^\alpha \phi \right) \right] .
\end{aligned}
\tag{26}
$$

Here the subscript $_{()}$ refers to a commutator among the indexes, while $_\phi$ and $_X$ are derivatives with respect to the scalar field ϕ and its kinetic term X respectively. The equation for the scalar field is obtained by varying the action (25) with respect to the scalar field:

$$
\begin{aligned}
&2G_{3\phi}\Box\phi + G_{3\phi\phi}(\nabla\phi)^2 + G_{3X\phi} \left[(\nabla\phi)^2 \Box\phi + 2\nabla_\mu\phi\nabla^\mu X \right] + G_{3X} \left[(\Box\phi)^2 - \nabla_\mu\nabla_\nu\phi\nabla^\mu\nabla^\nu\phi - R_{\mu\nu}\nabla^\mu\nabla^\nu\phi \right] \\
&+ G_{3XX} \left[\nabla_\mu\phi\nabla^\mu X + (\nabla X)^2 \right] = 0 ,
\end{aligned}
\tag{27}
$$

where recall that X is the kinetic term of the scalar field. As in the previous Lagrangian, a non-constant static scalar field, $\phi = \phi(x)$ is assumed. In order to show that the Nariai metric, expressed in the coordinates as in (11), may be a solution for the gravitational action (25), we use the $tt-$ and $xx-$ equations, which can be easily obtained from the field Equations (26) and yields:

$$
\begin{aligned}
tt- \qquad \frac{1}{\cosh^2 x} &= 8\pi G \phi'^2 \left[-G_{3\phi} + G_{3X}\Lambda^2 \left(\sinh x \cosh x \phi' + \cosh^2 x \phi'' \right) \right] , \\
xx- \qquad -\frac{1}{\cosh^2 x} &= 8\pi G \phi'^2 \left[-G_{3\phi} + G_{3X}\Lambda^2 \cosh x \sinh x \phi' \right] .
\end{aligned}
\tag{28}
$$

The $\theta\theta-$ and $\varphi\varphi-$ equations are just redundant, since the $tt-$ equation is reproduced up to proportional terms. In general, for an arbitrary $G_3(\phi, X)$, the system of Equation (28) has no solution $\phi(x)$, and consequently Nariai spacetime is not a solution for the gravitational Lagrangian (25). Nevertheless, Equation (28) can be used for reconstructing the appropriate $\mathcal{L}_3$ Lagrangian that reproduces the Nariai spacetime (11) when assuming a particular solution $\phi(x)$. As the corresponding partial derivatives $G_{3\phi}$ and G_{3X} are at the end functions of the coordinate x, we can express both of them in terms of the scalar field and its derivatives through Equation (28), which leads to:

$$
\begin{aligned}
G_{3\phi}(x) &= \frac{1}{8\pi G} \frac{2\tanh x\, \phi' + \phi''}{\phi'^2 \phi'' \cosh^2 x} , \\
G_{3X}(x) &= \frac{1}{4\pi G\Lambda} \frac{1}{\phi'^2 \phi'' \cosh^4 x} .
\end{aligned}
\tag{29}
$$

Hence, the corresponding Lagrangian (25) can be reconstructed as far as the expressions (29) are well defined for $\phi(x)$, such that the integrability condition holds $G_{3\phi X} = G_{3X\phi}$. Nevertheless, it is not straightforward to obtain an analytical and exact expression for the $\mathcal{L}_3$ Lagrangian, but we can consider a couple of ways that lead to an analytical reconstruction of the action.

Firstly, we may specify the form of the function $G_3(\phi, X)$, and reconstruct the corresponding action by using the system of Equation (28) and the integrability condition on $G_3(\phi, X)$. Let us consider the following $G_3(\phi, X)$:

$$G_3(\phi, X) = f_1(\phi) + f_2(X) . \tag{30}$$

The general kinetic term X is given by:

$$X = -\frac{1}{2}\Lambda \cosh^2 x \phi'^2 .$$ (31)

Then, by the partial derivative with respect to X in (29), we obtain:

$$G_{3X} = f_{2X} = \frac{\Lambda}{4\pi G}\frac{1}{X^2}\frac{\phi'^2}{\phi''} .$$ (32)

This equation together with the assumption (30) basically imposes that $\frac{\phi'^2}{\phi''} = g(X)$ must be expressed as a function of the kinetic term (31). As $g(X)$ is in principle arbitrary as far as providing a solution for the scalar field ϕ, we may assume $g(X) = X$ such that the scalar field becomes:

$$\phi(x) = -\frac{2\log{(\cosh x)}}{\Lambda}.$$ (33)

After integrating (32), the function $f_2(X)$ turns out:

$$f_2(X) = -\frac{\Lambda}{4\pi G}\frac{1}{X^2} .$$ (34)

While the partial derivative with respect to ϕ on G_3 leads to:

$$G_{3\phi} = f_{1\phi} = \frac{\Lambda^2}{32\pi G}\frac{1+2\log(\cosh x)}{\cosh^2 x} = \frac{1}{8\pi G}\frac{e^{\Lambda\phi}(1-\Lambda\phi)}{\phi^2} ,$$ (35)

which after integrating, provides the corresponding dependence on the scalar field ϕ:

$$f_1(\phi) = -\frac{1}{8\pi G}\frac{e^{\Lambda\phi}}{\phi} .$$ (36)

The full gravitational action $G_3(\phi, X)$ as given in (30) is reconstructed. Nevertheless, we may try to keep the form of G_3 arbitrary and consider a particular solution for the scalar field in order to reconstruct the action. For illustrative purposes, we consider the following solution:

$$\phi(x) = \phi_0 e^{\mu x} .$$ (37)

Then, by following the Equation (29), the following particular solutions are found in terms of the coordinate x:

$$\begin{aligned}G_{3\phi}(x) &= \frac{1}{8\pi G}\frac{\text{sech}^2 x(\mu+2\tanh x)e^{-2\mu x}}{\mu^3\phi_0^2} , \\ G_{3X}(x) &= \frac{1}{4\pi G}\frac{\text{sech}^4 x e^{-3\mu x}}{\mu^4\phi_0^3\Lambda} .\end{aligned}$$ (38)

The corresponding kinetic term $X = -\frac{1}{2}\partial_\mu\phi\partial^\mu\phi$ is given for this case by:

$$X = -\frac{1}{2}\phi_0^2\mu^2\Lambda\cosh^2 x e^{2\mu x} .$$ (39)

Hence, the partial derivative $G_{3X}(\phi, X)$ automatically leads to:

$$G_{3X}(\phi, X) = \frac{\Lambda}{16\pi G}\frac{\phi}{X^2} ,$$ (40)

After integrating, it leads to:

$$G_3(\phi, X) = -\frac{\Lambda}{16\pi G}\frac{\phi}{X} + f(\phi) ,$$ (41)

where $f(\phi)$ has to be computed by integrating the partial derivative $G_{3\phi}$, which is obtained by deriving expression (41) and equating to the expression in (38):

$$f_\phi = \frac{1}{4\pi G\phi_0^2\mu^3}\frac{\tanh x}{\cosh^2 x e^{2\mu x}} = \frac{1}{4\pi G\phi_0^2\mu^3}\frac{\tanh\left[\log\left(\frac{\phi}{\phi_0}\right)^{1/\mu}\right]}{\phi^2\cosh^2\left[\log\left(\frac{\phi}{\phi_0}\right)^{1/\mu}\right]}, \tag{42}$$

which after integrating, yields:

$$\begin{aligned}
f(\phi) &= \frac{1}{8\pi G\mu^2(\mu-2)\phi}\left\{F\left[1, 1-\mu/2, 2-\mu/2; -\left(\frac{\phi}{\phi_0}\right)^{2/\mu}\right] - (\mu-2)\mu F\left[1, -\mu/2, 1-\mu/2; -\left(\frac{\phi}{\phi_0}\right)^{2/\mu}\right]\right.\\
&+ \left. \text{sech}^2\left[\log\left(\frac{\phi}{\phi_0}\right)^{1/\mu}\right] + \mu\tanh\left[\log\left(\frac{\phi}{\phi_0}\right)^{1/\mu}\right]\right\}.
\end{aligned} \tag{43}$$

Here, $F(a, b, c; x)$ are hypergeometric functions, which can be computed analytically for some values of μ. For instance, $\mu = 1$ gives:

$$f(\phi) = -\frac{1}{4\pi G}\left[\frac{\phi^3 + 3\phi\phi_0^2}{(\phi^2 + \phi_0^2)^2} + \frac{\arctan\left(\frac{\phi}{\phi_0}\right)}{\phi_0}\right]. \tag{44}$$

Hence, the full reconstruction of the gravitational action (25) is explicitly shown for these two cases. The main conclusions can be obtained by analysing these two examples. As shown in the field equations, and by the expressions of $G_{3\phi}(x)$ and $G_{3X}(x)$, a constant scalar field $\phi(x) = \phi_0$ is not a solution for the Equation (28), at least whenever the Lagrangian (25) is considered as the sole action for gravity. In addition, the freedom of the function $G_3(\phi, X)$ implies that different Lagrangians can reproduce the Nariai metric, but leading to different solutions for the scalar field, as far as its partial derivatives (29) are well defined, as has been shown by these two examples.

3.3. Case $\mathcal{L}_4$

Let us now analyse the solutions when the Lagrangian $\mathcal{L}_4$ in (2) is considered as the sole gravitational action:

$$S_G = \int dx^4\sqrt{-g}\left[\frac{R}{16\pi G} + G_4(\phi, X)R + G_{4X}\left((\Box\phi)^2 - \nabla_\mu\nabla_\nu\phi\nabla^\mu\nabla^\nu\phi\right)\right]. \tag{45}$$

As usual, by varying the action (25) with respect to the metric $g_{\mu\nu}$, the corresponding field equations are obtained:

$$\begin{aligned}
\left(\frac{1}{16\pi G} + G_4\right)\left(R_{\mu\nu} - \frac{1}{2}g_{\mu\nu}R\right) - \nabla_\mu\nabla_\nu G_4 + g_{\mu\nu}\Box G_4 - \frac{1}{2}g_{\mu\nu}G_{4X}\left((\Box\phi)^2 - \nabla_\mu\nabla_\nu\phi\nabla^\mu\nabla^\nu\phi\right)\\
+... \text{ (second order terms)} = 0.
\end{aligned} \tag{46}$$

We can proceed as in the previous Lagrangian. However, the degree of freedom on the function $G_4(\phi, X)$ will lead to a set of infinite solutions for the scalar field, as shown above for $\mathcal{L}_3$, which does not provide any new insights on Nariai spacetime in Horndeski gravity, but just some similar features as in the previous case, i.e., for a given solution $\phi(x)$, one can in general reconstruct the appropriate action through $G_4(\phi, X)$, while the other way around, that is, given an arbitrary $G_4(\phi, X)$ function, the field Equation (46) does not have any solution for the scalar field in general, except for some special cases of the $G_4(\phi, X)$ function, as also shown for $G_3(\phi, X)$ above. In addition, note for the general Horndeski Lagrangian, the speed of gravitational waves is given by [47]:

$$c_{GW} = \frac{G_4 - X(\ddot{\phi}G_{5X} + G_{5\phi})}{G_4 - 2XG_{4X} - X(H\dot{\phi}G_{5X} - G_{5\phi})}, \tag{47}$$

where H is the Hubble parameter. Hence, by assuming $G_4(\phi, X) = G_4(\phi)$ and $G_5 = 0$, analogously to [48], the speed of propagation for GW's is kept as the speed of light $c_{GW} = 1$, satisfying the constraints obtained from the GW170817 detection [44,45]. Hence, we explore here the case where $G_4(\phi, X) = G_4(\phi)$, such that the field Equation (46) read:

$$
\begin{aligned}
tt- \quad & \left(\tfrac{1}{16\pi G} + G_4\right) \operatorname{sech}^2 x - \phi'^2 G_{4\phi\phi} - (\tanh x \phi' + \phi'') G_{4\phi} = 0 , \\
xx- \quad & -\left(\tfrac{1}{16\pi G} + G_4\right) \operatorname{sech}^2 x - \tanh x \phi' G_{4\phi} = 0 , \\
\theta\theta- \quad & -\left(\tfrac{1}{16\pi G} + G_4\right) \operatorname{sech}^2 x - \phi'^2 G_{4\phi\phi} + \phi'' G_{4\phi} = 0 .
\end{aligned}
\tag{48}
$$

By combining the $xx-$ and $\theta\theta-$ equations, it yields:

$$
\tanh x \, G_{4\phi} \phi' = 0 \quad \rightarrow \quad \phi = \text{constant} .
\tag{49}
$$

Hence, the only solution leads to a constant scalar field, similarly to $G_2(\phi, X)$, unless $G_{4\phi} = 0$, which together with other conditions is analysed in Appendix A. For this specific case, the only choice of G_4 that satisfies the equations of motion is:

$$
G_4(\phi) = -\frac{1}{16\pi G} ,
\tag{50}
$$

Nevertheless, for this choice the gravitational effective coupling constant in (45) becomes null and consequently the theory is ill defined in general. Then, for the particular case (45) with $G_4 = G_4(\phi)$, Nariai spacetime and consequently Schwarzschild-(A)dS is not reproduced by such Lagrangian. This is a natural consequence as Schwarzschild-(A)dS spacetime requires the presence of a cosmological constant, which cannot emerge from another term. However, such issue can be easily sorted out by adding a scalar potential in the action,

$$
S_G = \int dx^4 \sqrt{-g} \left[\frac{R}{16\pi G} + G_4(\phi)R - V(\phi) \right] .
\tag{51}
$$

The equations do not differ much from the ones above, but just up to a potential term,

$$
\begin{aligned}
tt- \quad & \left(\tfrac{1}{16\pi G} + G_4\right) \operatorname{sech}^2 x - \phi'^2 G_{4\phi\phi} - (\tanh x \phi' + \phi'') G_{4\phi} - \tfrac{\operatorname{sech}^2 x}{2\Lambda} V(\phi) = 0 , \\
xx- \quad & -\left(\tfrac{1}{16\pi G} + G_4\right) \operatorname{sech}^2 x - \tanh x \phi' G_{4\phi} + \tfrac{\operatorname{sech}^2 x}{2\Lambda} V(\phi) = 0 , \\
\theta\theta- \quad & -\left(\tfrac{1}{16\pi G} + G_4\right) \operatorname{sech}^2 x - \phi'^2 G_{4\phi\phi} + \phi'' G_{4\phi} + \tfrac{\operatorname{sech}^2 x}{2\Lambda} V(\phi) = 0 .
\end{aligned}
\tag{52}
$$

As in the previous case, by combining the $xx-$ and $\theta\theta-$ equations, the constraint Equation (49) is obtained, what leads to a constant scalar field $\phi(x) = \phi_0$, and by replacing in Equation (52), it leads to:

$$
-G_4(\phi_0) + \frac{V(\phi_0)}{2\Lambda} = \frac{1}{16\pi G} .
\tag{53}
$$

Hence, Nariai spacetime is a solution for the gravitational action (51) as long as the algebraic Equation (53) has at least a real solution.

Therefore, it is clear that Schwarzschild-(A)dS spacetime, and specifically Nariai spacetime is a solution for each of the Horndeski Lagrangians whereas some constraints are imposed on the Lagrangians $\mathcal{L}_i$. It is straightforward to show that the Nariai metric is also a solution of the full Horndeski Lagrangian as the degrees of freedom added by each $\mathcal{L}_i$ provides a way of reconstructing the corresponding gravitational action, what will imply an infinite number of choices on the G_i functions and a degenerate solution for the scalar field, as has been shown for some of the Lagrangians above, and which will also affect the full gravitational action due to the freedom of choosing the corresponding Lagrangians. In the next section, we analyse the stability of these extremal blackholes for those cases that the Nariai metric imposes real constraints on the Lagrangians.

4. Anti-Evaporation Regime in Horndeski Gravity

In this section, we analyse the stability of Nariai spacetime when perturbations around the background solution are introduced. To do so, we focus on the first four terms of the Horndeski Lagrangian:

$$S_G = \int dx^4 \sqrt{-g} \left[\frac{R}{16\pi G} + G_2(\phi, X) - G_3(\phi, X) \Box \phi + G_4(\phi) R \right] . \tag{54}$$

Note that (54) is the most general Horndeski Lagrangian that keeps the speed of gravitational waves (47) as the speed of light. As shown in the previous section, for a given solution $\phi(x)$ and the Nariai metric (11), one can reconstruct the corresponding Horndeski Lagrangian that reproduces such solution. Nevertheless, here we are assuming for simplicity while analysing the perturbations, a constant scalar field for the background $\phi(x, t) = \phi_0$, such that following the results from the above section, Nariai spacetime is a solution for the gravitational action (54) as long as the following constraint is satisfied:

$$\frac{G_{20}}{2\Lambda} + G_{40} = -\frac{1}{16\pi G} . \tag{55}$$

A useful way to define perturbations around the Nariai metric is:

$$ds^2 = e^{2\rho(x,t)} \left(-dt^2 + dx^2 \right) + e^{-2\varphi(x,t)} d\Omega_2^2, \tag{56}$$

in which $\rho(x, t)$ and $\varphi(x, t)$ at the background level are: $\rho = -\ln \sqrt{\Lambda} \cosh x$ and $\varphi = \ln \sqrt{\Lambda}$. The perturbations on the metric and the scalar field (with spherical symmetry) can be expressed as follows:

$$\begin{aligned} \phi &\rightarrow \phi_0 + \delta\phi(t, x) \\ \rho &\rightarrow -\ln \left[\sqrt{\Lambda} \cosh x \right] + \delta\rho \\ \varphi &\rightarrow \ln \sqrt{\Lambda} + \delta\varphi \end{aligned} \tag{57}$$

Let us show how the perturbations are transformed under a gauge transformation in order to construct gauge invariants that allow us to isolate the physical perturbations from gauge artifices. We can consider an infinitesimal transformation of coordinates, given by

$$x'^\mu = x^\mu + \delta x^\mu , \tag{58}$$

On any generic quantity F, this implies a transformation on its perturbation:

$$\delta F' = \delta F + \pounds_{\delta x} F_0 . \tag{59}$$

Here the prime denotes the quantity transformed in the new coordinates, F_0 is the background value and $\pounds_{\delta x}$ is the Lie derivate along the vector δx^μ. The corresponding perturbations on the metric are transformed as follows:

$$\begin{aligned} \delta\rho' &= \delta\rho + \pounds_{\delta x} \rho_0 , \\ \delta\varphi' &= \delta\varphi + \pounds_{\delta x} \varphi_0 = \delta\varphi . \end{aligned} \tag{60}$$

We are interested in the perturbation $\delta\varphi$, as the one that defines the perturbation on the radius of the horizon (see below). This is a gauge invariant quantity, such that we can work in an arbitrary gauge to solve the equations. Hence, introducing the perturbations (57) in the field equations, up to linear order leads to:

$$\left(\frac{1}{16\pi G} + G_4 \right) \delta G_{\mu\nu} + G_{\mu\nu} G_{4\phi} \delta\phi - G_{4\phi} \nabla_\mu \nabla_\nu \delta\phi + g_{\mu\nu} G_{4\phi} \Box \delta\phi - \frac{1}{2} \left(G_{2\phi} g_{\mu\nu} \delta\phi + G_2 \delta g_{\mu\nu} \right) = 0 . \tag{61}$$

Note that the functions G_i and their derivatives are evaluated at $\phi = \phi_0$ and expanded up to first order in perturbations as follows:

$$\begin{aligned}
G_2(\phi, X) &\rightarrow G_2(\phi_0, 0) + \left.\frac{\partial G_2(\phi, 0)}{\partial \phi}\right|_{\phi_0} \delta\phi \\
G_4(\phi) &\rightarrow G_4(\phi_0) + \left.\frac{\partial G_4(\phi)}{\partial \phi}\right|_{\phi_0} \delta\phi
\end{aligned} \tag{62}$$

Our next step will be the introduction of these perturbations into the field equations to study its evolution. The (tt), (xx) and (tx) perturbation equations are respectively:

$$\begin{aligned}
-2G_{20}\text{sech}^2x\delta\varphi + (G_{2\phi} + 2\Lambda G_{4\phi})\text{sech}^2x\delta\phi - 2G_{20}(\tanh x\delta\varphi' + \delta\varphi'') - 2G_{4\phi}\Lambda(\tanh x\delta\phi' + \delta\phi'') &= 0, \\
-2G_{20}\text{sech}^2x\delta\varphi + (G_{2\phi} + 2\Lambda G_{4\phi})\text{sech}^2x\delta\phi + 2G_{20}(\tanh x\delta\varphi' + \delta\ddot\varphi) + 2G_{4\phi}\Lambda(\tanh x\delta\phi' + \delta\ddot\phi) &= 0, \\
G_{20}(\tanh x\delta\dot\varphi + \delta\dot\varphi') + G_{4\phi}\Lambda(\tanh x\delta\dot\phi + \delta\dot\phi') &= 0,
\end{aligned} \tag{63}$$

The $(tx)-$ equation can be rewritten as follows:

$$\begin{aligned}
\frac{\partial}{\partial t}\left[G_{20}(\tanh x\delta\varphi + \delta\varphi') + G_{4\phi}\Lambda(\tanh x\delta\phi + \delta\phi')\right] &= 0, \\
\rightarrow \quad g(x, t)\tanh x + g'(x, t) &= h(x),
\end{aligned} \tag{64}$$

where $h(x)$ is an integration function to be determined, while $g(x, t) = G_{20}\delta\varphi + G_{40}\Lambda\delta\phi$, in which integrating the Equation (64) yields:

$$g(x, t) = G_{20}\delta\varphi + G_{40}\Lambda\delta\phi = f(t)\,\text{sech}x + \text{sech}x \int \cosh x\, h(x)dx. \tag{65}$$

Then, by combining the $tt-$ and $xx-$ equations, the functions $f(t)$ and $h(x)$ are determined,

$$\begin{aligned}
f(t) &= C_1 e^t + C_2 e^{-t}, \quad h(x) = C_3\tanh x + C_4, \\
\rightarrow \quad g(x, t) &= (C_1 e^t + C_2 e^{-t})\,\text{sech}x + C_3 + C_4\tanh x.
\end{aligned} \tag{66}$$

Here, C_i's are integration constants. Then, the expression for the metric perturbation $\delta\varphi$ can be easily obtained:

$$\delta\varphi = \frac{C_1 e^t + C_2 e^{-t}}{G_{20}}\text{sech}x + C_3\frac{G_{2\phi} + 2\Lambda G_{4\phi}}{G_{20}(G_{2\phi} + 4\Lambda G_{4\phi})} + C_4\tanh x. \tag{67}$$

We can now calculate how the horizon changes when considering the above perturbations on the metric. The horizon is a null hypersurface that can be defined as follows:

$$g^{\mu\nu}\nabla_\mu\varphi\nabla_\nu\varphi = 0, \tag{68}$$

By introducing (57) and (67) in (68), the following relation is obtained:

$$C_1^2 e^{4t} + C_2^2 - (C_4^2 + 2C_1 C_2\cosh 2x)e^{2t} + 2C_1 C_4 e^{3t}\sinh x + 2C_2 C_4 e^t\sinh x = 0, \tag{69}$$

which relates the $x-$coordinate and the $t-$coordinate at the horizon:

$$x = \log\left[\frac{C_4 + \sqrt{4C_1 C_2 + C_4^2}}{2C_1}e^{-t}\right]. \tag{70}$$

Hence, the perturbation (67) on the metric at the horizon leads to:

$$\delta\varphi_h = \frac{1}{G_{20}}\left[C_3\frac{G_{2\phi} + 2\Lambda G_{4\phi}}{G_{2\phi} + 4\Lambda G_{4\phi}} + \sqrt{4C_1 C_2 + C_4^2}\right] \tag{71}$$

Therefore, the perturbation at the horizon remains constant. By the Nariai metric (56), one can identify the radius of the horizon when it is perturbed as:

$$r_h = \frac{e^{-\delta\varphi_h}}{\sqrt{\Lambda}} = \frac{e^{-\frac{1}{G_{20}}\left[C_3 \frac{G_{2\phi}+2\Lambda G_{4\phi}}{G_{2\phi}+4\Lambda G_{4\phi}}+\sqrt{4C_1 C_2+C_4^2}\right]}}{\sqrt{\Lambda}}. \tag{72}$$

Note that this expression is time independent, such that no anti-evaporation effect arises when considering the restricted Horndeski Lagrangian (54) in Nariai spacetime. The only effect is a shift of the horizon, which may increase or decrease depending on the values of the integration constants (initial conditions) and on the functions G_i and their derivatives evaluated at ϕ_0 (Horndeski Lagrangian). In addition, if we set the integration constants to zero $C_i = 0$, the radius turns out $r_h = 1/\sqrt{\Lambda}$, i.e., the radius for the horizon in the Nariai spacetime. Moreover, by calculating the perturbation on the scalar field $\delta\phi$ through (65), it yields:

$$\delta\phi(x,t) = \frac{2C_3}{G_{2\phi}+4\Lambda G_{4\phi}}. \tag{73}$$

Hence, the scalar field perturbation does not propagate but just introduces a perturbation on the effective cosmological constant, which explains the absence of the anti-evaporation regime and the shift of the horizon radius when considering perturbations on Nariai spacetime in the framework of Horndeski gravity.

5. Conclusions

In the present paper we have analysed several aspects of Schwarzschild-de Sitter black holes, and particularly its extremal case when both horizons, the cosmological and the black hole ones coincide at the same hypersurface of the spacetime, the so-called Nariai metric. Focusing on the framework of Horndeski gravity, we have shown that the existence of such type solutions when Horndeski Lagrangians are considered can be easily achieved by the induction of an effective cosmological constant, which naturally arises when considering a constant scalar field for some of the Horndeski terms. In addition, we have found that not only a constant scalar field owns Nariai spacetime as a solution of the gravitational field equations but also non-constant scalar field can reproduce Schwarzschild-de Sitter extremal black holes when considering the appropriate functions on the gravitational Lagrangian. However, this result may not satisfy the generalised Birkhoff's theorem as for Brans–Dicke-like theories [63–65], since despite that Nariai spacetime is a static metric, the scalar field may become non-static, as explained in Appendix A.

By considering perturbations on the background scalar field, which is assumed constant, the induced perturbations on the metric turns out to be time dependent, which modifies the staticLK regime of the metric, inducing an exponential expansion, a natural solution when considering an effective cosmological constant. Nevertheless, the linear regime just induces a slight modification on the horizon radius, keeping it constant. Contrary to other frameworks where perturbations on the Nariai spacetime have been considered [8–11], which reproduces the so-called anti-evaporation regime, where the radius of the horizon may grow with time, this effect seems to be absent for the type of Horndeski Lagrangian analysed here. One obviously expects to find a non-constant scalar field perturbation by going beyond the linear regime, which will consequently induce the anti-evaporation regime. In addition, a non-constant scalar field for the background is also expected to produce such phenomena, as perturbations on its propagation will naturally induce effects on the horizon radius, making the Nariai metric unstable.

Author Contributions: Both authors contributed to this work equally in conceptualization, methodology, formal analysis, investigation, writing—original draft preparation, writing—review and editing. Both authors have read and agreed to the published version of the manuscript.

Acknowledgments: D.S.-C.G. is funded by the University of Valladolid (Spain). This article is based upon work from CANTATA COST (European Cooperation in Science and Technology) action CA15117, EU Framework Programme Horizon 2020. I.A. is funded by Fundação para a Ciência e a Tecnologia (FCT) grant number PD/BD/114435/2016 under the IDPASC PhD Program.

Appendix A

In Section 3 we found solutions to the equations for the $\mathcal{L}_2$, $\mathcal{L}_3$ and $\mathcal{L}_4$ cases. Nevertheless, the full set of solutions is not completely covered by the analysis above. Firstly, by analysing the equations for the $\mathcal{L}_2$ Lagrangian given in (13), a skillful reader could wonder about the special case in which the—in principle, non constant—coefficients of the equations become null. With the study of this case in mind, let us rewrite the Einstein tensor for an Einstein manifold, like the Nariai spacetime, as: $G_{\mu\nu} = -\Lambda g_{\mu\nu}$. Therefore, Equation (13) acquires the form:

$$0 = \left[G_2(\phi, X) + \frac{\Lambda}{8\pi G} \right] g_{\mu\nu} + \frac{\partial G_2(\phi, X)}{\partial X} \partial_\mu \phi \partial_\nu \phi. \tag{A1}$$

In addition, the scalar field equation is:

$$0 = 2G_{2\phi X}(\phi, X) X - G_{2XX}(\phi, X) \nabla^\mu X \nabla_\mu \phi - G_{2X}(\phi, X) \Box \phi - G_{2\phi}(\phi, X). \tag{A2}$$

So, as long as the G_2 function satisfies the following conditions:

$$\begin{aligned} G_2(\phi_0, X_0) &= -\frac{\Lambda}{8\pi G} \equiv -A \\ G_{2X}(\phi_0, X_0) = G_{2\phi}(\phi_0, X_0) = G_{2\phi X}(\phi_0, X_0) &= G_{2XX}(\phi_0, X_0) = 0 \quad , \end{aligned} \tag{A3}$$

the field equations above hold. A possible reconstruction of G_2 is given by:

$$G_2(\phi, X) = \sum_{n \geq 3} c_n \left[g(\phi, X) - C \right]^n - A \tag{A4}$$

in which the function $g(\phi, X)$ must satisfy that of $g(\phi_0, X_0) = C$, which becomes the field equation for the scalar field. Some examples of this can be found, but we think that the main issue here is the possibility of solutions with $\phi_0 \neq$ constant. In principle, this fact will affect the perturbations and change our conclusions because recall that we had considered a background scalar field as a constant to solve the system of the perturbations above. For this case, the perturbation equations are:

$$\delta \left(8\pi G \left[g_{\mu\nu} G_2(\phi, X) + \frac{\partial G_2(\phi, X)}{\partial X} \partial_\mu \phi \partial_\nu \phi \right] \right) = \delta G_{\mu\nu} , \tag{A5}$$

which under the constraints (A3) at first order yields:

$$\delta G_{\mu\nu} = -\Lambda \delta g_{\mu\nu} , \tag{A6}$$

As the perturbations on the Einstein tensor remains the same, the following system of equations for $\delta\varphi(x, t)$ is obtained:

$$\begin{array}{ll} \text{tt}- & \delta\varphi'' + \tanh x \delta\varphi' + \text{sech}^2 x \delta\varphi = 0 , \\ \text{xx}- & \delta\ddot{\varphi} + \tanh x \delta\varphi' - \text{sech}^2 x \delta\varphi = 0 , \\ \text{tx}- & \delta\dot{\varphi}' + \tanh x \delta\dot{\varphi} = 0 . \end{array} \tag{A7}$$

The general solution of this system of equations is:

$$\delta\varphi(x,t) = \left(C_1 e^t + C_1 2 e^{-t}\right)\operatorname{sech}x + \frac{1}{2}C_3 \tanh x \,, \tag{A8}$$

where C_i are integration constants. Hence, the perturbation on the horizon radius (72) is easily obtained:

$$r_H = \frac{e^{-\frac{1}{2}\sqrt{C_3^2 + 16 C_1 C_2}}}{\sqrt{\Lambda}} \,, \tag{A9}$$

which as in the case analysed throughout the paper, leads to a constant such that no instability occurs.

A similar analysis can be applied for the Lagrangians $\mathcal{L}_3$ and $\mathcal{L}_4$. For the former, the background equation can be expressed as:

$$\begin{aligned}
-\tfrac{1}{2}\Lambda g_{\mu\nu} =\ & 8\pi G\left[G_{3\phi}\left(g_{\mu\nu}\nabla_\alpha\phi\nabla^\alpha\phi - 2\nabla_\mu\phi\nabla_\nu\phi\right)\right.\\
& \left.+ G_{3X}\left(-\nabla_\mu\phi\nabla_\nu\phi\Box\phi - g_{\mu\nu}\nabla_\alpha\phi\nabla_\beta\phi\nabla^\alpha\phi\nabla^\beta\phi + 2\nabla^\alpha\phi\nabla_{(\mu}\phi\nabla_{\nu)}\nabla^\alpha\phi\right)\right].
\end{aligned} \tag{A10}$$

Then, we may impose to each term of the right hand side of this equation to be expressed as follows:

$$\begin{aligned}
G_{3\phi}\left(g_{\mu\nu}\nabla_\alpha\phi\nabla^\alpha\phi - 2\nabla_\mu\phi\nabla_\nu\phi\right) &= k_1 g_{\mu\nu}\,,\\
G_{3X}\left(-\nabla_\mu\phi\nabla_\nu\phi\Box\phi - g_{\mu\nu}\nabla_\alpha\phi\nabla_\beta\phi\nabla^\alpha\phi\nabla^\beta\phi + 2\nabla^\alpha\phi\nabla_{(\mu}\phi\nabla_{\nu)}\nabla^\alpha\phi\right) &= k_2 g_{\mu\nu}\,,
\end{aligned} \tag{A11}$$

where k_i are constants. Nevertheless, the first condition leads to:

$$\nabla_\mu\phi\nabla_\nu\phi \propto g_{\mu\nu}f(\mathbf{x})\,, \tag{A12}$$

with $f(\mathbf{x})$ being a function of the coordinates to be determined. By inspecting the $tt-$ and $xx-$ equations,

$$\dot\phi^2 = -\frac{1}{2}\frac{1}{\Lambda\cosh x}f(\mathbf{x})\,, \quad \phi'^2 = \frac{1}{2}\frac{1}{\Lambda\cosh x}f(\mathbf{x})\,. \tag{A13}$$

This does not guarantee a real solution for $\phi(x,t)$. Hence, we can not find general conditions on the function G_3 but the system of equations has to be analysed step by step, as done in Section 3.2.

The case for $\mathcal{L}_4$ is similar to $\mathcal{L}_2$. The background Equation (46) hold by imposing:

$$\begin{aligned}
G_4(\phi_0, X_0) &= -\tfrac{1}{16\pi G}\\
G_{4X}(\phi_0, X_0) &= G_{4\phi}(\phi_0, X_0) = G_{4\phi X}(\phi_0, X_0) = G_{4XX}(\phi_0, X_0) = 0\\
G_{4XXX}(\phi_0, X_0) &= G_{4XX\phi}(\phi_0, X_0) = G_{4X\phi\phi}(\phi_0, X_0) = G_{4\phi\phi\phi}(\phi_0, X_0) = 0\,.
\end{aligned} \tag{A14}$$

As above, this can be satisfied by:

$$G_4(\phi, X) = \sum_{n\geq 4} c_n\left[g(\phi, X) - C\right]^n - \frac{1}{16\pi G}\,, \tag{A15}$$

where $g(\phi_0, X_0) = C$. Nevertheless, all the coefficients of the perturbation equation become also null for any background solution ϕ_0, such that we have a degenerated equation that does not pose a well defined problem.

References

1. Gibbons, G.W.; Hawking, S.W. Cosmological event horizons, thermodynamics, and particle creation. *Phys. Rev. D* **1977**, *15*, 2738. [CrossRef]
2. Lake, K.; Roeder, R.C. Effects of a nonvanishing cosmological constant on the spherically symmetric vacuum manifold. *Phys. Rev. D* **1977**, *15*, 3513. [CrossRef]

3. Hawking, S.; Page, D.N. Thermodynamics of Black Holes in anti-De Sitter Space. *Commun. Math. Phys.* **1983**, *87*, 577. [CrossRef]

4. Nariai, H. On a New Cosmological Solution of Einstein's Field Equations of Gravitation. *Sci. Rep. Tohoku Univ. Ser. I* **1951**, *35* 62; reprinted in *Gen. Relativ. Gravit.* **1999**, *31*, 963–971. [CrossRef]

5. Podolsky, J. The Structure of the extreme Schwarzschild-de Sitter space-time. *Gen. Rel. Grav.* **1999**, *31*, 1703. [CrossRef]

6. van den Brink, A.M. Approach to the extremal limit of the Schwarzschild-de sitter black hole. *Phys. Rev. D* **2003**, *68*, 047501. [CrossRef]

7. Ginsparg, P.; Perry, M.J. Semiclassical Perdurance of de Sitter Space. *Nucl. Phys. B* **1983** *222*, 245. [CrossRef]

8. Bousso, R.; Hawking, S.W. (Anti)evaporation of Schwarzschild-de Sitter black holes. *Phys. Rev. D* **1998**, *57*, 2436. [CrossRef]

9. Nojiri, S.; Odintsov, S.D. Effective action for conformal scalars and anti-evaporation of black holes. *Int. J. Mod. Phys. A* **1999**, *14*, 1293. [CrossRef]

10. Nojiri, S.; Odintsov, S.D. Anti-Evaporation of Schwarzschild-de Sitter Black Holes in $F(R)$ gravity. *Class. Quant. Grav.* **2013**, *30*, 125003. [CrossRef]

11. Sebastiani, L.; Momeni, D.; Myrzakulov, R.; Odintsov, S.D. Instabilities and (anti)-evaporation of Schwarzschild–de Sitter black holes in modified gravity. *Phys. Rev. D* **2013**, *88*, 104022. [CrossRef]

12. Katsuragawa, T. Anti-Evaporation of Black Holes in Bigravity. *Universe* **2015**, *1*, 158–172. [CrossRef]

13. Katsuragawa, T.; Nojiri, S. Stability and antievaporation of the Schwarzschild–de Sitter black holes in bigravity. *Phys. Rev. D* **2015**, *91*, 084001. [CrossRef]

14. Nashed, G.G.L. Spherically symmetric black hole solution in mimetic gravity and anti-evaporation. *Int. J. Geom. Meth. Mod. Phys.* **2018**, *15*, 1850154. [CrossRef]

15. Bamba, K.; Capozziello, S.; Nojiri, S.; Odintsov, S.D. Dark energy cosmology: The equivalent description via different theoretical models and cosmography tests. *Astrophys. Space Sci.* **2012**, *342*, 155. [CrossRef]

16. Huterer, D.; Shafer, D.L. Dark energy two decades after: Observables, probes, consistency tests. *Rept. Prog. Phys.* **2018**, *81*, 016901. [CrossRef]

17. Frieman, J.; Turner, M.; Huterer, D. Dark Energy and the Accelerating Universe. *Ann. Rev. Astron. Astrophys.* **2008**, *46*, 385. [CrossRef]

18. Heisenberg, L. A systematic approach to generalisations of General Relativity and their cosmological implications. *Phys. Rept.* **2019**, *796*, 1–113. [CrossRef]

19. Joyce, A.; Jain, B.; Khoury, J.; Trodden, M. Beyond the Cosmological Standard Model. *Phys. Rept.* **2015**, *568*, 1–98. [CrossRef]

20. Nojiri, S.; Odintsov, S.D.; Sami, M. Dark energy cosmology from higher-order, string-inspired gravity and its reconstruction. *Phys. Rev. D* **2006**, *74*, 046004. [CrossRef]

21. Elizalde, E.; Nojiri, S.; Odintsov, S.D.; Saez-Gomez, D.; Faraoni, V. Reconstructing the universe history, from inflation to acceleration, with phantom and canonical scalar fields. *Phys. Rev. D* **2008**, *77*, 106005. [CrossRef]

22. Armendariz-Picon, C.; Mukhanov, V.F.; Steinhardt, P.J. A Dynamical solution to the problem of a small cosmological constant and late time cosmic acceleration. *Phys. Rev. Lett.* **2000**, *85*, 4438. [CrossRef]

23. Armendariz-Picon, C.; Mukhanov, V.F.; Steinhardt, P.J. Essentials of k essence. *Phys. Rev. D* **2001**, *63*, 103510. [CrossRef]

24. Nicolis, A.; Rattazzi, R.; Trincherini, E. The Galileon as a local modification of gravity. *Phys. Rev. D* **2009**, *79*, 064036. [CrossRef]

25. Horndeski, G.W. Second-order scalar-tensor field equations in a four-dimensional space. *Int. J. Theor. Phys.* **1974**, *10*, 363–384. [CrossRef]

26. Kobayashi, T. Horndeski theory and beyond: A review. *Rept. Prog. Phys.* **2019**, *82*, 086901. [CrossRef]

27. Deffayet, C.; Steer, D.A. A formal introduction to Horndeski and Galileon theories and their generalizations. *Class. Quant. Grav.* **2013**, *30*, 214006. [CrossRef]

28. Deffayet, C.; Esposito-Farese, G.; Vikman, A. Covariant Galileon. *Phys. Rev. D* **2009**, *79*, 084003. [CrossRef]

29. Deffayet, C.; Gao, X.; Steer, D.A.; Zahariade, G. From k-essence to generalised Galileons. *Phys. Rev. D* **2011**, *84*, 064039. [CrossRef]

30. Zumalacarregui, M.; Garcia-Bellido, J. Transforming gravity: From derivative couplings to matter to second-order scalar-tensor theories beyond the Horndeski Lagrangian. *Phys. Rev. D* **2014**, *89*, 064046. [CrossRef]

31. Achour, J.B.; Langlois, D.; Noui, K. Degenerate higher order scalar-tensor theories beyond Horndeski and disformal transformations. *Phys. Rev. D* **2016**, *93*, 124005. [CrossRef]

32. Gleyzes, J.; Langlois, D.; Piazza, F.; Vernizzi, F. Healthy theories beyond Horndeski. *Phys. Rev. Lett.* **2015**, *114*, 211101 . [CrossRef] [PubMed]

33. Kobayashi, T.; Yamaguchi, M.; Yokoyama, J. Generalized G-inflation: Inflation with the most general second-order field equations. *Prog. Theor. Phys.* **2011**, *126*, 511–529. [CrossRef]

34. Felice, A.D.; Tsujikawa, S. Inflationary non-Gaussianities in the most general second-order scalar-tensor theories. *Phys. Rev. D* **2011**, *84*, 083504. [CrossRef]

35. Gao, X.; Steer, D.A. Inflation and primordial non-Gaussianities of 'generalized Galileons'. *JCAP* **2011**, *12*, 019. [CrossRef]

36. Felice, A.D.; Kobayashi, T.; Tsujikawa, S. Effective gravitational couplings for cosmological perturbations in the most general scalar-tensor theories with second-order field equations. *Phys. Lett. B* **2011**, *706*, 123. [CrossRef]

37. Gao, X.; Kobayashi, T.; Yamaguchi, M.; Yokoyama, J. Primordial non-Gaussianities of gravitational waves in the most general single-field inflation model. *Phys. Rev. Lett.* **2011**, *107*, 211301. [CrossRef]

38. Charmousis, C.; Copeland, E.J.; Padilla, A.; Saffin, P.M. General second order scalar-tensor theory, self tuning, and the Fab Four. *Phys. Rev. Lett.* **2012**, *108*, 051101, [CrossRef]

39. Copeland, E.J.; Padilla, A.; Saffin, P.M. The cosmology of the Fab-Four. *JCAP* **2012**, *1212*, 026. [CrossRef]

40. Kobayashi, T. Generic instabilities of nonsingular cosmologies in Horndeski theory: A no-go theorem. *Phys. Rev. D* **2016**, *94*, 043511. [CrossRef]

41. Amendola, L.; Kunz, M.; Motta, M.; Saltas, I.D.; Sawicki, I. Observables and unobservables in dark energy cosmologies. *Phys. Rev. D* **2013**, *87*, 023501. [CrossRef]

42. Zumalacarregui, M.; Bellini, E.; Sawicki, I.; Lesgourgues, J.; Ferreira, P.G. hi_class: Horndeski in the Cosmic Linear Anisotropy Solving System. *JCAP* **2017**, *1708*, 019. [CrossRef]

43. Ezquiaga, J.M.; Zumalacarregui, M. Dark Energy After GW170817: Dead Ends and the Road Ahead. *Phys. Rev. Lett.* **2017**, *119*, 251304. [CrossRef] [PubMed]

44. Abbott, B.; Abbott, R.; Abbott, T.D.; Acernese, F.; Ackley, K.; Adams, C.; Adams, T.; Addesso, P.; Adhikari, R.X.; Adya, V.B.; et al. [LIGO Scientific and Virgo]. GW170817: Observation of Gravitational Waves from a Binary Neutron Star Inspiral. *Phys. Rev. Lett.* **2017**, *119*, 161101. [CrossRef]

45. Abbott, B.; Abbott, R.; Abbott, T.D.; Acernese, F.; Ackley, K.; Adams, C.; Adams, T.; Addesso, P.; Adhikari, R.X.; Adya, V.B.; et al. [LIGO Scientific, Virgo, Fermi-GBM and INTEGRAL]. Gravitational Waves and Gamma-rays from a Binary Neutron Star Merger: GW170817 and GRB 170817A. *Astrophys. J.* **2017**, *848*, L13. [CrossRef]

46. Deffayet, C.; Esposito-Farese, G.; Steer, D.A. Counting the degrees of freedom of generalized Galileons. *Phys. Rev. D* **2015**, *92*, 084013. [CrossRef]

47. Bettoni, D.; Ezquiaga, J.M.; Hinterbichler, K.; Zumalacarregui, M. Speed of Gravitational Waves and the Fate of Scalar-Tensor Gravity. *Phys. Rev. D* **2017**, *95*, 084029. [CrossRef]

48. Deffayet, C.; Pujolas, O.; Sawicki, I.; Vikman, A. Imperfect Dark Energy from Kinetic Gravity Braiding. *JCAP* **2010**, *10*, 026. [CrossRef]

49. Bartolo, N.; Karmakar, P.; Matarrese, S.; Scomparin, M. Cosmic structures and gravitational waves in ghost-free scalar-tensor theories of gravity. *JCAP* **2018**, *5*, 048. [CrossRef]

50. Kase, R.; Tsujikawa, S. Dark energy in Horndeski theories after GW170817: A review. *Int. J. Mod. Phys. D* **2019**, *28*, 1942005. [CrossRef]

51. de la Cruz-Dombriz, A.; Dobado, A.; Maroto, A. Black Holes in f(R) theories. *Phys. Rev. D* **2009**, *80*, 124011. [CrossRef]

52. Olmo, G.J.; Rubiera-Garcia, D. Palatini $f(R)$ Black Holes in Nonlinear Electrodynamics. *Phys. Rev. D* **2011**, *84*, 124059. [CrossRef]

53. Olmo, G.J.; Rubiera-Garcia, D.; Sanchis-Alepuz, H. Geonic black holes and remnants in Eddington-inspired Born-Infeld gravity. *Eur. Phys. J. C* **2014**, *74*, 2804. [CrossRef] [PubMed]

54. de la Cruz-Dombriz, A.d.; Saez-Gomez, D. Black holes, cosmological solutions, future singularities, and their thermodynamical properties in modified gravity theories. *Entropy* **2012**, *14*, 1717–1770. [CrossRef]

55. Sotiriou, T.P.; Zhou, S. Black hole hair in generalized scalar-tensor gravity. *Phys. Rev. Lett.* **2014**, *112*, 251102. [CrossRef]

56. Clifton, T. Spherically Symmetric Solutions to Fourth-Order Theories of Gravity. *Class. Quant. Grav.* **2006**, *23*, 7445. [CrossRef]

57. Yunes, N.; Sopuerta, C.F. Perturbations of Schwarzschild Black Holes in Chern-Simons Modified Gravity. *Phys. Rev. D* **2008**, *77*, 064007. [CrossRef]

58. Cardoso, V.; Gualtieri, L. Perturbations of Schwarzschild black holes in Dynamical Chern-Simons modified gravity. *Phys. Rev. D* **2009**, *80*, 064008. [CrossRef]

59. Olmo, G.J.; Rubiera-Garcia, D.; Sanchez-Puente, A. Geodesic completeness in a wormhole spacetime with horizons. *Phys. Rev. D* **2015**, *92*, 044047. [CrossRef]

60. Olmo, G.J.; Rubiera-Garcia, D.; Sanchez-Puente, A. Classical resolution of black hole singularities via wormholes. *Eur. Phys. J. C* **2016**, *76*, 143. [CrossRef]

61. Bejarano, C.; Olmo, G.J.; Rubiera-Garcia, D. What is a singular black hole beyond General Relativity? *Phys. Rev. D* **2017**, *95*, 064043. [CrossRef]

62. Nojiri, S.; Odintsov, S. Regular multihorizon black holes in modified gravity with nonlinear electrodynamics. *Phys. Rev. D* **2017**, *96*, 104008. [CrossRef]

63. Faraoni, V. The Jebsen-Birkhoff theorem in alternative gravity. *Phys. Rev. D* **2010**, *81*, 044002. [CrossRef]

64. Schleich, K.; Witt, D.M. A simple proof of Birkhoff's theorem for cosmological constant. *J. Math. Phys.* **2010**, *51*, 112502. [CrossRef]

65. Capozziello, S.; Saez-Gomez, D. Scalar-tensor representation of $f(R)$ gravity and Birkhoff's theorem. *Ann. Phys.* **2012**, *524*, 279–285. [CrossRef]

66. Moffat, J. Modified Gravity Black Holes and their Observable Shadows. *Eur. Phys. J. C* **2015**, *75*, 130. [CrossRef]

67. Davis, A.; Gregory, R.; Jha, R.; Muir, J. Astrophysical black holes in screened modified gravity. *JCAP* **2014**, *8*, 033. [CrossRef]

68. Guo, M.; Obers, N.A.; Yan, H. Observational signatures of near-extremal Kerr-like black holes in a modified gravity theory at the Event Horizon Telescope. *Phys. Rev. D* **2018**, *98*, 084063. [CrossRef]

69. Babichev, E.; Charmousis, C.; Lehebel, A. Asymptotically flat black holes in Horndeski theory and beyond. *JCAP* **2017**, *4*, 027. [CrossRef]

70. Tattersall, O.J.; Ferreira, P.G. Quasinormal modes of black holes in Horndeski gravity. *Phys. Rev. D* **2018**, *97*, 104047. [CrossRef]

71. Babichev, E.; Charmousis, C.; Lehebel, A. Black holes and stars in Horndeski theory. *Class. Quant. Grav.* **2016**, *33*, 154002. [CrossRef]

72. Babichev, E.; Charmousis, C.; Lehébel, A.; Moskalets, T. Black holes in a cubic Galileon universe. *JCAP* **2016**, *9*, 011. [CrossRef]

73. Tattersall, O.J.; Ferreira, P.G.; Lagos, M. Speed of gravitational waves and black hole hair. *Phys. Rev. D* **2018**, *97*, 084005. [CrossRef]

74. Babichev, E.; Charmousis, C.; Esposito-Farese, G.; Lehebel, A. Stability of Black Holes and the Speed of Gravitational Waves within Self-Tuning Cosmological Models. *Phys. Rev. Lett.* **2018**, *120*, 241101. [CrossRef] [PubMed]

75. Ganguly, A.; Gannouji, R.; Gonzalez-Espinoza, M.; Pizarro-Moya, C. Black hole stability under odd-parity perturbations in Horndeski gravity. *Class. Quant. Grav.* **2018**, *35*, 145008. [CrossRef]

76. Sakstein, J.; Babichev, E.; Koyama, K.; Langlois, D.; Saito, R. Towards Strong Field Tests of Beyond Horndeski Gravity Theories. *Phys. Rev. D* **2017**, *95*, 064013. [CrossRef]

77. Ogawa, H.; Kobayashi, T.; Suyama, T. Instability of hairy black holes in shift-symmetric Horndeski theories. *Phys. Rev. D* **2016**, *93*, 064078. [CrossRef]

78. Lehébel, A.; Babichev, E.; Charmousis, C. A no-hair theorem for stars in Horndeski theories. *JCAP* **2017**, *7*, 037. [CrossRef]

79. Kovacs, A.D.; Reall, H.S. Well-posed formulation of Lovelock and Horndeski theories. *Phys. Rev. D* **2020**, *101*, 124003. [CrossRef]

80. Ijjas, A.; Pretorius, F.; Steinhardt, P.J. Stability and the Gauge Problem in Non-Perturbative Cosmology. *JCAP* **2019**, *1*, 015. [CrossRef]

81. Ijjas, A. Space-time slicing in Horndeski theories and its implications for non-singular bouncing solutions. *JCAP* **2018**, *2*, 007. [CrossRef]

82.   Martel, K.; Poisson, E. Gravitational perturbations of the Schwarzschild spacetime: A Practical covariant and gauge-invariant formalism. *Phys. Rev. D* **2005**, *71*, 104003. [CrossRef]

83.   Kimura, M. Stability analysis of Schwarzschild black holes in dynamical Chern-Simons gravity. *Phys. Rev. D* **2018**, *98*, 024048. [CrossRef]

84.   Hung, P.K.; Keller, J.; Wang, M.T. Linear Stability of Schwarzschild Spacetime: Decay of Metric Coefficients. *arXiv* **2017**, arXiv:1702.02843.

2

Rotating Melvin-like Universes and Wormholes in General Relativity

Kirill A. Bronnikov [1,2,3,*] , **Vladimir G. Krechet** [4] **and Vadim B. Oshurko** [4]

[1] Center for Gravitation and Fundamental Metrology, VNIIMS, Ozyornaya ul. 46, 119361 Moscow, Russia
[2] Institute for Gravitation and Cosmology, Peoples' Friendship University of Russia (RUDN University), ul. Miklukho-Maklaya 6, 117198 Moscow, Russia
[3] Moscow Engineering Physics Institute, National Research Nuclear University "MEPhI", 105005 Moscow, Russia
[4] Department of Physics, Moscow State Technological University "Stankin", Vadkovsky per. 3A, 127055 Moscow, Russia
[*] Correspondence: kb20@yandex.ru

Abstract: We find a family of exact solutions to the Einstein–Maxwell equations for rotating cylindrically symmetric distributions of a perfect fluid with the equation of state $p = w\rho$ ($|w| < 1$), carrying a circular electric current in the angular direction. This current creates a magnetic field along the z axis. Some of the solutions describe geometries resembling that of Melvin's static magnetic universe and contain a regular symmetry axis, while some others (in the case $w > 0$) describe traversable wormhole geometries which do not contain a symmetry axis. Unlike Melvin's solution, those with rotation and a magnetic field cannot be vacuum and require a current. The wormhole solutions admit matching with flat-space regions on both sides of the throat, thus forming a cylindrical wormhole configuration potentially visible for distant observers residing in flat or weakly curved parts of space. The thin shells, located at junctions between the inner (wormhole) and outer (flat) regions, consist of matter satisfying the Weak Energy Condition under a proper choice of the free parameters of the model, which thus forms new examples of phantom-free wormhole models in general relativity. In the limit $w \to 1$, the magnetic field tends to zero, and the wormhole model tends to the one obtained previously, where the source of gravity is stiff matter with the equation of state $p = \rho$.

Keywords: exact solutions; cylindrical symmetry; rotation; perfect fluid; wormholes

1. Introduction

Cylindrical symmetry is the second (after the spherical one) simplest space-time symmetry making it possible to obtain numerous exact solutions in general relativity and its extensions, characterizing local strong gravitational field configurations. One of the motivations of studying cylindrically symmetric configurations is the possible existence of such linearly extended structures as cosmic strings as well as the observed cosmic jets. A large number of static cylindrically symmetric solutions have been obtained and studied since the advent of general relativity, including vacuum, electrovacuum, perfect fluid and others, see reviews in [1–3] and references therein.

Important arguments is favor of the studies of cylindrically symmetric and rotating configurations come from cosmological observations. Thus, for instance, Birch [4] has reported the discovery of polarization anisotropy in radio signals from extragalactic sources which could be a signature of a slow rotation of the Universe. This gave rise to the emergence of numerous cosmological models with rotation, most of which possess cylindrical symmetry; see [5,6] and references therein. There are indications of a distinguished direction in the Universe following from an analysis of the Cosmic Microwave Background [7] and the distribution of left-twirled and right-twirled spiral galaxies on the celestial sphere [8].

There are also reasons to try to include large-scale magnetic fields into cosmological models. A possible existence of a global magnetic field up to 10^{-15} G may be suspected due to the observed correlated orientations of quasars distant from each other [9]. Various possible manifestations of primordial magnetic fields are discussed in the literature; see, e.g., [10] for a review. Among numerous anisotropic cosmologies with a large-scale magnetic field, admitting late-time isotropization, one can mention Bianchi type I [11] and Kantowski-Sachs models [12], the latter appearing beyond the horizon of a regular black hole with a radial magnetic field and a phantom scalar field.

Melvin's famous solution to the Einstein–Maxwell equations, an "electric or magnetic geon" [13], is a completely regular static, cylindrically symmetric solution with a longitudinal electric or magnetic field as the only source of gravity. It is a special case from a large set of static cylindrically symmetric Einstein–Maxwell fields, see more details in [3,14,15].

An important distinguishing feature of cylindrical symmetry as compared to the spherical one is the possible inclusion of rotation, avoiding complications inherent to the more realistic axial symmetry, not to mention the general nonsymmetric space-times. Accordingly, a great number of exact stationary (assuming rotation) solutions to the Einstein equations are known, with various sources of gravity: the cosmological constant [16–20]; scalar fields with different self-interaction potentials [21–23]; rigidly or differentially rotating dust [24–26], dust with electric charge [27] or a scalar field [28], fluids with different equations of state, above all, perfect fluids with $p = w\rho$, $w = $ const (in usual notations) [29–33], some kinds of anisotropic fluids [34–37] etc., see also references therein and the reviews [1,3].

In this paper we obtain rotating counterparts of the static cylindrically symmetric solutions to the Einstein–Maxwell equations with a longitudinal magnetic field. It turns out that such a field cannot exist without a source in the form of an electric current, and we find solutions where such a source is a perfect fluid with $p = w\rho$. Many features of these solutions are quite different from those of the static ones, in particular, their common feature is the emergence of closed timelike curves at large radii. Also, there is a family of wormhole solutions that do not have a symmetry axis but contain a throat as a minimum of the circular radius. As in our previous studies [21,22,38,39], we try to make such wormholes potentially observable from spatial infinity by joining outer flat-space regions at some junction surfaces and verify the validity of the Weak Energy Condition for matter residing on these surfaces.

The structure of the paper is as follows. Section 2 briefly describes the general formalism. In Section 3, we find solutions of the field equations. In Section 4, we discuss the properties of Melvin-like solutions, and in Section 5, the wormhole family. Section 6 contains some concluding remarks.

2. Basic Relations

We consider stationary cylindrically symmetric space-times with the metric

$$ds^2 = e^{2\gamma(x)}[dt - E(x) e^{-2\gamma(x)} d\varphi]^2 - e^{2\alpha(x)}dx^2 - e^{2\mu(x)}dz^2 - e^{2\beta(x)}d\varphi^2, \qquad (1)$$

where $x^0 = t \in \mathbb{R}$, $x^1 = x$, $x^2 = z \in \mathbb{R}$ and $x^3 = \varphi \in [0, 2\pi)$ are the temporal, radial, longitudinal and angular (azimuthal) coordinates, respectively. The variable x is here specified up to a substitution $x \to f(x)$, therefore its range depends on both the geometry itself and the "gauge" (the coordinate condition). The off-diagonal component E describes rotation, or the vortex component of the gravitational field. In the general case, this vortex gravitational field is determined by the 4-curl of the orthonormal tetrad field e_m^μ (Greek and Latin letters are here assigned to world and tetrad indices, respectively) [40,41]:

$$\omega^\mu = \frac{1}{2}\varepsilon^{\mu\nu\rho\sigma}e_{m\mu}\partial_\rho e_\sigma^m. \qquad (2)$$

Kinematically, the axial vector ω^μ is the angular velocity of tetrad rotation, it determines the proper angular momentum density of the gravitational field,

$$S^\mu(g) = \omega^\mu/\kappa, \qquad \kappa = 8\pi G, \tag{3}$$

where G is the Newtonian gravitational constant. In space-times with the metric (1) we have

$$\omega^\mu = \frac{1}{2}\delta^{\mu 2}(E\,e^{-2\gamma})'\,e^{\gamma-\alpha-\beta-\mu} \tag{4}$$

(a prime stands for d/dx), and it appears sufficient to consider its absolute value $\omega(x) = \sqrt{\omega^\mu \omega_\mu}$ that has the meaning of the angular velocity of a congruence of timelike curves (vorticity) [21,40,41],

$$\omega = \frac{1}{2}(E\,e^{-2\gamma})'\,e^{\gamma-\beta-\alpha}. \tag{5}$$

Furthermore, in the reference frame comoving to matter as it rotates in the azimuthal (φ) direction, the stress-energy tensor (SET) component T_0^3 is zero, therefore due to the Einstein equations the Ricci tensor component $R_0^3 \sim (\omega\,e^{2\gamma+\mu})' = 0$, which leads to [21]

$$\omega = \omega_0\,e^{-\mu-2\gamma}, \qquad \omega_0 = \text{const.} \tag{6}$$

Then, according to (5),

$$E(x) = 2\omega_0\,e^{2\gamma(x)}\int e^{\alpha+\beta-\mu-3\gamma}dx. \tag{7}$$

Note that Equations (4)–(7) are valid for an arbitrary choice of the radial coordinate x. Preserving this arbitrariness, we can write the nonzero components of the Ricci (R_μ^ν) tensor as

$$\begin{aligned}
R_0^0 &= -e^{-2\alpha}[\gamma'' + \gamma'(\sigma' - \alpha')] - 2\omega^2, \\
R_1^1 &= -e^{-2\alpha}[\sigma'' + \sigma'^2 - 2U - \alpha'\sigma'] + 2\omega^2, \\
R_2^2 &= -e^{-2\alpha}[\mu'' + \mu'(\sigma' - \alpha')], \\
R_3^3 &= -e^{-2\alpha}[\beta'' + \beta'(\sigma' - \alpha')] + 2\omega^2, \\
R_3^0 &= G_3^0 = E\,e^{-2\gamma}(R_3^3 - R_0^0),
\end{aligned} \tag{8}$$

where we are using the notations

$$\sigma = \beta + \gamma + \mu, \qquad U = \beta'\gamma' + \beta'\mu' + \gamma'\mu'. \tag{9}$$

The Einstein equations may be written in two equivalent forms

$$G_\mu^\nu \equiv R_\mu^\nu - \tfrac{1}{2}\delta_\mu^\nu R = -\kappa T_\mu^\nu, \quad \text{or} \tag{10}$$

$$R_\mu^\nu = -\kappa\widetilde{T}_\mu^\nu \equiv -\kappa(T_\mu^\nu - \tfrac{1}{2}\delta_\mu^\nu T). \tag{11}$$

R being the Ricci scalar and T the SET trace. We will mostly use the form (11), of the equations, but it is also necessary to write the constraint equation from (10), which contains only first-order derivatives of the metric and represents a first integral of the other equations:

$$G_1^1 = e^{-2\alpha}U + \omega^2 = -\kappa T_1^1. \tag{12}$$

Owing to the last line of (8) and its analogue for T_μ^ν, in the Einstein equations it is sufficient to solve the diagonal components, and then their only off-diagonal component holds automatically [21].

As is evident from (8), the diagonal components of both the Ricci (R^ν_μ) and the Einstein (G^ν_μ) tensors split into the corresponding tensors for the static metric (the metric (1) with $E = 0$) plus a contribution containing ω [21]:

$$R^\nu_\mu = {}_sR^\nu_\mu + {}_\omega R^\nu_\mu, \qquad {}_\omega R^\nu_\mu = \omega^2 \, \mathrm{diag}(-2, 2, 0, 2), \tag{13}$$

$$G^\nu_\mu = {}_sG^\nu_\mu + {}_\omega G^\nu_\mu, \qquad {}_\omega G^\nu_\mu = \omega^2 \, \mathrm{diag}(-3, 1, -1, 1), \tag{14}$$

where ${}_sR^\nu_\mu$ and ${}_sG^\nu_\mu$ correspond to the static metric. It turns out that the tensors ${}_sG^\nu_\mu$ and ${}_\omega G^\nu_\mu$ (each separately) obey the conservation law $\nabla_\alpha G^\alpha_\mu = 0$ in terms of this static metric. Therefore, the tensor ${}_\omega G^\nu_\mu / \kappa$ may be interpreted as the SET of the vortex gravitational field. It possesses quite exotic properties (thus, the effective energy density is $-3\omega^2/\kappa < 0$), which favor the existence of wormholes, and indeed, a number of wormhole solutions with the metric (1) have already been obtained [21,22,39,41] with sources in the form of scalar fields, isotropic or anisotropic fluids. Further on we will obtain one more solution of this kind, now supported by a perfect fluid and a magnetic field due to an electric current. Let us mention that an alternative extension of static solutions to rotating ones, with a combination of electric and magnetic fields and a cosmological constant, was obtained in [42].

3. Solutions with A Perfect Fluid and A Magnetic Field

3.1. The Electromagnetic Field. A No-Go Theorem

Consider a longitudinal (z-directed) magnetic field, corresponding to the 4-potential

$$A_\mu = (0, 0, 0, \Phi(x)), \tag{15}$$

so that the only nonzero components of the Maxwell tensor $F_{\mu\nu} = \partial_\mu A_\nu - \partial_\nu A_\mu$ are $F_{13} = -F_{31} = \Phi'(x)$. The nonzero contravariant components $F^{\mu\nu}$ are

$$F^{13} = \mathrm{e}^{-2\alpha-2\beta}\Phi', \qquad F^{01} = E\,\mathrm{e}^{-2\alpha-2\beta-2\gamma}\Phi'. \tag{16}$$

The magnetic field magnitude (magnetic induction) B is determined by $B^2 = F^{13}F_{13}$, so that the electromagnetic field invariant is $F_{\mu\nu}F^{\mu\nu} = 2B^2$.

No-go theorem. It can be shown that a free longitudinal magnetic field is incompatible with a nonstatic ($\omega \neq 0$) metric (1). This follows from solving the Maxwell equations, which, for $F^{\mu\nu}$ of the form (16) read

$$(\sqrt{-g}F^{13})' = 0, \qquad (\sqrt{-g}F^{01})' = 0, \tag{17}$$

where

$$g = \det(g_{\mu\nu}), \qquad \sqrt{-g} = \mathrm{e}^{\alpha+\beta+\gamma+\mu}.$$

Equation (17) are integrated to give, respectively,

$$\Phi'\,\mathrm{e}^{-\alpha-\beta+\gamma+\mu} = h_1, \qquad E\Phi'\,\mathrm{e}^{-\alpha-\beta-\gamma+\mu} = h_2, \tag{18}$$

where $h_1, h_2 = \mathrm{const} \neq 0$. From (18), it follows $E\,\mathrm{e}^{-2\gamma} = h_2/h_1$, whence $(E\,\mathrm{e}^{-2\gamma})' = 0$, and according to (6), $\omega = 0$. We have shown that a free longitudinal magnetic field cannot support a vortex gravitational field with the metric (1).

Let us also note that in the case $E\,\mathrm{e}^{-2\gamma} = E_1 = \mathrm{const}$, the term $E\,\mathrm{e}^{-2\gamma}d\varphi$ is eliminated from (1) by introducing the new time coordinate $t' = t - E_1\varphi$, making the metric explicitly static.

3.2. The Fluid

To circumvent the above no-go theorem, that is, to avoid the relation $E\,e^{-2\gamma} = $ const, let us introduce a source of the magnetic field in the form of an electric current density $J^\mu = \rho_e u^\mu$, where ρ_e is the effective electric charge density (If we introduce a real nonzero charge density, it becomes necessary to consider, in addition, a Coulomb electric field, which will make the problem hardly tractable. We therefore consider an azimuthal electric current as if in a coil, in a neutral medium like a conductor with free electrons and ions at rest.), and u^μ is the 4-velocity satisfying the usual normalization condition $u_\mu u^\mu = 1$. We will assume that the effective charge distribution is at rest in our rotating reference frame, so that

$$u^\mu = (e^{-\gamma}, 0, 0, 0), \qquad J^\mu = (\rho_e\,e^{-\gamma}, 0, 0, 0). \tag{19}$$

As usual, the electric charge conservation equation $\nabla_\mu J^\mu$ holds automatically due to the Maxwell equations $\nabla_\nu F^{\mu\nu} = J^\mu$.

Two nontrivial Maxwell equations now read

$$(\sqrt{-g}F^{13})' = 0, \tag{20}$$

$$\frac{1}{\sqrt{-g}}(\sqrt{-g}F^{01})' = \rho_e\,e^{-\gamma}. \tag{21}$$

Integrating Equation (20), we obtain, as before,

$$\Phi' = h\,e^{\alpha+\beta-\gamma-\mu}, \qquad h = \text{const}, \tag{22}$$

and substituting this Φ' to Equation (21) with (16), we arrive at the following expression for ρ_e:

$$\rho_e = h\omega_0\,e^{-2\mu-3\gamma}. \tag{23}$$

On the other hand, the electric charges should have a material carrier, for which we will assume a perfect fluid with a barotropic equation of state and postulate a constant ratio of the effective charge density ρ_e to energy density ρ:

$$\rho_e/\rho = A = \text{const}; \qquad p/\rho = w = \text{const}, \tag{24}$$

p being the fluid pressure. We do not fix the value of w but later on we will obtain a restriction on it. The fluid must obey the conservation law $\nabla_\nu T^\nu_\mu = 0$, which gives in our comoving reference frame

$$p' + (p + \rho)\gamma' = 0, \tag{25}$$

which, for $w \neq 0$, leads to the expression

$$\rho = \rho_0\,e^{-\gamma(w+1)/w}, \qquad \rho_0 = \text{const}. \tag{26}$$

Comparing (23) and (26), taking into account the assumption $\rho_e/\rho = A = $ const, we obtain a relation between the metric coefficients $e^{2\gamma}$ and $e^{2\mu}$:

$$A\rho_0\,e^{2\mu} = h\omega_0\,e^{-\gamma(w+1)/(2w)}. \tag{27}$$

In the case $w = 0$ (zero pressure), the conservation law (25) simply leads to $\gamma = $ const.

3.3. *Solution of the Einstein Equations*

To address the Einstein equations, let us write the expressions for the SETs of the perfect fluid and the electromagnetic field. For the fluid we have

$$T_\mu^\nu[f] = \rho \, \text{diag}(1, -w, -w, -w). \tag{28}$$

For the electromagnetic field SET we have the standard expression

$$T_\mu^\nu[e] = \frac{1}{16\pi}\left(-F_{\mu\alpha}F^{\nu\alpha} + 4\delta_\mu^\nu F_{\alpha\beta}F^{\alpha\beta}\right),$$

which in our case leads to

$$T_\mu^\nu[e] = \frac{B^2}{8\pi}\, \text{diag}(1. - 1, 1, -1) \oplus T_3^0[e], \tag{29}$$

where $B^2 = h^2\,e^{-2\gamma-2\mu}$, and the only off-diagonal component

$$T_3^0[e] = -\frac{1}{4\pi}h\Phi'\,e^{-2\gamma} \tag{30}$$

does not affect the solution process, as mentioned in a remark after Equation (12).

Thus far all relations and expressions were written in terms of an arbitrary radial coordinate x. However, to solve the Einstein equations it is helpful, at last, to choose the "gauge", and by analogy with our previous studies we will use the harmonic radial coordinate corresponding to

$$\alpha = \beta + \gamma + \mu. \tag{31}$$

With the above expressions for the SET and Equation (8), the noncoinciding components of Equations (11) and (12), taking into account the expressions (6) for ω and (26) for ρ, may be written as

$$\gamma'' = -2\omega_0^2\,e^{2\beta-2\gamma} + \frac{1+3w}{2}K\,e^{2\beta-\gamma} + Gh^2\,e^{2\beta}, \tag{32}$$

$$\mu'' = \frac{w-1}{2}K\,e^{2\beta-\gamma} + Gh^2\,e^{2\beta}, \tag{33}$$

$$\beta'' = 2\omega_0^2\,e^{2\beta-2\gamma} + \frac{w-1}{2}K\,e^{2\beta-\gamma} - Gh^2\,e^{2\beta}, \tag{34}$$

$$\beta'\gamma' + \beta'\mu' + \gamma'\mu' = -\omega_0^2\,e^{2\beta-2\gamma} + wK\,e^{2\beta-\gamma} + Gh^2\,e^{2\beta}, \tag{35}$$

where $K = \kappa h\omega_0/A$ (recall that $\kappa = 8\pi G$).

Now, combining Equations (32) and (33) with (27), we arrive at the algebraic equation for γ with constant coefficients,

$$4\omega_0^2(1-2w)\,e^{-2\gamma} + (8w^2 - 3w - 1)K\,e^{-\gamma} + 2(4w-1)Gh^2 = 0, \tag{36}$$

from which it follows $\gamma = \text{const}$, and we can without loss of generality put $\gamma \equiv 0$ by choosing a time scale. Then Equation (27) implies $\mu = \text{const}$ which also allows us to put $\mu \equiv 0$ by choosing the scale along z; from (27) we then obtain a relation among the constants: $h\omega_0 = A\rho_0$, so that, in particular, $K = \kappa\rho_0$.

With $e^\gamma = e^\mu = 1$, from Equations (32) and (33) we find

$$Gh^2 = \frac{1}{2}(1-w)K, \qquad \omega_0^2 = \frac{1}{2}(1+w)K, \tag{37}$$

which leads to a conclusion on the range of w:

$$-1 < w < 1. \tag{38}$$

With (37) it is directly verified that Equations (35) and (36) also hold. All our constant parameters may be expressed in terms of two of them, for example, ω_0 and w:

$$Gh^2 = \frac{1-w}{1+w}\omega_0^2, \qquad \kappa\rho_0 = \frac{2\omega_0^2}{1+w}, \qquad A = \frac{\rho_e}{\rho} = 4\pi\sqrt{G(1-w^2)}. \tag{39}$$

We see that in our system not only $\mu = \gamma = 0$, but also both densities ρ and ρ_e as well as the angular velocity ω are constant. It is also of interest that the two limiting cases of the equation of state, $w = 1$ (maximum stiffness compatible with causality) and $w = -1$ (the cosmological constant) are excluded in the present system. In both these cases, static and stationary cylindrically symmetric solutions without an electromagnetic field are well known [3,17,18,21,23,37,39].

The remaining differential Equation (34) has the Liouville form

$$\beta'' = \frac{4w}{w+1}\omega_0^2\, e^{2\beta}. \tag{40}$$

and has the first integral

$$\beta'^2 = \frac{4w}{w+1}\omega_0^2\, e^{2\beta} + k^2\,\mathrm{sign}\,k, \tag{41}$$

with $k = \mathrm{const}$. The further integration depends on the signs of w and k:

$$\textbf{1.}\ w < 0,\ k > 0, \qquad e^\beta = \frac{k}{m\cosh(kx)};. \tag{42}$$

$$\textbf{2.}\ w > 0,\ k > 0: \qquad e^\beta = \frac{k}{m\sinh(kx)};. \tag{43}$$

$$\textbf{3.}\ w > 0,\ k = 0: \qquad e^\beta = \frac{1}{mx};. \tag{44}$$

$$\textbf{4.}\ w > 0,\ k < 0: \qquad e^\beta = \frac{|k|}{m\cos(|k|x)};. \tag{45}$$

where we have denoted $m = \left(\dfrac{4|w|\omega_0^2}{w+1}\right)^{1/2}$.

In the previously excluded case $w = 0$ (dustlike matter), the equality $\gamma = \mathrm{const}$ is immediately obtained from (25), $\mu = \mathrm{const}$ then follows from (27), and as before, without loss of generality, we can put $\mu = \gamma = 0$. Instead of (40), we obtain $\beta'' = 0$ whence we can write

$$e^\beta = r_0\, e^{kx}, \qquad r_0, k = \mathrm{const}. \tag{46}$$

In all cases the off-diagonal metric function E is easily obtained as

$$E(x) = 2\omega_0 \int e^{2\beta}dx. \tag{47}$$

4. Melvin-Like Universes

Melvin's electric or magnetic geon [13] is among the most well-known static, cylindrically symmetric solutions to the Einstein–Maxwell equations; it is a special solution from a large class of static, cylindrically symmetric solutions with radial, azimuthal and longitudinal electric and/or magnetic fields; see, e.g., [1,3,14]. Its metric may be written in the form [3,14]

$$ds^2 = (1 + q^2x^2)^2(dt^2 - dx^2 - dz^2) - \frac{x^2}{(1+q^2x^2)^2}d\varphi^2, \tag{48}$$

where $x \geq 0$, and the magnetic (let us take it for certainty) field magnitude is

$$B = B_z = 2q(1 + q^2 x^2)^{-2},$$

with $q = \text{const}$ characterizing the effective current that might be its source. However, this solution describes a purely field configuration existing without any massive matter, electric charges or currents. Both the metric and the magnetic field are regular on the axis $x = 0$. The other "end", $x \to \infty$, is infinitely far away (the distance $\int \sqrt{-g_{xx}} dx$ diverges), the magnetic field vanishes there, and the circular radius $r = \sqrt{-g_{\varphi\varphi}}$ also tends to zero, so that the whole configuration is closed in nature, without spatial infinity, and with finite total magnetic field energy per unit length along the z axis.

As we saw in Section 3.1, such a free magnetic field cannot support a rotating counterpart of Melvin's solution, but Einstein–Maxwell solutions with a longitudinal magnetic field are obtained in the presence of perfect fluids with electric currents. Let us briefly discuss their main features.

In all cases under consideration, the magnetic field is directed along the z axis and has the constant magnitude $B = h$, while the metric has the form

$$ds^2 = (dt - E d\varphi)^2 - e^{2\beta} dx^2 - dz^2 - e^{2\beta} d\varphi^2, \tag{49}$$

and E is determined by Equation (47). Note that both φ and x are dimensionless while t, z and e^β have the dimension of length.

Dustlike Matter, Equation (46)

Let us begin with the case $w = 0$. For $E(x)$ we find

$$E = E_0 + \frac{\omega_0 r_0^2}{k} e^{2kx} = E_0 + \frac{\omega_0}{k} r^2,$$
$$r = r_0 e^{kx}, \qquad E_0 = \text{const}, \tag{50}$$

where E_0 is an integration constant. In terms of the coordinate r, the metric reads

$$ds^2 = \left(dt - E d\varphi^2\right)^2 - k^{-2} dr^2 - dz^2 - r^2 d\varphi^2. \tag{51}$$

The symmetry axis $r = 0$ is regular in the case $E_0 = 0$, $k = 1$ (The axis regularity conditions require [1,3,43] finite values of the curvature invariants plus local flatness (sometimes also called "elementary flatness") as a correct circumference to radius ratio for small circles around the axis, which in our case leads to the condition $e^{-2\alpha} R'^2 \to 1$, where $R = \sqrt{-g_{33}}$.). Also, in this case

$$g_{33} = E^2 - r^2 = r^2(-1 + \omega_0^2 r^2) \tag{52}$$

changes its sign at $r = \omega^{-1}$, and at larger r the lines of constant t, r, z (coordinate circles) are timelike, thus being closed timelike curves (CTCs) violating causality.

Solution 1, Equation (42)

For $w < 0$, with (42), for $E(x)$ we calculate

$$E = E_0 + \frac{2k\omega_0}{m^2} \tanh kx, \tag{53}$$

The metric has the form

$$ds^2 = (dt - E d\varphi)^2 - dz^2 - \frac{k^2}{m^2 \cosh^2(kx)} (dx^2 + d\varphi^2), \tag{54}$$

where, for convenience, we have rearranged the terms with dz^2 and dx^2 as compared to (49).

For g_{33}, similarly to (52), again putting $E_0 = 0$ and recalling the definition of m, we obtain

$$g_{33} = -\frac{k^2}{m^2}\left[1 - \frac{1}{|w|}\tanh^2 kx\right]. \tag{55}$$

In this solution $x \in \mathbb{R}$, and at both extremes $x \to \pm\infty$ we have $r = e^\beta \to 0$, i.e., these are two centers of symmetry (or poles) on the (x, φ) 2-surface, or two symmetry axes from the viewpoint of 3-dimensional space. However, as follows from (55), g_{33} is positive (hence contains CTCs) where $|\tanh kx| > |w|$, that is, at large enough $|x|$, in circular regions around the two poles.

By choosing another value of the integration constant E_0 one can make one of the poles free from CTCs, at the expense of enlarging the CTC region around the other pole. One of the poles can even be made regular by a proper choice of the parameters. For example, choosing $E_0 = 2k\omega_0/m^2$, we obtain $E = 0$ at $x = -\infty$, and it is easy to verify that the pole $x = -\infty$ is then regular under the condition $k = 1$.

Solution 2, Equation (43)

For $w > 0, k > 0$, with (43), for $E(x)$ we find

$$E = E_0 - \frac{2k\omega_0}{m^2}\coth kx. \tag{56}$$

The metric takes the form

$$ds^2 = (dt - Ed\varphi)^2 - dz^2 - \frac{k^2}{m^2\sinh^2(kx)}(dx^2 + d\varphi^2), \tag{57}$$

It is convenient to introduce the new coordinate y by substituting

$$e^{-2kx} = 1 - \frac{2k}{y}, \tag{58}$$

after which we obtain

$$r^2 = e^{2\beta} = \frac{y}{m^2}(y - 2k). \qquad E = E_0 - \frac{2\omega_0}{m^2}y. \tag{59}$$

The range $x > 0$ is converted to $y \geq 2k$, where $y = 2k$ is the axis of symmetry. The metric now has the form

$$ds^2 = (dt - Ed\varphi)^2 - \frac{dy^2}{my(y - 2k)} - dz^2 - \frac{y}{m^2}(y - 2k)d\varphi^2. \tag{60}$$

Assuming $E_0 = 0$, for g_{33} it is then easy to obtain the expression

$$g_{33} = \frac{y}{m^2}\left(\frac{y}{w} + 2k\right) > 0, \tag{61}$$

which means that CTCs are present everywhere, and actually this space-time has an incorrect signature, $(+ - -+)$ instead of $(+ - --)$.

However, with nonzero values of E_0 it becomes possible to get rid of CTCs in some part of space. Thus, choosing E_0 in such a way that $E = 0$ at some $y_0 > 2k$, we will obtain the normal sign $g_{33} < 0$ in some range of y around y_0.

Solution 3, Equation (44)

In the case $w > 0$, $k = 0$, with (44), it is convenient to use the coordinate $r = 1/(mx)$, and then we obtain

$$E = E_0 + \frac{2\omega_0}{r}, \qquad ds^2 = (dt - E d\varphi)^2 - \frac{dr^2}{m^2 r^2} - dz^2 - r^2 d\varphi^2, \tag{62}$$

and assuming $E_0 = 0$, we arrive at

$$g_{33} = -r^2 \left(1 - \frac{2\omega_0^2}{m^2}\right) = r^2 \frac{1 - w}{2w} > 0. \tag{63}$$

We again obtain a configuration with an incorrect signature, possessing CTCs at all r. However, as in the previous case, by choosing E_0 so that $E = 0$ at some $r = r_0$ we can provide a CTC-free region in a thick layer around $r = r_0$.

5. Wormholes

With the solution (45) for $r = e^\beta$, the range of x is $x \in (-\pi/(2\bar{k}), \pi/(2\bar{k}))$, where $\bar{k} = -k > 0$, and we see that $r \to \infty$ on both ends, confirming the wormhole nature of this configuration, where $x = 0$ is the wormhole throat (minimum of r). Substituting $y = \bar{k} \tan \bar{k}x$, we obtain the metric in the form

$$ds^2 = (dt - E d\varphi)^2 - \frac{dy^2}{m^2 (\bar{k}^2 + y^2)} - dz^2 - \frac{\bar{k}^2 + y^2}{m^2} d\varphi^2, \tag{64}$$

where $y \in \mathbb{R}$, and $y = 0$ is the throat; furthermore,

$$E = E_0 + \frac{2\omega_0}{m^2} y, \tag{65}$$

and for g_{33} in the case $E_0 = 0$ (which makes the solution symmetric with respect to $y = 0$) it follows

$$g_{33} = -\frac{\bar{k}^2}{m^2} + \frac{1 - w}{2w} \frac{y^2}{m^2}. \tag{66}$$

The expression (66) shows that CTCs are absent around the throat, at $y^2 < 2\bar{k}^2 w/(1 - w)$, while at larger $|y|$ the CTCs emerge.

Let us note that in the limit $w \to 1$, so that the fluid EoS tends to that of maximally stiff matter, the magnetic field disappears ($h \to 0$ according to (37)), and the whole solution tends to the one obtained in [39] for a cylindrical wormhole with stiff matter.

As always with rotating cylindrical wormhole solutions, these wormholes do not have a flat-space asymptotic behavior at large $|x|$, which makes it impossible to interpret them as objects that can be observed from regions with small curvature. To overcome this problem, it has been suggested [21] to cut out of the obtained wormhole solution a regular region, containing a throat, and to place it between two flat regions, thus making the whole system manifestly asymptotically flat. However, to interpret such a "sandwich" as a single space-time, it is necessary to identify the internal and external metrics on the junction surfaces Σ_+ and Σ_-, which should be common for these regions. The internal region will be described in the present case by (64), (65)). Furthermore, since the internal metric contains rotation, the external Minkowski metric should also be taken in a rotating reference frame.

Thus we take the Minkowski metric in cylindrical coordinates, $ds_M^2 = dt^2 - dX^2 - dz^2 - X^2 d\varphi^2$, and convert it to a rotating reference frame with angular velocity $\Omega = \text{const}$ by substituting $\varphi \to \varphi + \Omega t$, so that

$$ds_M^2 = dt^2 - dX^2 - dz^2 - X^2 (d\varphi + \Omega dt)^2. \tag{67}$$

In the notations of (1), the relevant quantities in (67) are

$$e^{2\gamma} = 1 - \Omega^2 X^2, \qquad e^{2\beta} = \frac{X^2}{1 - \Omega^2 X^2},$$

$$E = \Omega X^2, \qquad \omega = \frac{\Omega}{1 - \Omega^2 X^2}. \tag{68}$$

This stationary metric admits matching with the internal metric at any $|X| < 1/|\Omega|$, inside the "light cylinder" $|X| = 1/|\Omega|$ on which the linear rotational velocity coincides with the speed of light.

Making use of the symmetry of (64), let us assume that the internal region is $-y_* < y < y_*$, so that the junction surfaces $\Sigma_\pm$ are situated at $y = \pm y_*$, to be identified with $X_\pm = \pm X_*$ in Minkowski space, respectively, so that the external flat regions are $X < -X_*$ and $X > X_+$. Matching is achieved if we identify there the two metrics, so that

$$[\beta] = 0, \quad [\mu] = 0, \quad [\gamma] = 0, \quad [E] = 0, \tag{69}$$

where, as usual, the brackets $[f]$ denote a discontinuity of any function f across the surface. Under the conditions (69), we can suppose that the coordinates t, z, ϕ are the same in the whole space. At the same time, there is no need to adjust the choice of radial coordinates on different sides of the junction surfaces since the quantities involved in all matching conditions are insensitive to possible reparametrizations of y or X.

Having identified the metrics, we certainly did not adjust their normal derivatives, whose jumps are well known to determine the properties of matter filling a junction surface Σ and forming there a thin shell. The SET S_a^b of such a thin shell is calculated using the Darmois–Israel formalism [44–46], and in the present case of a timelike surface, S_a^b is related to the extrinsic curvature K_a^b of Σ as

$$S_a^b = -(8\pi G)^{-1}[\tilde{K}_a^b], \quad \tilde{K}_a^b := K_a^b - \delta_a^b K_c^c, \tag{70}$$

where the indices $a, b, c = 0, 2, 3$. The general expressions for nonzero components of $\tilde{K}_a^b$ for surfaces $x = \text{const}$ in the metric (1) are [39]

$$\begin{aligned}
\tilde{K}_{00} &= -e^{-\alpha + 2\gamma}(\beta' + \mu'), \\
\tilde{K}_{03} &= -\tfrac{1}{2}e^{-\alpha}E' + Ee^{-\alpha}(\beta' + \gamma' + \mu'), \\
\tilde{K}_{22} &= e^{-\alpha + 2\mu}(\beta' + \gamma'), \\
\tilde{K}_{33} &= e^{-\alpha + 2\beta}(\gamma' + \mu') + e^{-\alpha - 2\gamma}[EE' - E^2(\beta' + 2\gamma' + \mu')].
\end{aligned} \tag{71}$$

From (71) it is straightforward to find S_b^a on the surfaces $\Sigma_\pm$. However, our interest is not in finding these quantities themselves but, instead, a verification of whether or not the resulting SET S_b^a satisfies the WEC. Let us use for this purpose the necessary and sufficient conditions obtained in a general form in [38], see also a detailed description in [39]. These conditions are

$$a + c + \sqrt{(a - c)^2 + 4d^2} \geq 0, \tag{72}$$

$$a + c + \sqrt{(a - c)^2 + 4d^2} + 2b \geq 0, \tag{73}$$

$$a + c \geq 0, \tag{74}$$

where

$$\begin{aligned}
a &= -[e^{-\alpha}(\beta' + \mu')], \quad b = [e^{-\alpha}(\beta' + \gamma')], \\
c &= [e^{-\alpha}(\gamma' + \mu')], \quad d = -[\omega]
\end{aligned} \tag{75}$$

Let us discuss, for certainty, the conditions on $\Sigma_+ : y = y_*$, $X = X_*$ with our metrics (64), (65) and (67). Among the matching conditions (69), $[\mu] = 0$ holds automatically, while to satisfy the condition $[\gamma] = 0$ we will rescale the time coordinate in the internal region according to

$$t = \sqrt{P}\tilde{t}, \qquad P := 1 - \Omega^2 X^2 \tag{76}$$

and use the new coordinate $\tilde{t}$, with which, instead of E, we must use $E\sqrt{P}$ in all formulas. The remaining two conditions (69) yield

$$\Omega X^2 = \frac{2\omega_0}{m} y \sqrt{P}, \qquad \frac{k^2 + y^2}{m^2} = \frac{X^2}{P}, \tag{77}$$

where, without risk of confusion, we omit the asterisk at X and y. With these conditions, there are four independent parameters of the system, for example, we can choose as such parameters

$$X, \; y, \; P, \; n = \frac{2\omega_0^2}{m} = \frac{w+1}{2w}. \tag{78}$$

The other parameters are expressed in their terms as

$$\Omega = \frac{\sqrt{1-P}}{X}, \qquad \omega_0 = \frac{ny\sqrt{P}}{X\sqrt{1-P}}, \qquad k^2 = \frac{y^2(2n-1+P)}{1-P}. \tag{79}$$

Now we can calculate the quantities (75), with $[f] = f_{\text{out}} - f_{\text{in}}$ on Σ_+:

$$a = \frac{yP^{3/2} - 1}{PX}, \qquad b = \frac{1 - y\sqrt{P}}{X}, \qquad c = -\frac{1-P}{PX}, \qquad d = \frac{nyP^{3/2} - 1 + P}{PX\sqrt{1-P}}. \tag{80}$$

It can be easily verified that the conditions (72)–(74) are satisfied as long as

$$y \geq \frac{2-P}{P^{3/2}}, \tag{81}$$

in full analogy with the corresponding calculation in [39].

We have shown that under the condition (81) the WEC holds on Σ_+. Now, what changes on the surface Σ_- specified by $X = -X_* < 0$ and $y = -y_* < 0$, where we must take $[f] = f_{\text{out}} - f_{\text{in}}$ for any function f? As in [39], it can be verified that the parameters a, b, c do not change from (80) if we replace X with $|X|$ (we denote, as before, $y = y_* > 0$). For $d = -[\omega]$ there will be another expression since, according to (69), $\Omega(\Sigma_-) = -\Omega(\Sigma_+)$, while in the internal solution $\omega(\Sigma_-) = \omega(\Sigma_+)$, hence on Σ_-

$$d \mapsto d_- = -\frac{n|y|P^{3/2} + 1 - P}{|X|P\sqrt{1-P}},$$

so that $|d_-| > |d|$, making it even easier to satisfy the WEC requirements. As a result, the WEC holds under the same condition (81), providing a wormhole model which is completely phantom-free.

We can also notice that from (79) it follows $y_*^2 < k^2$, therefore, $y^2 < k^2$ in the whole internal region, which is thus free from CTCs.

There is one more point to bear in mind: since there is a z-directed magnetic field in the internal region, we must suppose that there are some surface currents on $\Sigma_\pm$ in the φ direction. Their values can be easily calculated using the Maxwell equations $\nabla_\nu F^{\mu\nu} = J^\mu$. Indeed, say, Σ_+ ($x = x_*$) separates the region where $F^{\mu\nu} = 0$ from the one with nonzero $F^{\mu\nu}$, therefore at their junction we have $J^\mu = -\delta(x - x_*)J^\mu(x_* - 0)$, so that the surface current is $J^a = -J^\mu(x_* - 0)\big|_{\mu=a}$. Similarly, on Σ_- ($x = -x_*$) we obtain $J^a = J^\mu(-x_* + 0)\big|_{\mu=a}$. In our wormhole configurations we obtain, according to (23), (24), (27) and taking into account that $\gamma = \mu \equiv 0$,

$$J^a = (J^0, 0, 0), \quad J^0(\Sigma_\pm) = \mp h\omega_0. \tag{82}$$

Thus the surface currents on $\Sigma_\pm$ have only the temporal component, i.e., they are comoving to the matter and current in the internal region.

As is the case with the internal wormhole solution, in the limit $w \to 1$ (hence $n \to 1$) the whole twice asymptotically flat construction tends to the one obtained in [39] with a stiff matter source.

6. Concluding Remarks

We have obtained a family of stationary cylindrically symmetric solutions to the Einstein–Maxwell equations in the presence of perfect fluids with $p = w\rho$, $|w| < 1$. Some of them (Solutions 1–3) contain a symmetry axis which can be made regular by properly choosing the solution parameters. The only geometry of closed type belongs to Solution 1, Equations (42) and (53)–(55). Unlike Melvin's solution and like all other solutions with rotation, it inevitably contains a region where $g_{33} > 0$, so that the coordinate circles parametrized by the angle φ are timelike, violating causality.

The wormhole models discussed here are of interest as new examples of phantom-free wormholes in general relativity, respecting the WEC. As in other known examples [38,39], such a result is achieved owing to the exotic properties of vortex gravitational fields with cylindrical symmetry, and their asymptotic behavior making them potentially observable from flat or weakly curved regions of space is provided by joining flat regions on both sides of the throat. Such a complex structure is necessary because asymptotic flatness at large circular radii cannot be achieved in any cylindrical solutions with rotation. The present family of models with a magnetic field, parametrized by the equation-of-state parameter $w < 1$, tends to the one obtained in [39] in the limit $w \to 1$, in which the magnetic field vanishes.

Let us mention that other static or stationary wormhole models with proper asymptotic behavior and matter sources respecting the WEC have been obtained in extensions of general relativity, such as the Einstein–Cartan theory [47,48], Einstein–Gauss–Bonnet gravity [49], multidimensional gravity including brane worlds [50,51], theories with nonmetricity [52], Horndeski theories [53], etc.

One can also notice that the same trick as was used with wormhole models, that is, joining a flat region taken in a rotating reference frame, can be used as well with solutions possessing a symmetry axis. It is important that in all such cases the surface to be used as a junction should not contain CTCs (in other words, there should be, as usual, $g_{33} < 0$) because $g_{33} < 0$ in the admissible part of flat space, while g_{33} taken from the external and internal regions should be identified at the junction. In this way one can obtain completely CTC-free models of extended cosmic strings with rotation.

A possible observer can be located far from such an extended string or wormhole configuration and be at rest in a nonrotating frame in flat space, other than the one used for the object construction. A question of interest is that of their observational appearance. If such a stringlike object does not emit radiation of its own, it can undoubtedly manifest itself by gravitational lensing in the same way as is discussed for cosmic strings (certainly if there is the corresponding angular deficit in the external, locally flat region); see, e.g., [54–56] and references therein. Moreover, possible signals scattered in the strong field region can carry information of interest on the nature and motion of rotating matter that forms such objects.

An evident further development of this study can be a search for other rotating configurations with electromagnetic fields, possibly including radiation in different directions in the spirit of [57], where radial, azimuthal and longitudinal radiation flows were considered as sources of gravity in space-times with the metric (1). Another set of problems concerns electrostatics in the fields of extended strings or wormholes with sources including electromagnetism. As follows from [58], even in simpler, partly conical cylindrical geometries with thin shells electrostatics turns out to be rather interesting and complex.

Author Contributions: All authors contributed equally to the manuscript. All authors have read and agreed to the published version of the manuscript.

Funding: This publication was supported by the RUDN University program 5-100 and by the Russian Foundation for Basic Research Project 19-02-00346. The work of K.B. was also performed within the framework of the Center FRPP supported by MEPhI Academic Excellence Project (contract No. 02.a03.21.0005, 27.08.2013). The work of V.K. and V.O. was supported by the Ministry of Education and Science of Russia, Grant FSFS-2020-0025.

References

1. Stephani, H.; Kramer, D.; MacCallum, M.A.H.; Hoenselaers, C.; Herlt, E. Exact solutions of Einstein's field equations. In *Cambridge Monographs on Mathematical Physics*; Cambridge University Press: Cambridge, UK, 2009.

2. Griffiths, J.B.; Podolsky, J. *Exact Space-Times in Einstein's General Relativity*; Cambride University Press: Cambridge, UK, 2009.

3. Bronnikov, K.A.; Santos, N.O.; Wang, A. Cylindrical systems in general relativity. *Class. Quantum Gravity* **2020**, *37*, 113002. [CrossRef]

4. Birch, P. Is the Universe Rotating? *Nature* **1982**, *298*, 451–454. [CrossRef]

5. Panov, V.F.; Pavelkin, V.N.; Kuvshinova, E.V.; Sandakova, O.V. *Cosmology with Rotation*; Perm University Press: Perm, Russia, 2016.

6. Sandakova, O.V.; Yanishevsky, D.M.; Panov, V.F. A cosmological scenario with rotation. *Grav. Cosmol.* **2019**, *25*, 362. [CrossRef]

7. Land, K.; Magueijo, J. The axis of evil. *Phys. Rev. Lett.* **2005**, *95*, 071301. [CrossRef]

8. Longo, M.J. Detection of a dipole in the handedness of spiral galaxies with redshifts $z \sim 0.04$. *Phys. Lett. B* **2011**, *699*, 224–229. [CrossRef]

9. Poltis, R.; Stojkovic, D. Can primordial magnetic fields seeded by electroweak strings cause an alignment of quasar axes on cosmological scales? *Phys. Rev. Lett.* **2010**, *105*, 161301. [CrossRef]

10. Yamazaki, D.G.; Kajino, T.; Mathews, G.J.; Ichiki, K. The search for a primordial magnetic field. *Phys. Rep.* **2012**, *517*, 141–167. [CrossRef]

11. Di Gioia, F.; Montani, G. Linear perturbations of an anisotropic Bianchi I model with a uniform magnetic field. *Eur. Phys. J. C* **2019**, *79*, 921. [CrossRef]

12. Bolokhov, S.V.; Bronnikov, K.A.; Skvortsova, M.V. Magnetic black universes and wormholes with a phantom scalar. *Class. Quantum Grav.* **2012**, *29*, 245006. [CrossRef]

13. Melvin, M.A. Pure magnetic and electric geons. *Phys. Lett.* **1964**, *8*, 65–68. [CrossRef]

14. Bronnikov, K.A. Static, cylindrically symmetric Einstein–Maxwell fields. In *Problems in Gravitation Theory and Particle Theory (PGTPT)*; 10th issue; Staniukovich, K.P., Ed.; Atomizdat: Moscow, Russia, 1979; pp. 37–50.

15. Bronnikov, K.A. Inverted black holes and anisotropic collapse. *Soviet Phys. J.* **1979**, *22*, 594–600. [CrossRef]

16. Lanczos, C. Ueber eine stationäre Kosmologie in Sinne der Einsteinischen Gravitationstheories. *Z. Physik* **1924**, *21*, 73. [CrossRef]

17. Lewis, T. Some special solutions of the equations of axially symmetric gravitational fields. *Proc. R. Soc. A* **1932**, *136*, 176.

18. Santos, N.O. Solution of the vacuum Einstein equations with nonzero cosmological constant for a stationary cylindrically symmetric spacetime. *Class. Quantum Grav.* **1993**, *10*, 2401. [CrossRef]

19. Krasinski, A. Solutions of the Einstein field equations for a rotating perfect fluid II: properties of the flow-stationary and and vortex-homogeneous solutions. *Acta Phys. Pol.* **1975**, *6*, 223.

20. MacCallum, M.A.H.; Santos, N.O. Stationary and static cylindrically symmetric Einstein spaces of the Lewis form. *Class. Quantum Grav.* **1998**, *15*, 1627. [CrossRef]

21. Bronnikov, K.A.; Krechet, V.G.; Lemos, P.S. Rotating cylindrical wormholes. *Phys. Rev. D* **2013**, *87*, 084060. [CrossRef]

22. Bronnikov, K.A.; Krechet, V.G. Rotating cylindrical wormholes and energy conditions. *Int. J. Mod. Phys. A* **2016**, *31*, 1641022. [CrossRef]

23. Erices, C.; Martinez, C. Stationary cylindrically symmetric spacetimes with a massless scalar field and a nonpositive cosmological constant. *Phys. Rev. D* **2015**, *92*, 044051. [CrossRef]

24. Van Stockum, W.J. The gravitational field of a distribution of particles rotating about an axis of symmetry. *Proc. R. Soc. Edinburgh A* **1937**, *57*, 135. [CrossRef]

25. Bonnor, W.B.; Steadman, B.R. A vacuum exterior to Maitra's cylindrical dust solution. *Gen. Rel. Grav.* **2009**, *41*, 1381. [CrossRef]

26. Ivanov, B.V. The General Double-dust Solution. *arXiv* **2002**, arXiv:gr-qc/0209032.

27. Ivanov, B.V. Rigidly rotating cylinders of charged dust. *Class. Quantum Gravity* **2002**, *19*, 5131. [CrossRef]

28. Santos, N.O.; Mondaini, R.P. Rigidly rotating relativistic generalized dust cylinder. *Nuovo Cim. B.* **1982**, *72*, 13. [CrossRef]

29. Hoenselaers, C.; Vishveshwara, C.V. A relativistically rotating fluid cylinder. *Nuovo Cim. B* **1979**, *10*, 43–51. [CrossRef]

30. Davidson, W. Barotropic perfect fluid in steady cylindrically symmetric rotation. *Class. Quantum Gravity* **1997**, *14*, 119. [CrossRef]

31. Davidson, W. A cylindrically symmetric stationary solution of Einstein's equations describing a perfect fluid of finite radius. *Class. Quantum Gravity* **2000**, *17*, 2499. [CrossRef]

32. Sklavenites, D. Stationary perfect fluid cylinders. *Class. Quantum Gravity* **1999**, *16*, 2753. [CrossRef]

33. Ivanov, B.V. On rigidly rotating perfect fluid cylinders. *Class. Quantum Gravity* **2002**, *19*, 3851. [CrossRef]

34. Letelier, P.S.; Verdaguer, E. Anisotropic fluid with SU(2) type structure in general relativity: A model of localized matter. *Int. Math. Phys.* **1987**, *28*, 2431. [CrossRef]

35. Herrera, L.; Denmat, G.L.; Marcilhacy, G.; Santos, N.O. Static cylindrical symmetry and conformal flatness. *Int. J. Mod. Phys. D* **2005**, *14*, 657. [CrossRef]

36. Debbasch, F.; Herrera, L.; Pereira, P.R.C.T.; Santos, N.O. Stationary cylindrical anisotropic fluid. *Gen. Rel. Grav.* **2006**, *38*, 1825. [CrossRef]

37. Bolokhov, S.V.; Bronnikov, K.A.; Skvortsova, M.V. Rotating cylinders with anisotropic fluids in general relativity. *Gravit. Cosmol.* **2019**, *25*, 122. [CrossRef]

38. Bronnikov, K.A.; Krechet, V.G. Potentially observable cylindrical wormholes without exotic matter in GR. *Phys. Rev. D* **2019**, *99*, 084051. [CrossRef]

39. Bronnikov, K.A.; Bolokhov, S.V.; Skvortsova, M.V. Cylindrical wormholes: a search for viable phantom-free models in GR. *Int. J. Mod. Phys. D* **2019**, *28*, 1941008. [CrossRef]

40. Krechet, V.G. Topological and physical effects of rotation and spin in the general relativistic theory of gravitation. *Russ. Phys. J.* **2007**, *50*, 1021. [CrossRef]

41. Krechet, V.G.; Sadovnikov, D.V. Spin-spin interaction in general relativity and induced geometries with nontrivial topology. *Gravit. Cosmol.* **2009**, *15*, 337. [CrossRef]

42. Astorino, M. Charging axisymmetric space-times with cosmological constant. *J. High Energy Phys.* **2012**, *6*, 86. [CrossRef]

43. Bronnikov, K.A.; Rubin, S.G. *Black Holes, Cosmology and Extra Dimensions*; World Scientific: Singapore, 2013.

44. Darmois, G. Les équations de la gravitation einsteinienne. In *Mémorial des Sciences Mathematiques*; Gauthier-Villars: Paris, France, 1927; Volume 25.

45. Israel, W. Singular hypersurfaces and thin shells in general relativity. *Nuovo Cim. B* **1967**, *48*, 463. [CrossRef]

46. Berezin, V.A.; Kuzmin, V.A.; Tkachev, I.I. Dynamics of bubbles in general relativity. *Phys. Rev. D* **1987**, *36*, 2919. [CrossRef]

47. Bronnikov, K.A.; Galiakhmetov, A.M. Wormholes without exotic matter in Einstein-Cartan theory. *Gravit. Cosmol.* **2015**, *21*, 283. [CrossRef]

48. Bronnikov, K.A.; Galiakhmetov, A.M. Wormholes and black universes without phantom fields in Einstein-Cartan theory. *Phys. Rev. D* **2016**, *94*, 124006. [CrossRef]

49. Maeda, H.; Nozawa, M. Static and symmetric wormholes respecting energy conditions in Einstein-Gauss-Bonnet gravity. *Phys. Rev. D* **2008**, *78*, 024005. [CrossRef]

50. Bronnikov, K.A.; Kim, S.-W. Possible wormholes in a brane world. *Phys. Rev. D* **2003**, *67*, 064027. [CrossRef]

51. Bronnikov, K.A.; Skvortsova, M.V. Wormholes leading to extra dimensions. *Grav. Cosmol.* **2016**, *22*, 316. [CrossRef]

52. Krechet, V.G.; Oshurko, V.B.; Lodi, M.N. Induced nonlinearities of the scalar field and wormholes in the metric-affine theory of gravity. *Grav. Cosmol.* **2018**, *24*, 186–190. [CrossRef]

53. Sushkov, S.V.; Korolev, R.V. Scalar wormholes with nonminimal derivative coupling. *Class. Quantum Grav.* **2012**, *29*, 085008. [CrossRef]

54. Huterer, D.; Vachaspati, T. Gravitational lensing by cosmic strings in the era of wide-field surveys. *Phys. Rev. D* **2003**, *68*, 041301. [CrossRef]

55. Sazhin, M.V.; Khovanskaya, O.S.; Capaccioli, M.; Longo, G.; Paolillo, M.; Covone, G.; Grogin, N.A.; Schreier, E.J. Gravitational lensing by cosmic strings: what we learn from the CSL-1 case. *Mon. Not. R. Astron. Soc.* **2007**, *376*, 1731–1739. [CrossRef]

56. Fernández-Núñez, I.; Bulashenko, O. Emergence of Fresnel diffraction zones in gravitational lensing by a cosmic string. *Phys. Lett. A* **2017**, *381*, 1764–1772. [CrossRef]

57. Bronnikov, K.A. String clouds and radiation flows as sources of gravity in static or rotating cylinders. *Int. J. Mod. Phys. A* **2020**, *35*, 2040004. [CrossRef]

58. Rubín de Celis, E.; Simeone, C. Electrostatics and self-force in asymptotically flat cylindrical wormholes. *Eur. Phys. J. C* **2020**, *80*, 501. [CrossRef]

3

Tests of Lorentz Symmetry in the Gravitational Sector

Aurélien Hees [1,*], Quentin G. Bailey [2], Adrien Bourgoin [3], Hélène Pihan-Le Bars [3], Christine Guerlin [3,4] and Christophe Le Poncin-Lafitte [3]

[1] Department of Physics and Astronomy, University of California, Los Angeles, CA 90095, USA
[2] Physics Department, Embry-Riddle Aeronautical University, Prescott, AZ 86301, USA; baileyq@erau.edu
[3] SYRTE, Observatoire de Paris, PSL Research University, CNRS, Sorbonne Universités, UPMC Univ. Paris 06, LNE, 61 avenue de l'Observatoire, 75014 Paris, France; adrien.bourgoin@obspm.fr (A.B.); helene.pihan-lebars@obspm.fr (H.P.-L.B.); christine.guerlin@obspm.fr (C.G.); christophe.leponcin@obspm.fr (C.L.P.-L.)
[4] Laboratoire Kastler Brossel, ENS-PSL Research University, CNRS, UPMC-Sorbonne Universités, Collège de France, 75005 Paris, France
[*] Correspondence: ahees@astro.ucla.edu.

Academic Editors: Lorenzo Iorio and Elias C. Vagenas

Abstract: Lorentz symmetry is one of the pillars of both General Relativity and the Standard Model of particle physics. Motivated by ideas about quantum gravity, unification theories and violations of CPT symmetry, a significant effort has been put the last decades into testing Lorentz symmetry. This review focuses on Lorentz symmetry tests performed in the gravitational sector. We briefly review the basics of the pure gravitational sector of the Standard-Model Extension (SME) framework, a formalism developed in order to systematically parametrize hypothetical violations of the Lorentz invariance. Furthermore, we discuss the latest constraints obtained within this formalism including analyses of the following measurements: atomic gravimetry, Lunar Laser Ranging, Very Long Baseline Interferometry, planetary ephemerides, Gravity Probe B, binary pulsars, high energy cosmic rays, ... In addition, we propose a combined analysis of all these results. We also discuss possible improvements on current analyses and present some sensitivity analyses for future observations.

Keywords: experimental tests of gravitational theories; Lorentz and Poincaré invariance; modified theories of gravity; celestial mechanics; atom interferometry; binary pulsars

1. Introduction

The year 2015 was the centenary of the theory of General Relativity (GR), the current paradigm for describing the gravitational interaction (see e.g., the Editorial of this special issue [1]). Since its creation, this theory has passed all experimental tests with flying colors [2,3] ; the last recent success was the discovery of gravitational waves [4], summarized in [5]. On the other hand, the three other fundamental interactions of Nature are described within the Standard Model of particle physics, a framework based on relativistic quantum field theory. Although very successful so far, it is commonly admitted that these two theories are not the ultimate description of Nature but rather some effective theories. This assumption is motivated by the construction of a quantum theory of gravitation that has not been successfully developed so far and by the development of a theory that would unify all the fundamental interactions. Moreover, observations requiring the introduction of Dark Matter and Dark Energy also challenge GR and the Standard Model of particle physics since they cannot be explained by these two paradigms altogether [6]. It is therefore extremely important to test our current description of the four fundamental interactions [7].

Lorentz invariance is one of the fundamental symmetry of relativity, one of the corner stones of both GR and the Standard Model of particle physics. It states that the outcome of any local experiment is independent of the velocity and of the orientation of the laboratory in which the experiment is performed. If one considers non-gravitational experiments, Lorentz symmetry is part of the Einstein Equivalence Principle (EEP). A breaking of Lorentz symmetry implies that the equations of motion, the particle thresholds, etc. . . may be different when the experiment is boosted or rotated with respect to a background field [8]. More precisely, it is related to a violation of the invariance under "particle Lorentz transformations" [8] which are the boosts and rotations that relate the properties of two systems within a specific oriented inertial frame (or in other words they are boosts and rotations on localized fields but not on background fields). On the other hand, the invariance under coordinates transformations known as "observer Lorentz transformations" [8] which relate observations made in two inertial frames with different orientations and velocities is always preserved. Considering the broad field of applicability of this symmetry, searches for Lorentz symmetry breaking provide a powerful test of fundamental physics. Moreover, it has been suggested that Lorentz symmetry may not be a fundamental symmetry of Nature and may be broken at some level. While some early motivations came from string theories [9–11], breaking of Lorentz symmetry also appears in loop quantum gravity [12–15], non commutative geometry [16,17], multiverses [18], brane-world scenarios [19–21] and others (see for example [22,23]).

Tests of Lorentz symmetry have been performed since the time of Einstein but the last decades have seen the number of tests increased significantly [24] in all fields of physics. In particular, a dedicated effective field theory has been developed in order to systematically consider all hypothetical violations of the Lorentz invariance. This framework is known as the Standard-Model Extension (SME) [8,25] and covers all fields of physics. It contains the Standard Model of particle physics, GR and all possible Lorentz-violating terms that can be constructed at the level of the Lagrangian, introducing a large numbers of new coefficients that can be constrained experimentally.

In this review, we focus on the gravitational sector of the SME which parametrizes deviations from GR. GR is built upon two principles [2,26,27]: (i) the EEP; and (ii) the Einstein field equations that derive from the Einstein-Hilbert action. The EEP gives a geometric nature to gravitation allowing this interaction to be described by spacetime curvature. From a theoretical point of view, the EEP implies the existence of a spacetime metric to which all matter minimally couples [28]. A modification of the matter part of the action will lead to a breaking of the EEP. In SME, such a breaking of the EEP is parametrized (amongst others) by the matter-gravity coupling coefficients $\bar{a}_\mu$ and $\bar{c}_{\mu\nu}$ [29,30]. From a phenomenological point of view, the EEP states that [2,27]: (i) the universality of free fall (also known as the weak equivalence principle) is valid; (ii) the outcome of any local non-gravitational experiment is independent of the velocity of the free-falling reference frame in which it is performed; and (iii) the outcome of any local non-gravitational experiment is independent of where and when in the universe it is performed. The second part of Einstein theory concerns the purely gravitational part of the action (the Einstein-Hilbert action) which is modified in SME to introduce hypothetical Lorentz violations in the gravitational sector. This review focuses exclusively on this kind of Lorentz violations and not on breaking of the EEP.

A lot of tests of GR have been performed in the last decades (see [2] for a review). These tests rely mainly on two formalisms: the parametrized post-Newtonian (PPN) framework and the fifth force formalism. In the former one, the weak gravitational field spacetime metric is parametrized by 10 dimensionless coefficients [27] that encode deviations from GR. This formalism therefore provides a nice interface between theory and experiments. The PPN parameters have been constrained by a lot of different observations [2] confirming the validity of GR. In particular, three PPN parameters encode violations of the Lorentz symmetry: the $\alpha_{1,2,3}$ PPN coefficients. In the fifth force formalism, one is looking for a deviation from Newtonian gravity where the gravitational potential takes the form of a Yukawa potential characterized by a length λ and a strength α of interaction [31–34]. These two parameters are very well constrained as well except at very small and large distances (see [35]).

The gravitational sector of SME offers a new framework to test GR by parametrizing deviations from GR at the level of the action, introducing new terms that are breaking Lorentz symmetry. The idea is to extend the standard Einstein-Hilbert action by including Lorentz-violating terms constructed by contracting new fields with some operators built from curvature tensors and covariant derivatives with increasing mass dimension [36]. The lower mass dimension (dimension 4) term is known as the minimal SME and its related new fields can be split into a scalar part u, a symmetric trace free part $s^{\mu\nu}$ and a traceless piece $t^{\kappa\lambda\mu\nu}$. In order to avoid conflicts with the underlying Riemann geometry, the Lorentz violating coefficients can be assumed to be dynamical fields and the Lorentz violation to arise from a spontaneous symmetry breaking [37–42]. The Lorentz violating fields therefore acquire a non-vanishing vacuum expectation value (denoted by a bar). It has been shown that in the linearized gravity limit the fluctuations around the vacuum values can be integrated out so that only the vacuum expectation values of the SME coefficients influence observations [39]. In the minimal SME, the coefficient $\bar{u}$ corresponds to a rescaling of the gravitational constant and is therefore unobservable and the coefficients $\bar{t}^{\kappa\lambda\mu\nu}$ do not play any role at the post-Newtonian level, a surprising phenomenon known as the t-puzzle [43,44]. The $\bar{s}^{\mu\nu}$ coefficients lead to modifications from GR that have thoroughly been investigated in [39]. In particular, the SME framework extends standard frameworks such as the PPN or fifth force formalisms meaning that "standard" tests of GR cannot directly be translated into this formalism.

In the last decade, several measurements have been analyzed within the gravitational sector of the minimal SME framework: Lunar Laser Ranging (LLR) analysis [45,46], atom interferometry [47,48], planetary ephemerides analysis [49,50], short-range gravity [51], Gravity Probe B (GPB) analysis [52], binary pulsars timing [53,54], Very Long Baseline Interferometry (VLBI) analysis [55] and Čerenkov radiation [56]. In addition to the minimal SME, there exist some higher order Lorentz-violating curvature couplings in the gravity sector [43] that are constrained by short-range experiments [57–59], Čerenkov radiation [30,56] and gravitational waves analysis [60,61]. Finally, some SME experiments have been used to derive bounds on spacetime torsion [62,63]. A review for these measurements can be found in [30]. The classic idea to search for or to constrain Lorentz violations in the gravitational sector is to search for orientation or boost dependence of an observation. Typically, one will take advantage of modulations that will occur through an orientation dependence of the observations due to the Earth's rotation, the motion of satellites around Earth (the Moon or artificial satellites), the motion of the Earth (or other planets) around the Sun, the motion of binary pulsars, …The main goal of this communication is to review all the current analyses performed in order to constrain Lorentz violation in the pure gravitational sector.

Two distinct procedures have been used to analyze data within the SME framework. The first procedure consists in deriving analytically the signatures produced by the SME coefficients on some observations. Then, the idea is to fit these signatures within residuals obtained by a data analysis performed in pure GR. This approach has the advantage to be relatively easy and fast to perform. Nevertheless, when using this postfit approach, correlations with other parameters fitted in the data reduction are completely neglected and may lead to overoptimistic results. A second way to analyze data consists in introducing the Lorentz violating terms directly in the modeling of observables and in the global data reduction. In this review, we highlight the differences between the two approaches.

In this communication, a brief theoretical review of the SME framework in the gravitational sector is presented in Section 2. The two different approaches to analyze data within the SME framework (postfit analysis versus full modeling of observables within the SME framework) are discussed and compared in Section 3. Section 4 is devoted to a discussion of the current measurements analyzed within the SME framework. This discussion includes a general presentation of the measurements, a brief review of the effects of Lorentz violation on each of them, the current analyses performed with real data and a critical discussion. A "grand fit" combining all existing analyses is also presented. In Section 5, some future measurements that are expected to improve the current analyses are developed. Finally, our conclusion is presented in Section 6.

2. The Standard-Model Extension in the Gravitational Sector

Many of the tests of Lorentz and CPT symmetry have been analyzed within an effective field theory framework which generically describes possible deviations from exact Lorentz and CPT invariance [8,25] and contains some traditional test frameworks as limiting cases [64,65]. This framework is called, for historical reasons, the Standard-Model Extension (SME). One part of the activity has been a resurgence of interest in tests of relativity in the Minkowski spacetime context, where global Lorentz symmetry is the key ingredient. Numerous experimental and observational constraints have been obtained on many different types of hypothetical Lorentz and CPT symmetry violations involving matter [24]. Another part, which has been developed more recently, has seen the SME framework extended to include the curved spacetime regime [37]. Recent work shows that there are many ways in which the spacetime symmetry foundations of GR can be tested [29,39].

In the context of effective field theory in curved spacetime, violations of these types can be described by an action that contains the usual Einstein-Hilbert term of GR, a matter action, plus a series of terms describing Lorentz violation for gravity and matter in a generic way. While the fully general coordinate invariant version of this action has been studied in the literature, we focus on a limiting case that is valid for weak-field gravity and can be compactly displayed. Using an expansion of the spacetime metric around flat spacetime, $g_{\mu\nu} = \eta_{\mu\nu} + h_{\mu\nu}$, the effective Lagrange density to quadratic order in $h_{\mu\nu}$ can be written in a compact form as

$$\mathcal{L} = \mathcal{L}_{\text{EH}} + \frac{c^3}{32\pi G} h^{\mu\nu} \bar{s}^{\alpha\beta} \mathcal{G}_{\alpha\mu\nu\beta} + ..., \tag{1}$$

where $\mathcal{L}_{\text{EH}}$ is the standard Einstein-Hilbert term, $\mathcal{G}_{\alpha\mu\nu\beta}$ is the double dual of the Einstein tensor linearized in $h_{\mu\nu}$, G the bare Newton constant and c the speed of light in a vacuum. The Lorentz-violating effects in this expression are controlled by the 9 independent coefficients in the traceless and dimensionless $\bar{s}^{\mu\nu}$ [39]. These coefficients are treated as constants in asymptotically flat cartesian coordinates. The ellipses represent additional terms in a series including terms that break CPT symmetry for gravity; such terms are detailed elsewhere [43,56,60] and are part of the so-called nonminimal SME expansion. Note that the process by which one arrives at the effective quadratic Lagrangian (1) is consistent with the assumption of the spontaneous breaking of local Lorentz symmetry, which is discussed below.

Also of interest are the matter-gravity couplings. This form of Lorentz violation can be realized in the classical point-mass limit of the matter sector. In the minimal SME the point-particle action can be written as

$$S_{\text{Matter}} = \int d\lambda \, c \left(-m\sqrt{-(g_{\mu\nu} + 2c_{\mu\nu})u^\mu u^\nu} - a_\mu u^\mu \right), \tag{2}$$

where the particle's worldline tangent is $u^\mu = dx^\mu/d\lambda$ [29]. The coefficients controlling local Lorentz violation for matter are $c_{\mu\nu}$ and a_μ. In contrast to $\bar{s}^{\mu\nu}$, these coefficients depend on the type of point mass (particle species) and so they can also violate the EEP. When the coefficients $\bar{s}_{\mu\nu}$, $c_{\mu\nu}$, and a_μ vanish perfect local Lorentz symmetry for gravity and matter is restored. It is also interesting to mention that this action with fixed (but not necessarily constant) a_μ and $c_{\mu\nu}$ represents motion in a Finsler geometry [66,67].

It has been shown that explicit local Lorentz violation is generically incompatible with Riemann geometry [37]. One natural way around this is assumption of spontaneous Lorentz-symmetry breaking. In this scenario, the tensor fields in the underlying theory acquire vacuum expectation values through a dynamical process. Much of the literature has been devoted to studying this possibility in the last decades [9,38,68–78], including some original work on spontaneous Lorentz-symmetry breaking in string field theory [10,11]. For the matter-gravity couplings in Equation (2), the coefficient fields $c_{\mu\nu}$, and a_μ are then expanded around their background (or vacuum) values $\bar{c}_{\mu\nu}$, and $\bar{a}_\mu$. Both a modified spacetime metric $g_{\mu\nu}$ and modified point-particle equations of motion result from the spontaneous breaking of Lorentz symmetry. In the linearized gravity limit these results rely only on the vacuum values $\bar{c}_{\mu\nu}$, and $\bar{a}_\mu$. The dominant signals for Lorentz violation controlled by these coefficients are revealed in the calculation of observables in the post-Newtonian limit.

Several novel features of the post-Newtonian limit arise in the SME framework. It was shown in Ref. [39] that a subset of the $\bar{s}^{\mu\nu}$ coefficients can be matched to the PPN formalism [2,27], but others lie outside it. For example, a dynamical model of spontaneous Lorentz symmetry breaking can be constructed from an antisymmetric tensor field $B_{\mu\nu}$ that produces $\bar{s}^{\mu\nu}$ coefficients that cannot be reduced to an isotropic diagonal form in any coordinate system, thus lying outside the PPN assumptions [78]. We can therefore see that the SME framework has a partial overlap with the PPN framework, revealing new directions to explore in analysis via the $\bar{s}^{\mu\nu}$, $\bar{c}_{\mu\nu}$, and $\bar{a}_\mu$ coefficients. The equations of motion for matter are modified by the matter-gravity coefficients for Lorentz violation $\bar{c}_{\mu\nu}$ and $\bar{a}_\mu$, which can depend on particle species, thus implying that these coefficients also control EEP violations. One potentially important class of experiments from the action (2) concerns the Universality of Free Fall of antimatter whose predictions are discussed in [29,79]. In addition, the post-Newtonian metric itself receives contributions from the matter coefficients $\bar{c}_{\mu\nu}$ and $\bar{a}_\mu$. So for example, two (chargeless) sources with the same total mass but differing composition will yield gravitational fields of different strength.

For solar-system gravity tests, the primary effects due to the nine coefficients $\bar{s}^{\mu\nu}$ can be obtained from the post-Newtonian metric and the geodesic equation for test bodies. A variety of ground-based and space-based tests can measure these coefficients [80–82]. Such tests include Earth-laboratory tests with gravimeters, lunar and satellite laser ranging, studies of the secular precession of orbital elements in the solar system, and orbiting gyroscope experiments, and also classic effects such as the time delay and bending of light around the Sun and Jupiter. Furthermore, some effects described by the Lagrangian (1) can be probed by analyzing data from binary pulsars and measurements of cosmic rays [56].

For the matter-gravity coefficients $\bar{c}_{\mu\nu}$ and $\bar{a}_\mu$, which break Lorentz symmetry and EEP, several experiments can be used for analysis in addition to the ones already mentioned above including ground-based gravimeter and WEP experiments. Dedicated satellite EEP tests are among the most sensitive where the relative acceleration of two test bodies of different composition is the observable of interest. Upon relating the satellite frame coefficients to the standard Sun-centered frame used for the SME, oscillations in the acceleration of the two masses occur at a number of different harmonics of the satellite orbital and rotational frequencies, as well as the Earth's orbital frequency. Future tests of particular interest include the currently flying MicroSCOPE experiment [83,84].

While the focus of the discussion to follow are the results for the minimal SME coefficients $\bar{s}^{\mu\nu}$, recent work has also involved the nonminimal SME coefficients in the pure-gravity sector associated with mass dimension 5 and 6 operators. One promising testing ground for these coefficients is sensitive short-range gravity experiments. The Newtonian force between two test masses becomes modified in the presence of local Lorentz violation by an anisotropic quartic force that is controlled by a subset of coefficients from the Lagrangian organized as the totally symmetric $(\bar{k}_{\text{eff}})_{jklm}$, which has dimensions of length squared [43]. This contains 14 measurable quantities and any one short-range experiment is sensitive to 8 of them. Two key experiments, from Indiana University and Huazhong University of Science and Technology, have both reported analysis in the literature [57,58] . A recent work combines the two analyses to place new limits on all 14, a priori independent, $(\bar{k}_{\text{eff}})_{jklm}$ coefficients [59]. Other higher mass dimension coefficients play a role in gravitational wave propagation [60] and gravitational Čerenkov radiation [56].

To conclude this section, we ask: what can be said about the possible sizes of the coefficients for Lorentz violation? A broad class of hypothetical effects is described by the SME effective field theory framework, but it is a test framework and as such does not make specific predictions concerning the sizes of these coefficients. One intriguing suggestion is that there is room in nature for violations of spacetime symmetry that are large compared to other sectors due to the intrinsic weakness of gravity. Considering the current status of the coefficients $\bar{s}^{\mu\nu}$, the best laboratory limits are at the 10^{-10}–10^{-11} level, with improvements of four orders of magnitude in astrophysical tests on these coefficients [56]. However, the limits are at the $10^{-8}\,\text{m}^2$ level for the mass dimension 6 coefficients

$(\bar{k}_{\text{eff}})_{jklm}$ mentioned above. Comparing this to the Planck length 10^{-35} m, we see that symmetry breaking effects could still have escaped detection that are not Planck suppressed. This kind of "countershading" was first pointed out for the $\bar{a}_\mu$ coefficients [85], which, having dimensions of mass, can still be as large as a fraction of the electron mass and still lie within current limits.

In addition, any action-based model that breaks local Lorentz symmetry either explicitly or spontaneously can be matched to a subset of the SME coefficients. Therefore, constraints on SME coefficients can directly constrain these models. Matches between various toy models and coefficients in the SME have been achieved for models that produce effective $\bar{s}^{\mu\nu}$, $\bar{c}_{\mu\nu}$, $\bar{a}_\mu$, and other coefficients. This includes vector and tensor field models of spontaneous Lorentz-symmetry breaking [29,39,75–78], models of quantum gravity [12,65] and noncommutative quantum field theory [17]. Furthermore, Lorentz violations may also arise in the context of string field theory models [86].

3. Postfit Analysis Versus Full Modeling

Since the last decade, several studies aimed to find upper limits on SME coefficients in the gravitational sector. A lot of these studies are based on the search of possible signals in post-fit residuals of experiments. This was done with LLR [45], GPB [52], binary pulsars [53,54] or Solar System planetary motions [49,50]. However, two new works focused on a direct fit to data with LLR [46] and VLBI [55], which are more satisfactory.

Indeed, in the case of a post-fit analysis, a simple modeling of extra terms containing SME coefficients are least square fitted in the residuals, attempting to constrain the SME coefficients of a testing function in residual noise obtained from a pure GR analysis, where of course Lorentz symmetry is assumed. It comes out correlations between SME coefficients and other global parameters previously fitted (masses, position and velocity...) cannot be assessed in a proper way. In others words, searching hypothetical SME signals in residuals, i.e., in noise, can lead to an overestimated formal error on SME coefficients, as illustrated in the case of VLBI [55], and without any chance to learn something about correlations with other parameters, as for example demonstrated in the case of LLR [46]. Let us consider the VLBI example to illustrate this fact. The VLBI analysis is described in Section 4.2. Including the SME contribution within the full VLBI modeling and estimating the SME coefficient $\bar{s}^{TT}$ altogether with the other parameters fitted in standard VLBI data reduction leads to the estimate $\bar{s}^{TT} = (-5 \pm 8) \times 10^{-5}$. A postfit analysis performed by fitting the SME contribution within the VLBI residuals obtained after a pure GR analysis leads to $\bar{s}^{TT} = (-0.6 \pm 2.1) \times 10^{-8}$ [55]. This example shows that a postfit analysis can lead to results with overoptimistic uncertainties and one needs to be extremely careful when using such results.

4. Data Analysis

In this section, we will review the different measurements that have already been used in order to constrain the SME coefficients. The different analyses are based on quite different types of observations. In order to compare all the corresponding results, we need to report them in a canonical inertial frame. The standard canonical frame used in the SME framework is a Sun-centered celestial equatorial frame [64], which is approximately inertial over the time scales of most observations. This frame is asymptotically flat and comoving with the rest frame of the Solar System. The cartesian coordinates related to this frame are denoted by capital letters

$$X^\Xi = (cT, X^J) = (cT, X, Y, Z).\tag{3}$$

The Z axis is aligned with the rotation axis of the Earth, while the X axis points along the direction from the Earth to the Sun at vernal equinox. The origin of the coordinate time T is given by the time when the Earth crosses the Sun-centered X axis at the vernal equinox. These conventions are depicted in Figure 2 from [39].

In the following subsections, we will present the different measurements used to constrain the SME coefficients. Each subsection contains a brief description of the principle of the experiment, how it can be used to search for Lorentz symmetry violations, what are the current best constraints obtained with such measurements and eventually how it can be improved in the future.

4.1. Atomic Gravimetry

The most sensitive experiments on Earth searching for Lorentz Invariance Violation (LIV) in the minimal SME gravity sector are gravimeter tests. As Earth rotates, the signal recorded in a gravimeter, i.e., the apparent local gravitational acceleration g of a laboratory test body, would be modulated in the presence of LIV in gravity. This was first noted by Nordtvedt and Will in 1972 [87] and used soon after with gravimeter data to constrain preferred-frame effects in the PPN formalism [88,89] at the level of 10^{-3}.

This test used a superconducting gravimeter, based on a force comparison (the gravitational force is counter-balanced by an electromagnetic force maintaining the test mass at rest). While superconducting gravimeters nowadays reach the best sensitivity on Earth, force comparison gravimeters intrinsically suffer from drifts of their calibration factor (with e.g., aging of the system). Development of other types of gravimeters has evaded this drawback: free fall gravimeters. Monitoring the motion of a freely falling test mass, they provide an absolute measurement of g. State-of-the art free fall gravimeters use light to monitor the mass free fall. Beyond classical gravimeters that drop a corner cube, the development of atom cooling and trapping techniques and atom interferometry has led to a new generation of free fall gravimeters, based on a quantum measurement: atomic gravimeters.

Atomic gravimeters use atoms in gaseous phase as a test mass. The atoms are initially trapped with magneto-optical fields in vacuum, and laser cooled (down to 100 nK) in order to control their initial velocity (down to a few mm/s). The resulting cold atom gas, containing typically a million atoms, is then launched or dropped for a free fall measurement. Manipulating the electronic and motional state of the atoms with two counterpropagating lasers, it is possible to measure, using atom interferometry, their free fall acceleration with respect to the local frame defined by the two lasers [90]. This sensitive direction is aligned to be along the local gravitational acceleration noted $\hat{z}$; the atom interferometer then measures the phase $\varphi = ka^{\hat{z}}T^2$, where T is half the interrogation time, $k \simeq 2(2\pi/\lambda)$ with λ the laser wavelength, and $a^{\hat{z}}$ is the free fall acceleration along the laser direction. The free fall time is typically on the order of 500 ms, corresponding to a free fall distance of about a meter. A new "atom preparation—free fall—detection" cycle is repeated every few seconds. Each measurement is affected by white noise, but averaging leads to a typical sensitivity on the order of or below $10^{-9}\,g$ [91–93].

Such an interferometer has been used by H. Müller et al. in [47] and K. Y. Chung et al. in [48] for testing Lorentz invariance in the gravitational sector with Caesium atoms, leading to the best terrestrial constraints on the $\bar{s}^{\mu\nu}$ coefficients. The analysis uses three data sets of respectively 2.5 days for the first two and 10 days for the third, stretched over 4 years, which allows one to observe sidereal and annual LIV signatures. The gravitational SME model used for this analysis can be found in [39,47,48]; its derivation will be summarized hereunder. Since the atoms in free fall are sensitive to the local phase of the lasers, LIV in the interferometer observable could also come from the pure electromagnetic sector. This contribution has been included in the experimental analysis in [48]. Focusing here on the gravitational part of SME, we ignore it in the following.

The gravitational LIV model adjusted in this test restricts to modifications of the Earth-atom two-body gravitational interaction. The Lagrangian describing the dynamics of a test particle at a point on the Earth's surface can be approximated by a post-Newtonian series as developed in [39]. At the Newtonian approximation, the two bodies Lagrangian is given by

$$\mathcal{L} = \frac{1}{2}mV^2 + G_N\frac{Mm}{R}\left(1 + \frac{1}{2}\bar{s}_t^{JK}\hat{R}^J\hat{R}^K - \frac{3}{2}\bar{s}^{TJ}\frac{V^J}{c} - \bar{s}^{TJ}\hat{R}^J\frac{V^K}{c}\hat{R}^k\right), \tag{4}$$

where $\boldsymbol{R}$ and $\boldsymbol{V}$ are the position and velocity expressed in the standard SME Sun-centered frame and $\hat{\boldsymbol{R}} = \boldsymbol{R}/R$ with $R = |\boldsymbol{R}|$. In addition, we have introduced G_N the observed Newton constant measured by considering the orbital motion of bodies and defined by (see also [39,50] or Section IV of [52])

$$G_N = G \left(1 + \frac{5}{3}\bar{s}^{TT}\right) ,\tag{5}$$

and the 3-dimensional traceless tensor

$$\bar{s}_t^{JK} = \bar{s}^{JK} - \frac{1}{3}\bar{s}^{TT}\delta^{JK} .\tag{6}$$

From this Lagrangian one can derive the equations of motion of the free fall mass in a laboratory frame (see the procedure in Section V.C.1. from [39]). It leads to the modified local acceleration in the presence of LIV [39] given by

$$a^{\hat{z}} = g \left(1 - \frac{1}{6}i_4\bar{s}^{TT} + \frac{1}{2}i_4\bar{s}^{\hat{z}\hat{z}}\right) - \omega_\oplus^2 R_\oplus \sin^2\chi - gi_4\bar{s}^{T\hat{z}}\beta_\oplus^{\hat{z}} - 3gi_1\bar{s}^{TJ}\beta_\oplus^J ,\tag{7}$$

where $g = G_N M_\oplus / R_\oplus^2$, $\omega_\oplus$ is the Earth's angular velocity, $\beta_\oplus = \frac{V_\oplus}{c} \sim 10^{-4}$ is the Earth's boost, $R_\oplus$ is the Earth radius, $M_\oplus$ is the Earth mass and χ the colatitude of the lab whose reference frame's $\hat{z}$ direction is the sensitive axis of the instrument as previously defined here. This model includes the shape of the Earth through its spherical moment of inertia $I_\oplus$ which appears in $i_\oplus = \frac{I_\oplus}{M_\oplus R_\oplus^2}$, $i_1 = 1 + \frac{1}{3}i_\oplus$ and $i_4 = 1 - 3i_\oplus$. In [48], Earth has been approximated as spherical and homogeneous leading to $i_\oplus = \frac{1}{2}$, $i_1 = \frac{7}{6}$ and $i_4 = -\frac{1}{2}$.

The sensing direction of the experiment precesses around the Earth rotation axis with sidereal period, and the lab velocity varies with sidereal period and annual period. At first order in $V_\oplus$ and $\omega_\oplus$ and as a function of the SME coefficients, the LIV signal takes the form of a harmonic series with sidereal and annual base frequencies (denoted resp. $\omega_\oplus$ and Ω) together with first harmonics. The time dependence of the measured acceleration $a^{\hat{z}}$ from Equation (7) arises from the terms involving the $\hat{z}$ indices. It can be decomposed in frequency according to [39]

$$\frac{\delta a^{\hat{z}}}{a^{\hat{z}}} = \sum_l C_l \cos\left(\omega_l t + \phi_l\right) + D_l \sin\left(\omega_l t + \phi_l\right) .\tag{8}$$

The model contains seven frequencies $l \in \{\Omega, \omega_\oplus, 2\omega_\oplus, \omega_\oplus \pm \Omega, 2\omega_\oplus \pm \Omega\}$. The 14 amplitudes C_l and D_l are linear combinations of 7 $\bar{s}^{\mu\nu}$ components: $\bar{s}^{JK}$, $\bar{s}^{TJ}$ and $\bar{s}^{XX} - \bar{s}^{YY}$ which can be found in Table 1 of [48] or Table IV from [39].

In order to look for tiny departures from the constant Earth-atom gravitational interaction, a tidal model for $a^{\hat{z}}$ variations due to celestial bodies is removed from the data before fitting to Equation (8). This tidal model consists of two parts. One part is based on a numerical calculation of the Newtonian tide-generating potential from the Moon and the Sun at Earth's surface based on ephemerides. It uses here the Tamura tidal catalog [94] which gives the frequency, amplitude and phase of 1200 harmonics of the tidal potential. These arguments are used by a software (ETGTAB) that calculates the time variation of the local acceleration in the lab and includes the elastic response of Earth's shape to the tides, called "solid Earth tides", also described analytically e.g., by the DDW model [95]. A previous SME analysis of the atom gravimeter data using only this analytical tidal correction had been done, but it led to a degraded sensitivity of the SME test [47]. Indeed, a non-negligible contribution to $a^{\hat{z}}$ is not covered by this non-empirical tidal model: oceanic tide effects such as ocean loading, for which good global analytical models do not exist. They consequently need to be adjusted from measurements. For the second analysis, reported here, additional local tidal corrections fitted on altimetric data have been removed [96] allowing to improve the statistical uncertainty of the SME test by one order of magnitude.

After tidal subtraction, signal components are extracted from the data using a numerical Fourier transform (NFT). Due to the finite data length, Fourier components overlap, but the linear combinations of spectral lines that the NFT estimates can be expressed analytically. Since the annual component $\omega_l = \Omega$ has not been included in this analysis, the fit provides 12 measurements. From there, individual constraints on the 7 SME coefficients and their associated correlation coefficients can be estimated by a least square adjustment. The results obtained are presented in Table 1.

Table 1. Atom-interferometry limits on Lorentz violation in gravity from [48]. The correlation coefficients can be derived from Table III of [48].

Coefficient	
$\bar{s}^{TX}$	$(-3.1 \pm 5.1) \times 10^{-5}$
$\bar{s}^{TY}$	$(0.1 \pm 5.4) \times 10^{-5}$
$\bar{s}^{TZ}$	$(1.4 \pm 6.6) \times 10^{-5}$
$\bar{s}^{XX} - \bar{s}^{YY}$	$(4.4 \pm 11) \times 10^{-9}$
$\bar{s}^{XY}$	$(0.2 \pm 3.9) \times 10^{-9}$
$\bar{s}^{XZ}$	$(-2.6 \pm 4.4) \times 10^{-9}$
$\bar{s}^{YZ}$	$(-0.3 \pm 4.5) \times 10^{-9}$

Correlation Coefficients						
1						
0.05	1					
0.11	−0.16	1				
−0.82	0.34	−0.16	1			
−0.38	−0.86	0.10	−0.01	1		
−0.41	0.13	−0.89	0.38	0.02	1	
−0.12	−0.19	−0.89	0.04	0.20	0.80	1

All results obtained are compatible with null Lorentz violation. As expected from boost suppressions in Equation (7) and from the measurement uncertainty, on the order of a few 10^{-9} g [97], typical limits obtained are in the 10^{-9} range for purely spatial $\bar{s}^{\mu\nu}$ components and 4 orders of magnitude weaker for the spatio-temporal components $\bar{s}^{TJ}$. It can be seen e.g., with the purely spatial components that these constraints do not reach the intrinsic limit of acceleration resolution of the instrument (which has a short term stability of 11×10^{-9} $g/\sqrt{\text{Hz}}$) because the coefficients are still correlated. Their marginalized uncertainty is broadened by their correlation.

Consequently, improving the uncertainty could be reached through a better decorrelation, by analyzing longer data series. In parallel, the resolution of these instruments keeps increasing and has nowadays improved by about a factor 10 since this experiment. However, increasing the instrument's resolution brings back to the question of possible accidental cancelling in treating "postfit" data. Indeed, it should be recalled here that local tidal corrections subtracted prior to analysis are based on adjusting a model of ocean surface from altimetry data. In principle, this observable would as well be affected by gravity LIV; fitting to these observations thus might remove part of SME signatures from the atom gravimeter data. This was mentioned in the first atom gravimeter SME analysis [47]. The adjustment process used to assess local corrections in gravimeters is not made directly on the instrument itself, but it always involves a form of tidal measurement (here altimetry data, or gravimetry data from another instrument in [98]). All LIV frequencies match to the main tidal frequencies. Further progress on SME analysis with atom gravimeters would thus benefit from addressing in more details the question of possible signal cancelling.

4.2. Very Long Baseline Interferometry

VLBI is a geometric technique measuring the time difference in the arrival of a radio wavefront, emitted by a distant quasar, between at least two Earth-based radio-telescopes. VLBI observations are done daily since 1979 and the database contains nowadays almost 6000 24 h sessions, corresponding to 10 millions group-delay observations, with a present precision of a few picoseconds. One of the principal goals of VLBI observations is the kinematical monitoring of Earth rotation with respect to a global inertial frame realized by a set of defining quasars, the International Celestial Reference Frame [99], as defined by the International Astronomical Union [100]. The International VLBI Service for Geodesy and Astrometry (IVS) organizes sessions of observation, storage of data and distribution of products, in particular the Earth Orientation parameters. Because of this precision, VLBI is also a very interesting tool to test gravitation in the Solar System. Indeed, the gravitational fields of the Sun and the planets are responsible of relativistic effects on the quasar light beam through the propagation of the signal to the observing station and VLBI is able to detect these effects very accurately. By using the complete VLBI observations database, it was possible to obtain a constraint on the γ PPN parameter at the level of 1.2×10^{-4} [101,102]. In its minimal gravitational sector, SME can also be investigated with VLBI and obtaining a constrain on the $\bar{s}^{TT}$ coefficient is possible.

Indeed, the propagation time of a photon emitted at the event $(cT_e, \mathbf{X}_e)$ and received at the position $\mathbf{X}_r$ can be computed in the SME formalism using the time transfer function formalism [103–107] and is given by [39,80]

$$\mathcal{T}(\mathbf{X}_e, T_e, \mathbf{X}_r) = T_r - T_e = \frac{R_{er}}{c} + 2\frac{G_N M}{c^3}\left[1 - \frac{2}{3}\bar{s}^{TT} - \bar{s}^{TJ}N_{er}^J\right]\ln\frac{R_e - N_{er}.\mathbf{X}_e}{R_r - N_{er}.\mathbf{X}_r}$$

$$+ \frac{G_N M}{c^3}\left(\bar{s}^{TJ}P_{er}^J - \bar{s}^{JK}N_{er}^J P_{er}^K\right)\frac{R_e - R_r}{R_e R_r} + \frac{G_N M}{c^3}\left[\bar{s}^{TJ}N_{er}^J + \bar{s}^{JK}\hat{P}_{er}^J \hat{P}_{er}^K - \bar{s}^{TT}\right](N_r.N_{er} - N_e.N_{er})$$

$$(9)$$

where the terms a_1 and a_2 from [80] are taken as unity (which corresponds to using the harmonic gauge, which is the one used for VLBI data reduction), $R_e = |\mathbf{X}_e|$, $R_r = |\mathbf{X}_r|$, $R_{er} = |\mathbf{X}_r - \mathbf{X}_e|$ with the central body located at the origin and where we introduce the following vectors

$$\mathbf{K} = \frac{\mathbf{X}_e}{R_e}, \quad \mathbf{N}_{ij} \equiv \frac{\mathbf{X}_{ij}}{R_{ij}} = \frac{\mathbf{X}_j - \mathbf{X}_i}{|\mathbf{X}_{ij}|}, \quad \mathbf{N}_i = \frac{\mathbf{X}_i}{|\mathbf{X}_i|}, \quad \mathbf{P}_{er} = \mathbf{N}_{er} \times (\mathbf{X}_r \times \mathbf{N}_{er}), \quad \text{and} \quad \hat{\mathbf{P}}_{er} = \frac{\mathbf{P}_{er}}{|\mathbf{P}_{er}|}, \quad (10)$$

and where G_N is the observed Newton constant measured by considering the orbital motion of bodies and is defined in Equation (5). This equation is the generalization of the well-known Shapiro time delay including Lorentz violation. The VLBI is actually measuring the difference of the time of arrival of a signal received by two different stations. This observable is therefore sensitive to a differential time delay (see [108] for a calculation in GR). Assuming a radio-signal emitted by a quasar at event $(T_e, \mathbf{X}_e)$ and received by two different VLBI stations at events $(T_1, \mathbf{X}_1)$ and $(T_2, \mathbf{X}_2)$ (all quantities being expressed in a barycentric reference frame), respectively, the VLBI group-delay $\Delta\tau_{(SME)}$ in SME formalism can be written [55]

$$\Delta\tau_{(SME)} = 2\frac{G_N M}{c^3}\left(1 - \frac{2}{3}\bar{s}^{TT}\right)\ln\frac{R_1 + \mathbf{K}.\mathbf{X}_1}{R_2 + \mathbf{K}.\mathbf{X}_2} + \frac{2}{3}\frac{G_N M}{c^3}\bar{s}^{TT}(\mathbf{N}_2.\mathbf{K} - \mathbf{N}_1.\mathbf{K}), \quad (11)$$

where we only kept the $\bar{s}^{TT}$ contribution (see Equation (7) from [55] for the full expression) and we use the same notations as in [108] by introducing three unit vectors

$$\mathbf{K} = \frac{\mathbf{X}_e}{|\mathbf{X}_e|}, \quad \mathbf{N}_1 = \frac{\mathbf{X}_1}{|\mathbf{X}_1|}, \quad \text{and} \quad \mathbf{N}_2 = \frac{\mathbf{X}_2}{|\mathbf{X}_2|}. \quad (12)$$

Ten million VLBI delay observations between August 1979 and mid-2015 have been used to estimate the $\bar{s}^{TT}$ coefficient. First, VLBI observations are corrected from delay due to the radio wave crossing of dispersive media by using 2 GHz and 8 GHz recordings. Then, we used only the 8 GHz delays and the Calc/Solve geodetic VLBI analysis software, developed at NASA Goddard Space Flight Center and coherent with the latest standards of the International Earth Rotation and Reference Systems Service [109]. We added the partial derivative of the VLBI delay with respect to $\bar{s}^{TT}$ from Equation (11) to the software package using the USERPART module of Calc/Solve. We turned to a global solution in which we estimated $\bar{s}^{TT}$ as a global parameter together with radio source coordinates. We obtained

$$\bar{s}^{TT} = (-5 \pm 8) \times 10^{-5}, \tag{13}$$

with a postfit root mean square of 28 picoseconds and a χ^2 per degree of freedom of 1.15. Correlations between radio source coordinates and $\bar{s}^{TT}$ are lower than 0.02, the global estimate being consistent with the mean value obtained with the session-wise solution with a slightly lower error.

In conclusion, VLBI is an incredible tool to test Lorentz symmetry, especially the $\bar{s}^{TT}$ coefficient. This coefficient has an isotropic impact on the propagation speed of gravitational waves as can be noticed from Equation (27) below (or see Equation (9) from [56] or Equation (11) from [60]). The analysis performed in [55] includes the SME contribution in the modeling of VLBI observations and includes the $\bar{s}^{TT}$ parameter in the global fit with other parameters. It is therefore a robust analysis that produces the current best estimate on the $\bar{s}^{TT}$ parameter. In the future, the accumulation of VLBI data in the framework of the permanent geodetic monitoring program leads us expect improvement of this constraint.

4.3. Lunar Laser Ranging

On 20 August 1969, after ranging to the lunar retro-reflector placed during the Apollo 11 mission, the first LLR echo was detected at the McDonald Observatory in Texas. Currently, there are five stations spread over the world which have realized laser shots on five lunar retro-reflectors. Among these stations four are still operating: Mc Donald Observatory in Texas, Observatoire de la Côte d'Azur in France, Apache point Observatory in New Mexico and Matera in Italy while one on Maui, Hawaii has stopped lunar ranging since 1990. Concerning the lunar retro-reflectors three are located at sites of the Apollo missions 11, 14 and 15 and two are French-built array operating on the Soviet roving vehicle Lunakhod 1 and 2.

LLR is used to conduct high precision measurements of the light travel time of short laser pulses emitted at time t_1 by a LLR station, reflected at time t_2 by a lunar retro-reflector and finally received at time t_3 at a station receiver. The data are presented as normal points which combine time series of measured light travel time of photons, averaged over several minutes to achieve a higher signal-to-noise ratio measurement of the lunar range at some characteristic epoch. Each normal-point is characterized by one emission time (t_1 in universal time coordinate—UTC), one time delay (Δt_c in international atomic time—TAI) and some additional observational parameters as laser wavelength, atmospheric temperature and pressure *etc*. According to [110], the theoretical pendent of the observed time delay ($\Delta t_c = t_3 - t_1$ in TAI) is defined as

$$\Delta t_c = \left[T_3 - \Delta \tau_t(T_3) \right] - \left[T_1 - \Delta \tau_t(T_1) \right], \tag{14}$$

where T_1 is the emission time expressed in barycentric dynamical time (TDB) and $\Delta \tau_t$ is a relativistic correction between the TDB and the terrestrial time (TT) at the level of the station. The reception time T_3 expressed in TDB is defined by the following two relations

$$T_3 = T_2 + \frac{1}{c} \left\| X_{o'}(T_3) - X_r(T_2) \right\| + \Delta \mathcal{T}_{(\mathrm{grav})} + \Delta \tau_a, \tag{15a}$$

$$T_2 = T_1 + \frac{1}{c} \left\| X_r(T_2) - X_o(T_1) \right\| + \Delta \mathcal{T}_{(\mathrm{grav})} + \Delta \tau_a, \tag{15b}$$

with T_2 the time in TDB at the reflection point X_o and $X_{o'}$ are respectively the barycentric position vector at the emitter and the reception point, X_r is the barycentric position vector at the reflection point, $\Delta \mathcal{T}_{(\mathrm{grav})}$ is the one way gravitational time delay correction and $\Delta \tau_a$ is the one way tropospheric correction.

LLR measurements are used to produce the Lunar ephemeris but also provide a unique opportunity to study the Moon's rotation, the Moon's tidal acceleration, the lunar rotational dissipation, etc. [111]. In addition, LLR measurements have turn the Earth-Moon system into a laboratory to study fundamental physics and to conduct tests of the gravitation theory. Nordtvedt was the first to suggest that LLR can be used to test GR by testing one of its pillar: the Strong Equivalence Principle [112–114]. He showed that precise laser ranging to the Moon would be capable of measuring precisely the ratio of gravitational mass to inertial mass of the Earth to an accuracy sufficient to constrain a hypothetical dependence of this ratio on the gravitational self-energy. He concluded that such a measurement could be used to test Einstein's theory of gravity and others alternative theories as scalar tensor theories. The best current test of the Strong Equivalence Principle is provided by a combination of torsion balance measurements with LLR analysis and is given by [115–117]

$$\eta = (4.4 \pm 4.5) \times 10^{-4}, \tag{16}$$

where η is the Nordtvedt parameter that is defined as $m_G/m_I = 1 + \eta U/mc^2$ with m_G the gravitational mass, m_I the inertial mass and U the gravitational self-energy of the body. Using the Cassini constraint on the γ PPN parameter [118] and the relation $\eta = 4\beta - \gamma - 3$ leads to a constraint on β PPN parameter at the level $\beta - 1 = (1.2 \pm 1.1) \times 10^{-4}$ [116].

In addition to tests of the Strong Equivalence Principle, many other tests of fundamental physics were performed with LLR analysis. For instance, LLR data can be used to search for a temporal evolution of the gravitational constant $\dot{G}/G$ [115] and to constrain the fifth force parameters [119]. In addition, LLR has been used to constrain violation of the Lorentz symmetry in the PPN framework. Müller et al. [119] deduced from LLR data analysis constraints on the preferred frame parameters α_1 and α_2 at the level $\alpha_1 = (-7 \pm 9) \times 10^{-5}$ and $\alpha_2 = (1.8 \pm 2.5) \times 10^{-5}$.

Considering all the successful GR tests performed with LLR observations, it is quite natural to use them to search for Lorentz violations in the gravitation sector. In the SME framework, Battat et al. [45] used the lunar orbit to provide estimates on the SME coefficients. Using a perturbative approach, the main signatures produced by SME on the lunar orbit have analytically been computed in [39]. These computations give a first idea of the amplitude of the signatures produced by a breaking of Lorentz symmetry. Nevertheless, these analytical signatures have been computed assuming the lunar orbit to be circular and fixed (i.e., neglecting the precession of the nodes for example). These analytical signatures have been fitted to LLR residuals obtained from a data reduction performed in pure GR [45]. They determined a *"realistic"* error on their estimates from a similar postfit analysis performed in the PPN framework. The results obtained by this analysis are presented in Table 2. It is important to note that this analysis uses projections of the SME coefficients into the lunar orbital plane $\bar{s}^{11}, \bar{s}^{22}, \bar{s}^{0i}$ (see Section V.B.2 of [39]) while the standard SME analyses uses coefficients defined in a Sun-centered equatorial frame (and denoted by capital letter $\bar{s}^{IJ}$).

Table 2. Estimation of Standard-Model Extension (SME) coefficients from Lunar Laser Ranging (LLR) postfit data analysis from [45]. No correlations coefficients have been derived in this analysis. The coefficients $\bar{s}^{ij}$ are projections of the $\bar{s}^{IJ}$ into the lunar orbital plane (see Equation (107) from [39]) while the linear combinations $\bar{s}_{\Omega_\oplus c}$ and $\bar{s}_{\Omega_\oplus s}$ are given by Equation (108) from [39].

	Coefficient
$\bar{s}^{11} - \bar{s}^{22}$	$(1.3 \pm 0.9) \times 10^{-10}$
$\bar{s}^{12}$	$(6.9 \pm 4.5) \times 10^{-11}$
$\bar{s}^{01}$	$(-0.8 \pm 1.1) \times 10^{-6}$
$\bar{s}^{02}$	$(-5.2 \pm 4.8) \times 10^{-7}$
$\bar{s}_{\Omega_\oplus c}$	$(0.2 \pm 3.9) \times 10^{-7}$
$\bar{s}_{\Omega_\oplus s}$	$(-1.3 \pm 4.1) \times 10^{-7}$

However, as discussed in Section 3 and in [46,55], a postfit search for SME signatures into residuals of a data reduction previously performed in pure GR is not fully satisfactory. First of all, the uncertainties obtained by a postfit analysis based on a GR data reduction can be underestimated by up to two orders of magnitude. This is mainly due to correlations between SME coefficients and others global parameters (masses, positions and velocities, . . .) that are neglected in this kind of approach. In addition, in the case of LLR data analysis, the oscillating signatures derived in [39] and used in [45] to determine pseudo-constraints are computed only accounting for short periodic oscillations, typically at the order of magnitude of the mean motion of the Moon around the Earth. Therefore, this analytic solution remains only valid for few years while LLR data spans over 45 years (see also the discussions in footnote 2 from [50] and page 22 from [39]).

Regarding LLR data analysis, a more robust strategy consists in including the SME modeling in the complete data analysis and to estimate the SME coefficients in a global fit along with others parameters by taking into account short and long period terms and also correlations (see [46]). In order to perform such an analysis, a new numerical lunar ephemeris named "Éphéméride Lunaire Parisienne Numérique" (ELPN) has been developed within the SME framework. The dynamical model of ELPN is similar to the DE430 one [120] but includes the Lorentz symmetry breaking effects arising on the orbital motion of the Moon. The SME contribution to the lunar equation of motion has been derived in [39] and is given by

$$
\begin{aligned}
a_{\mathrm{SME}}^{J} &= \frac{G_N M}{r^3}\left[\bar{s}_t^{JK} r^K - \frac{3}{2}\bar{s}_t^{KL}\hat{r}^K\hat{r}^L r^J + 2\frac{\delta m}{M}\left(\bar{s}^{TK}\hat{\partial}^K r^J - \bar{s}^{TJ}\hat{\partial}^K r^K \right) \right. \\
&\quad \left. + 3\bar{s}^{TK}\hat{V}^K r^J - \bar{s}^{TJ}\hat{V}^K r^K - \bar{s}^{TK}\hat{V}^J r^K + 3\bar{s}^{TL}\hat{V}^K\hat{r}^K\hat{r}^L r^J \right],
\end{aligned}
\tag{17}
$$

where G_N is the observed Newtonian constant defined by Equation (5), M is the mass of the Earth-Moon barycenter, δm is the difference between the Earth and the lunar masses; $\hat{r}^J$ being the unit position vector of the Moon with respect to the Earth; $\hat{v}^J = v^J/c$ with v^J being the relative velocity vector of the Moon with respect to the Earth; $\hat{V}^J = V^J/c$ with V^J being the Heliocentric velocity vector of the Earth-Moon barycenter and the 3-dimensional traceless tensor defined by Equation (6). These equations of motion as well as their partial derivatives are integrated numerically in ELPN.

In addition to the orbital motion, effects of a violation of Lorentz symmetry on the light travel time of photons is also considered. More precisely, the gravitational time delay $\Delta \mathcal{T}_{(\mathrm{grav})}$ appearing in Equation (14) is given by the gravitational part of Equation (9) [80].

Estimates on the SME coefficients are obtained by a standard chi-squared minimization: the LLR residuals are minimized by an iterative weighted least squares fit using partial derivatives previously computed from variational equations in ELPN. After an adjustment of 82 parameters including the SME coefficients a careful analysis of the covariance matrix shows that LLR data does not allow to estimate independently all the SME coefficients but that they are sensitive to the following three linear combinations:

$$
\bar{s}^{XX} - \bar{s}^{YY}, \qquad \bar{s}^{TY} + 0.43\bar{s}^{TZ}, \qquad \bar{s}^{XX} + \bar{s}^{YY} - 2\bar{s}^{ZZ} - 4.5\bar{s}^{YZ}.
\tag{18}
$$

The estimations on the 6 SME coefficients derived in [46] is summarized in Table 3. In particular, it is worth emphasizing that the quoted uncertainties are the sum of the statistical uncertainties obtained from the least-square fit with estimations of systematics uncertainties obtained with a Jackknife resampling method [121,122].

Table 3. Estimation of SME coefficients from a full LLR data analysis from [46] and associated correlation coefficients.

Coefficient	Estimates
$\bar{s}^{TX}$	$(-0.9 \pm 1.0) \times 10^{-8}$
$\bar{s}^{XY}$	$(-5.7 \pm 7.7) \times 10^{-12}$
$\bar{s}^{XZ}$	$(-2.2 \pm 5.9) \times 10^{-12}$
$\bar{s}^{XX} - \bar{s}^{YY}$	$(0.6 \pm 4.2) \times 10^{-11}$
$\bar{s}^{TY} + 0.43\,\bar{s}^{TZ}$	$(6.2 \pm 7.9) \times 10^{-9}$
$\bar{s}^{XX} + \bar{s}^{YY} - 2\bar{s}^{ZZ} - 4.5\,\bar{s}^{YZ}$	$(2.3 \pm 4.5) \times 10^{-11}$

Correlation Coefficients					
1					
−0.06	1				
−0.04	0.29	1			
0.58	−0.12	−0.16	1		
0.16	−0.01	−0.09	0.25	1	
0.07	−0.10	−0.13	−0.10	0.03	1

In summary, LLR is a powerful experiment to constrain gravitation theory and in particular hypothetical violation of the Lorentz symmetry. A first analysis based on a postfit estimations of the SME coefficients have been performed [45] which is not satisfactory regarding the neglected correlations with other global parameters as explained in Section 3. A full analysis including the integration of the SME equations of motion and the SME contribution to the gravitational time delay has been done in [46]. The resulting estimates on some SME coefficients are presented in Table 3. In addition, some SME coefficients are still correlated with parameters appearing in the rotational motion of the Moon as the principal moment of inertia, the quadrupole moment, the potential Stockes coefficient C_{22} and the polar component of the velocity vector of the fluid core [46]. A very interesting improvement regarding this analysis would be to produce a joint GRAIL (Gravity Recovery And Interior Laboratory) [123–125] and LLR data analysis that would help in decorrelating the SME parameters from the lunar potential Stockes coefficients of degree 2 and therefore improve marginalized estimations of the SME coefficients. Finally, in [45,46], the effects of SME on the translational lunar equations of motion are considered and used to derive constraints on the SME coefficients. It would be also interesting to extend these analyses by considering the modifications due to SME on the rotation of the Moon. A first attempt has been proposed in Section V. A. 2. of [39] but needs to be extended.

4.4. Planetary Ephemerides

The analysis of the motion of the planet Mercury around the Sun was historically the first evidence in favor of GR with the explanation of the famous advance of the perihelion in 1915. From there, planetary ephemerides have always been a very powerful tool to constrain GR and alternative theories of gravitation. Currently, three groups in the world are producing planetary ephemerides: the NASA Jet Propulsion Laboratory with the DE ephemerides [120,126–131], the French INPOP (Intégrateur Numérique Planétaire de l'Observatoire de Paris) ephemerides [132–137] and the Russian EPM ephemerides [138–142]. These analyses use an impressive number of different observations to produce high accurate planetary and asteroid trajectories. The observations used to produce ephemerides comprise radioscience observations of spacecraft that orbited around Mercury, Venus, Mars and Saturn, flyby tracking of spacecraft close to Mercury, Jupiter, Uranus and Neptune and optical observations of all planets. This huge set of observations have been used to constrain the γ and β post-Newtonian parameter at the level of 10^{-5} [136,137,141–143], the fifth force interaction (see [32] and Figure 31 from [143]), the quantity of Dark Matter in our Solar System [144], the Modified Newtonian Dynamics [131,145–147], . . .

A violation of Lorentz symmetry within the gravity sector of SME induces different types of effects that can have implications on planetary ephemerides analysis: effects on the orbital dynamics and effects on the light propagation. Simulations using the Time Transfer Formalism [104,106,107] based on the software presented in [148] have shown that only the $\bar{s}^{TT}$ coefficients produce a non-negligible effect on the light propagation (while it has impact only at the next post-Newtonian level on the orbital dynamics [29,39]). On the other hand, the other coefficients produce non-negligible effects on the orbital dynamics [39] and can therefore be constrained using planetary ephemerides data. In the linearized gravity limit, the contribution from SME to the 2-body equations of motion within the gravitational sector of SME are given by the first line of Equation (17) (i.e., for a vanishing V^k). The coefficient $\bar{s}^{TT}$ is completely unobservable in this context since absorbed in a rescaling of the gravitational constant (see the discussion in [39,52]).

Ideally, in order to perform a solid estimation of the SME coefficients using planetary ephemerides, one should include the full SME equations in the integration of the planets motion and fit them simultaneously with the other estimated parameters (positions and velocities of planets, J_2 of the Sun, ...). This solid analysis within the SME formalism has not been performed so far.

As a first step, a postfit analysis has been performed [49,50]. The idea of this analysis is to derive the analytical expression for the secular evolution of the orbital elements produced by the SME contribution to the equations of motion. Using the Gauss equations, secular perturbations induced by SME on the orbital elements have been computed in [39] (see also [49] for a similar calculations done for the $\bar{s}^{TJ}$ coefficients only). In particular, the secular evolution of the longitude of the ascending node Ω and the argument of the perihelion ω is given by

$$\left\langle \frac{d\Omega}{dt} \right\rangle = \frac{n}{\sin i(1-e^2)^{1/2}} \left[\frac{\varepsilon}{e^2}\bar{s}_{kP}\sin\omega + \frac{(e^2-\varepsilon)}{e^2}\bar{s}_{kQ}\cos\omega - \frac{\delta m}{M}\frac{2na\varepsilon}{ec}\bar{s}^k\cos\omega \right], \quad (19a)$$

$$\left\langle \frac{d\omega}{dt} \right\rangle = -\cos i \left\langle \frac{d\Omega}{dt} \right\rangle - n \left[-\frac{\varepsilon^2}{2e^4}(\bar{s}_{PP} - \bar{s}_{QQ}) + \frac{\delta m}{M}\frac{2na(e^2-\varepsilon)}{ce^3(1-e^2)^{1/2}}\bar{s}^Q \right], \quad (19b)$$

where a is the semimajor axis, e the eccentricity, i the orbit inclination (with respect to the ecliptic), $n = (G_N m_\odot/a^3)^{1/2}$ is the mean motion, $\varepsilon = 1 - (1-e^2)^{1/2}$, δm the difference between the two masses and M their sum (in the cases of planets orbiting the Sun, one has $M \approx \delta m$). In all these expressions, the coefficients for Lorentz violation with subscripts P, Q, and k are understood to be appropriate projections of $\bar{s}^{\mu\nu}$ along the unit vectors P, Q, and k, respectively. For example, $\bar{s}^k = k^i\bar{s}^{Ti}$, $\bar{s}_{PP} = P^iP^j\bar{s}^{ij}$. The unit vectors P, Q and k define the orbital plane (see [39] or Equation (8) from [50]).

Instead of including the SME equations of motion in planetary ephemerides, the postfit analysis uses estimations of supplementary advances of perihelia and nodes derived from ephemerides analysis [135,140,144] to fit the SME coefficients through Equation (19). In [50], estimations of supplementary advances of perihelia and longitude of nodes from INPOP (see Table 5 from [135]) are used to fit a posteriori the SME coefficients. This analysis suffers from large correlations due to the fact that the planetary orbits are very similar to each other: nearly eccentric orbit and very low inclination orbital planes. In order to deal properly with these correlations a Bayesian Monte Carlo inference has been used [50]. The posterior probability distribution function can be found on Figure 1 from [50]. The intervals corresponding to the 68% Bayesian confidence levels are given in Table 4 as well as the correlation matrix. It is interesting to mention that a decomposition of the normal matrix in eigenvectors allows one to find linear combinations of SME coefficients that are uncorrelated with the planetary ephemerides analysis (see Equation (15) and Table IV from [50]).

Table 4. Estimations of the SME coefficients from a postfit data analysis based on planetary ephemerides from [50]. The uncertainties correspond to the 68% Bayesian confidence levels of the marginal posterior probability distribution function. The associated correlation coefficients can be found in Table III from [50].

Coefficient	
$\bar{s}^{XX} - \bar{s}^{YY}$	$(-0.8 \pm 2.0) \times 10^{-10}$
$\bar{s}^{XX} + \bar{s}^{YY} - 2\,\bar{s}^{ZZ}$	$(-0.8 \pm 2.7) \times 10^{-10}$
$\bar{s}^{XY}$	$(-0.3 \pm 1.1) \times 10^{-10}$
$\bar{s}^{XZ}$	$(-1.0 \pm 3.5) \times 10^{-11}$
$\bar{s}^{YZ}$	$(5.5 \pm 5.2) \times 10^{-12}$
$\bar{s}^{TX}$	$(-2.9 \pm 8.3) \times 10^{-9}$
$\bar{s}^{TY}$	$(0.3 \pm 1.4) \times 10^{-8}$
$\bar{s}^{TZ}$	$(-0.2 \pm 5.0) \times 10^{-8}$

Correlation coefficients							
1							
0.99	1						
0.99	0.99	1					
0.98	0.98	0.99	1				
−0.32	−0.24	−0.26	−0.26	1			
0.99	0.98	0.98	0.98	−0.32	1		
0.62	0.67	0.62	0.59	0.36	0.60	1	
−0.83	−0.86	−0.83	−0.81	−0.14	−0.82	−0.95	1

In summary, planetary ephemerides offer a great opportunity to constrain hypothetical violations of Lorentz symmetry. So far, only postfit estimations of the SME coefficients have been performed [49,50]. In this analysis, estimations of secular advances of perihelia and longitude of nodes obtained with the INPOP planetary ephemerides [135] are used to fit a posteriori the SME coefficients using the Equations (19). The 68% marginalized confidence intervals are given in Table 4. This analysis suffers highly from correlations due to the fact that the planetary orbits are very similar. A very interesting improvement regarding this analysis would be to perform a full analysis by integrating the planetary equations of motion directly within the SME framework and by fitting the SME coefficients simultaneously with the other parameters fitted during the ephemerides data reduction.

4.5. Gravity Probe B

In GR, a gyroscope in orbit around a central body undergoes two relativistic precessions with respect to a distant inertial frame: (i) a geodetic drift in the orbital plane due to the motion of the gyroscope in the curved spacetime [149]; and (ii) a frame-dragging due to the spin of the central body [150]. In GR, the spin of a gyroscope is parallel transported, which at the post-Newtonian approximation gives the relativistic drift

$$\boldsymbol{R} = \frac{d\hat{\boldsymbol{S}}}{dt} = \boldsymbol{\Omega}_{GR} \times \boldsymbol{S}, \tag{20a}$$

$$\boldsymbol{\Omega}_{GR} = \frac{3GM}{2c^2r^3}\boldsymbol{r} \times \boldsymbol{v} + \frac{3\hat{\boldsymbol{r}}(\hat{\boldsymbol{r}}.\boldsymbol{J}) - \boldsymbol{J}}{c^2r^3}, \tag{20b}$$

where $\hat{\boldsymbol{S}}$ is the unit vector pointing in the direction of the spin $\boldsymbol{S}$ of the gyroscope, $\boldsymbol{r}$ and $\boldsymbol{v}$ are the position and velocity of the gyroscope, $\hat{\boldsymbol{r}} = \boldsymbol{r}/r$ and $\boldsymbol{J}$ is the angular momentum of the central body. In 1960, it has been suggested to use these two effects to perform a new test of GR [151,152]. In April 2004, GPB, a satellite carrying 4 cryogenic gyroscopes was launched in order to measure these two precessions. GPB was orbiting Earth on a polar orbit such that the two relativistic drifts are orthogonal to each other [153]: the geodetic effect is directed along the NS direction (North-South, i.e., parallel to the satellite motion) while the frame-dragging effect is directed on the WE direction (West-East, see [52,153] for further details about the axes conventions in the GPB data reduction).

A year of data gives the following measurements of the relativistic drift: (i) the geodetic drift $R_{NS} = -6601.8 \pm 18.3$ mas/yr (milliarcsecond per year) to be compared to the GR prediction of -6606.1 mas/yr; and (ii) the frame-dragging drift $R_{WE} = -37.2 \pm 7.2$ mas/yr to be compared with the GR prediction of -39.2 mas/yr. In other word, the GPB results can be written as a measurement of a deviation from GR given by

$$\Delta R_{NS} = 4.3 \pm 18.3 \text{ mas/yr} \qquad \text{and} \qquad \Delta R_{WE} = 2 \pm 7.2 \text{ mas/yr}. \tag{21}$$

Within the SME framework, if one considers only the $\bar{s}^{\mu\nu}$ coefficients, the equation of parallel transport in term of the spacetime metric is not modified (see Equation (143) from [39]). Nevertheless, the expression of the spacetime metric is modified leading to a modification of the relativistic drift given by Equation (150) from [39]. In order to focus only on the dominant secular part of the evolution of the spin orientation, the relativistic drift equation has been averaged over a period. The SME contribution to the precession can be written as [39]

$$\Delta \Omega^J = \frac{G_N M}{r^2} v \left[\left(-\frac{4}{3} \bar{s}^{TT} - \frac{9}{8} \tilde{i}_{(-5/3)} \bar{s}_t^{JK} \hat{\sigma}^J \hat{\sigma}^K \right) \hat{\sigma}^J + \frac{5}{4} \tilde{i}_{(-3/5)} \bar{s}_t^{JK} \hat{\sigma}^K \right], \tag{22}$$

where G_N is the effective gravitational constant defined by Equation (5), the coefficients $\tilde{i}$ are defined by $\tilde{i}_{(\beta)} = 1 + \beta I_{\oplus} / (M_{\oplus} r^2)$, $\hat{\sigma}^J$ is a unit vector normal to the gyroscope orbital plane, r and v are the norm of the position and velocity of the gyroscope and $\bar{s}_t^{JK}$ is the traceless part of $\bar{s}^{JK}$ as defined by Equation (6). Using the geometry of GPB into the last equation and using Equation (20a), one finds that the gyroscope anomalous drift is given by

$$\Delta R_{NS} = 5872\bar{s}^{TT} + 794 \left(\bar{s}^{XX} - \bar{s}^{YY} \right) - 317 \left(\bar{s}^{XX} + \bar{s}^{YY} - 2\bar{s}^{ZZ} \right) - 1050\bar{s}^{XY}, \tag{23a}$$

$$\Delta R_{WE} = -368(\bar{s}^{XX} - \bar{s}^{YY}) - 1112\bar{s}^{XY} + 1269\bar{s}^{XZ} + 4219\bar{s}^{YZ}, \tag{23b}$$

where the units are mas/yr. These are the SME modifications to the relativistic drift arising from the modification of the equations of evolution of the gyroscope axis (i.e., modification of the parallel transport equation due to the modification of the underlying spacetime metric).

In addition to modifying the evolution of the spin axis, a breaking of Lorentz symmetry will impact the orbital motion of the gyroscope. As a result, the position and velocity of the gyroscope will depend on the SME coefficients and therefore also impact the evolution of the spin axis through the GR contribution given by Equation (20b). The best way to deal with this effect is to use the GPB tracking measurements (GPS) in order to constrain the gyroscope orbital motion and eventually constrain the SME coefficients through the equations of motion. In [52], these tracking observations are not used and only the gyroscope drift is used in order to constrain the SME contributions coming from both the modification of the parallel transport and from the modification of GPB orbital motion. In order to do this, the contribution of SME on the evolution of the orbital elements given by Equations (19) and (26) are used, averaged over a period and in the low eccentricity approximation. This secular evolution for the osculating elements is introduced in the relativistic drift equation for the gyroscope from Equation (20b) and averaged over the measurement time using Equation (20a). Using the GPB geometry, this contribution to the relativistic drift is given by

$$\Delta R'_{NS} = 5.7 \times 10^6 (\bar{s}^{XX} - \bar{s}^{YY}) + 1.7 \times 10^7 \bar{s}^{XY} - 1.9 \times 10^7 \bar{s}^{XZ} - 6.6 \times 10^7 \bar{s}^{YZ}, \tag{24a}$$

$$\Delta R'_{WE} = -1.89 \times 10^7 (\bar{s}^{XX} - \bar{s}^{YY}) - 5.71 \times 10^7 \bar{s}^{XY} - 5.96 \times 10^6 \bar{s}^{XZ} - 1.98 \times 10^7 \bar{s}^{YZ}, \tag{24b}$$

with units of mas/yr.

The sum of the two SME contributions to the gyroscope relativistic drift given by Equations (23) and (24) can be compared to the GPB estimations given by Equation (21). The result is given in Table 5. The main advantage of GPB comes from the fact that it is sensitive to the $\bar{s}^{TT}$ coefficient. The constraint on this coefficient is at the level of 10^{-3}, a little bit less good than the one obtained with VLBI or with binary pulsars but relying on a totally different type of observations. The constraints on the spatial part of the SME coefficients $(\bar{s}^{IJ})$ are at the level of 10^{-7} and are superseded by the other measurements. The constraints on these coefficients come mainly from the contribution arising from the orbital dynamics of GPB and not from a direct modification of the spin evolution. Constraining the orbital motion from GPB by using the gyroscope observations only is not optimal and tracking observations may help to improve the corresponding constraints (in this case, a dedicated satellite may be more appropriate as discussed in Section 5.3).

Table 5. Estimations of the SME coefficients from a postfit data analysis based on Gravity Probe B (GPB) [52].

Coefficient		
$\bar{s}_{\mathrm{GPB}}^{(1)}$	$= \bar{s}^{TT} + 970\left(\bar{s}^{XX} - \bar{s}^{YY}\right) - 0.05\left(\bar{s}^{XX} + \bar{s}^{YY} - 2\bar{s}^{ZZ}\right)$ $+2895\,\bar{s}^{XY} - 3235\,\bar{s}^{XZ} - 11\,240\,\bar{s}^{YZ}$	$(0.7 \pm 3.1) \times 10^{-3}$
$\bar{s}_{\mathrm{GPB}}^{(2)}$	$= \bar{s}^{XX} - \bar{s}^{YY} + 3.02\,\bar{s}^{XY} + 0.32\,\bar{s}^{XZ} + 1.05\,\bar{s}^{YZ}$	$(-1.1 \pm 3.8) \times 10^{-7}$

In summary, the GPB measurement of a gyroscope relativistic drifts due to geodetic precession or frame-dragging can be used to search for a breaking of Lorentz symmetry. The main advantage of this technique comes from its sensitivity to $\bar{s}^{TT}$. As already mentioned, this coefficient has an isotropic impact on the propagation velocity of gravitational waves as can be noticed from Equation (27) below (see also Equation (9) from [56] or Equation (11) from [60]). A preliminary result based on a post-fit analysis performed after a GR data reduction of GPB measurements gives a constraint on $\bar{s}^{TT}$ at the level of 10^{-3} [52]. This should be investigated further since the Earth's quadrupole moment has been neglected and Lorentz-violating effects on the aberration terms can also change slightly the results. In addition, impacts from Lorentz violations on frame-dragging arising in other contexts such as satellite laser ranging (see Section 5.3) or signals from accretion disks around collapsed stars [154] would also be interesting to consider.

4.6. Binary Pulsars

The discovery of the first binary pulsars PSR 1913+16 by Hulse and Taylor in 1975 [155] has opened a new window to test the theory of gravitation. Observations of this pulsar have allowed one to measure the relativistic advance of the periastron [156] and more importantly to measure the rate of orbital decay due to gravitational radiation [157]. Pulsars are rotating neutron stars that are emitting very strong radiation. The periods of pulsars are very stable which allows us to consider them as "clocks" that are moving in an external gravitational field (typically in the gravitational field generated by a companion). The measurements of the pulse time of arrivals can be used to infer several parameters by fitting an appropriate timing model (see for example Section 6.1 from [2]): (i) non-orbital parameter such as the pulsar period and its rate of change; (ii) five Keplerian parameters; and (iii) some post-Keplerian parameters [158]. In GR, the expressions of these post-Keplerian parameters are related to the masses of the two bodies and to the Keplerian parameters. If more than 2 of these post-Keplerian parameters can be determined, they can be used to test GR [159]. Nowadays, more than 70 binary pulsars have been observed [160]. A description of the most interesting binary pulsars in order to test the gravitation theory can be found in Section 6.2 from [2] or in the supplemental material from [53].

The model fitted to the observations is based on a post-Newtonian analytical solution to the 2 body equations of motion [161] (see also [162]) and includes contribution from the Einstein time delay (i.e., the transformation between proper and coordinate time), the Shapiro time delay, the Roemer time delay [158]. The model also corrects for several systematics like atmospheric delay, Solar system dispersion, interstellar dispersion, motion of the Earth and the Solar System, ... (see for example [163]).

Pulsars observations provide some of the best current constraints on alternative theories of gravitation (for a review, see [164,165]). In addition to the Hulse and Taylor pulsar, the double pulsar [166] now provides the best measurement of the pulsar orbital rate of change [165]. In addition, the post-Keplerian modeling has been fully derived in tensor-scalar theories [167–169] such that pulsars observations have provided some of the best constraints on this class of theory [165,170,171]. It is important to mention that non perturbative strong field effects may arise in binary pulsars system and needs to be taken into account [169,172].

In addition, binary pulsars have also been successfully used to test Lorentz symmetry. For example, analyses of the pulses time of arrivals provide a constraint on the $\alpha_{1,2,3}$ PPN parameters. Since non perturbative strong field effects may arise in binary pulsars system (see for example [173] for strong field effects in Einstein-Aether theory), the obtained constraints are interpreted as strong field version of the PPN parameters denoted by $\hat{\alpha}_i$. Estimates of these parameters should be compared carefully to the standard weak field constraints since they may depend on the gravitational binding energy of the neutron star. The best current constraint on $\hat{\alpha}_1 = -0.4^{+3.7}_{-3.1} \times 10^{-5}$ is obtained by considering the orbital dynamics of the binary pulsars PSR J1738+0333 [174,175]. The best current constraint on $\hat{\alpha}_2$ takes advantage from the fact that this parameter produces a precession of the spin axis of massive bodies [176]. The combination of observations of two solitary pulsars lead to the best current constraints on $|\hat{\alpha}_2| < 1.6 \times 10^{-9}$ [177]. Finally, the parameter $\hat{\alpha}_3$ produces a violation of the momentum conservation in addition to a violation of the Lorentz symmetry. This parameter will induce a self-acceleration for rotating body that can be constrained using binary pulsars [178]. The best current constraint uses a set of 5 pulsars (4 binary pulsars and one solitary pulsar) and is given by $\hat{\alpha}_3 < 5.5 \times 10^{-20}$ [179].

Furthermore, specific Lorentz violating theories have also been constrained with binary pulsars. In [72,73], binary pulsars observations are used to constrain Einstein-Aether and khronometric theory. In these theories, the low-energy limit Lorentz violations can be parametrized by four parameters: the α_1 and α_2 PPN parameters and two other parameters. It has been shown [72,73,173] that the orbital period decay depends on these four parameters. Assuming the solar system constraints on α_1 and α_2 [2], measurements of the rate of change of the orbital period of binary pulsars have been used to constrain the two other parameters (see for example Figure 2 from [72]). In this work, strong field effects have been taken into account by solving numerically the field equations in order to determine the neutron stars sensitivity [73].

Finally, binary pulsars have been used in order to derive constraints on the SME coefficients. As in the PPN formalism, constraints obtained from binary pulsars need to be considered as constraints on strong-field version of the SME coefficients that may include non perturbative effects. Two different types of effects have been used to determine estimates on the SME coefficients: (i) tests using the spin precession of solitary pulsars and (ii) tests using effects on the orbital dynamics of binary pulsars [53]. The SME contribution to the precession rate of an isolated spinning body has been derived in [39] and is given by

$$\Omega^k_{\text{SME}} = \frac{\pi}{P} \bar{s}^{kj} \hat{S}^j \, , \tag{25}$$

where P is the spin period and $\hat{S}^j$ is the unit vector pointing along the spin direction. The effects from the pulsar spin precession on the pulse width can be found in [177,180]. Two solitary pulsars have been used to constrain the SME coefficients with this effect. The second type of tests come from the orbital dynamics of binary pulsars. As mentioned in Sections 4.3 and 4.4, the SME will modify the two-body equations of motion by including the term from Equation (17). At first order in the SME coefficients, this will produce several secular effects that have been computed in [39]. In particular, an additional advance in the argument of periastron and of the longitude of the nodes has been mentioned in Equation (19) and used to constrain the SME with planetary ephemerides. For binary pulsars, it is possible to constrain a secular evolution of two other orbital elements: the eccentricity and the projected semi-major axis x. The secular SME contributions to these quantities have been computed in [39,53,54] and are given by

$$\left\langle \frac{de}{dt} \right\rangle = -n\sqrt{1-e^2}\left[\frac{\varepsilon^2}{e^3}\bar{s}_{PQ} - 2\frac{\delta m}{M}\frac{na\varepsilon}{e^2}\bar{s}^P\right], \tag{26a}$$

$$\left\langle \frac{dx}{dt} \right\rangle = n\frac{m_C}{m_P+m_C}a\cos i\frac{\varepsilon}{e^2\sqrt{1-e^2}}\left[\bar{s}_{kP}\cos\omega - \sqrt{1-e^2}\bar{s}_{kQ}\sin\omega + 2\frac{\delta m}{M}na e\bar{s}^k\cos\omega\right], \tag{26b}$$

where m_P is the mass of the pulsar, m_C is the mass of the companion and all others quantities have been introduced after Eqs. (19). For each binary pulsar, in principle 3 tests can be constructed by using $\dot\omega, \dot e, \dot x$. In [53], 13 pulsars have been used to derive estimates on the SME coefficients. The combination of the observations from the solitary pulsars and from the 13 binary pulsars are reported in Table 6. Both orbital dynamics and spin precession are completely independent of $\bar{s}^{TT}$ whose constraint will be discussed later.

Table 6. Estimation of SME coefficients from binary pulsars data analysis from [53,54]. No correlations coefficients have been derived in this analysis. These estimates should be considered as estimates on the strong field version of the SME coefficients that may include non perturbative strong field effects due to the gravitational binding energy.

	Coefficient
$\|\bar{s}^{TT}\|$	$< 2.8 \times 10^{-4}$
$\bar{s}^{XX} - \bar{s}^{YY}$	$(0.2 \pm 9.9) \times 10^{-11}$
$\bar{s}^{XX} + \bar{s}^{YY} - 2\bar{s}^{ZZ}$	$(-0.05 \pm 12.25) \times 10^{-11}$
$\bar{s}^{XY}$	$(0.05 \pm 3.55) \times 10^{-11}$
$\bar{s}^{XZ}$	$(0.0 \pm 2.0) \times 10^{-11}$
$\bar{s}^{YZ}$	$(0.0 \pm 3.3) \times 10^{-11}$
$\bar{s}^{TX}$	$(0.05 \pm 5.25) \times 10^{-9}$
$\bar{s}^{TY}$	$(0.5 \pm 8.0) \times 10^{-9}$
$\bar{s}^{TZ}$	$(-0.05 \pm 5.85) \times 10^{-9}$

Several comments can be made about this analysis. First of all, it can be considered as a postfit analysis done after an initial fit performed in GR (or within the post-Keplerian formalism). In particular, correlations between the SME coefficients and other parameters (e.g., orbital parameters) are neglected. Secondly, for most of the pulsars, $\dot x$ $\dot\omega$ and $\dot e$ are not directly measured from the pulse time of arrivals but rather estimated from the uncertainties on x, ω and e divided by the time span of the observations. Further, it is important to mention that effects of Lorentz violations have been considered only for the orbital dynamics but never on the Einstein delay or on the Shapiro time delay in this analysis. The full timing model within SME can be found in Section V.E.3 from [39] (see also [181] for a similar derivation with the matter-gravity couplings). In addition, some parameters are not measured like for example the longitude of the ascending node Ω or the azimuthal angle of the spin. These parameters have been marginalized by using Monte Carlo simulations. It is unclear what type of prior probability distribution function has been used in this analysis and what is the impact of this choice. Nevertheless, the results obtained by this analysis (which does not include the $\bar{s}^{TT}$ parameter) are amongst the best ones currently available demonstrating the power of pulsars observations. The main advantages of using binary pulsars come from the fact that their orbital orientation vary which allows one to disentangle the different SME coefficients and to end up with low correlations. Furthermore, they are so far the only constraints on the strong field version of the SME coefficients.

In addition, a different analysis has been performed to constrain the parameter $\bar{s}^{TT}$ alone [54]. While the orbital dynamics and the spin precession is completely independent of $\bar{s}^{00}$ (i.e., the time component of $\bar{s}^{\mu\nu}$ in a local frame), the boost between the Solar System and the binary pulsar frame makes appear explicitly the $\bar{s}^{TT}$ coefficient. In [54], the assumption that there exists a preferred frame where the $\bar{s}^{\mu\nu}$ tensor is isotropic is made, which makes the results specific to that case (although the analysis can be done without this assumption). The analysis requires the knowledge of the pulsar velocity with respect to the preferred frame as well as the velocity of the Solar System with respect to the same frame. Three pulsars have their radial velocity measured, which combined with proper motion in the sky can be used to determine their velocity. The velocity of the Solar System is taken as its velocity with respect to the Cosmic Microwave Background (CMB) frame $w_\odot$ (with $|w_\odot| = 369$ km/s). The analysis is completely similar to the ones performed for the other SME coefficients (see the discussion in the previous paragraph). It is known that $\bar{s}^{TT}$ has a strong effect on the propagation of the light neglected in [54], which may impact the result. In addition, all correlations between $\bar{s}^{TT}$ and the other SME coefficients are neglected. Finally, two different scenarios have been considered regarding the preferred frame: (i) a scenario where the preferred frame is assumed to be the CMB frame and (ii) a scenario where the orientation of the preferred frame is left free and is marginalized over but the magnitude of the velocity of the Solar System with respect to that frame is still assumed to be the 369 km/s. The general case corresponding to a completely free preferred frame has not been considered. If the CMB frame is assumed to be the preferred frame, the constraint on $\bar{s}^{TT}$ is given by $|\bar{s}^{TT}| < 1.6 \times 10^{-5}$ which is a bit better than the one obtained with VLBI (see Equation (13)) although the VLBI analysis does not assume any preferred frame. The scenario where the orientation of the preferred frame is left as a free parameter leads to an upper bound on $|\bar{s}^{TT}| < 2.8 \times 10^{-4}$.

In summary, observations of binary pulsars are an incredible tool to test the gravitation theory. These tests are of the same order of magnitude (and sometimes better) than the ones performed in the Solar System. Moreover, observations of binary pulsars are sensitive to strong field effects. Observations of the pulse arrival times have been used to search for a breaking of Lorentz violation within the PPN framework by constraining the strong field version of the α_i parameters. The parameter $\hat{\alpha}_1$ is constrained at the level of 10^{-5}, $\hat{\alpha}_2$ at the level of 10^{-9} and $\hat{\alpha}_3$ at the level of 10^{-20} [164]. In addition, constraints on Einstein-Aether and khronometric theory have also been done by combining Solar System constraints with binary pulsars observations [72,73]. Finally, within the SME framework, a postfit analysis has been done by considering the spin precession of solitary pulsars and the orbital dynamics of binary pulsars. The obtained results are given in Table 6 and constrain the strong field version of the SME coefficients. The main advantage of using binary pulsars comes from the fact that they proved an estimate of all the SME coefficients with reasonable correlations. It has to be noted that the modification of the orbital period due to gravitational waves emission has not been computed so far in the SME formalism. In addition, the constraint on $\bar{s}^{TT}$ suffers from the assumption of the existence of a preferred frame. Moreover, the corresponding analysis has neglected all effects on the timing delay that may also impact the results and has neglected the other SME coefficients that may also impact this constraint.

4.7. Čerenkov Radiation

Gravitational Čerenkov radiation is an effect that occurs when the velocity of a particle exceeds the phase velocity of gravity. In this case, the particle will emit gravitational radiation until the particle loses enough energy to drop below the gravity speed [56]. In modified theory of gravity, the speed of gravity in a vacuum may be different from the speed of light and Čerenkov radiation may occur and produces energy losses for particles traveling over long distances. Observations of high energy cosmic rays that have not lost all their energy through Čerenkov radiation can be used to put constraints on models of gravitation that predicts gravitational waves that are propagating slower than light. This effect has been used to constrain some alternative gravitation theories [182,183]: a class of tensor-vector theories [184], a class of tensor-scalar theories [185], extended theories of gravitation [186] and some ghost-free bigravity [187].

The propagation of gravitational waves within the SME framework has been derived in [56,60] (including nonminimal SME contributions). In particular, in the minimal SME, the dispersion relation for the gravitational waves is given by [56]

$$l_0^2 = |\boldsymbol{l}|^2 + \bar{s}^{\mu\nu} l_\mu l_\nu \, , \tag{27}$$

where l^α is the 4-momentum of the gravitational wave. A similar expression including nonminimal higher order SME terms can be found in [56,60]. If the minimal SME produces dispersion-free propagation, the higher order terms lead to dispersion and birefringence [60]. As can be directly inferred from the last equation, gravitational Čerenkov radiation can arise when the effective refractive index n is

$$n^2 = 1 - \bar{s}^{\mu\nu} \hat{l}_\mu \hat{l}_\nu > 1 \, , \tag{28}$$

where $\hat{l}_\mu = l_\mu / |\boldsymbol{l}|$. The expression for the energy loss rate due to Lorentz-violating gravitational Čerenkov emission has been calculated from tree-level graviton emission for photons, fermions and scalar particles and is given by [56]

$$\frac{dE}{dt} = -F^w(d) G \left(\bar{s}^{(d)}(\hat{\boldsymbol{p}}) \right)^2 |\boldsymbol{p}|^{2d-4} \, , \tag{29}$$

where d is the dimension of the Lorentz violating operator ($d = 4$ for the minimal SME), $F^w(d)$ is a dimensionless factor depending on the flavor w of the particle emitting the radiation, $\boldsymbol{p}$ is the particle incoming momentum (with $\hat{\boldsymbol{p}} = \boldsymbol{p} / |\boldsymbol{p}|$) and $\bar{s}^{(d)}$ is a direction-dependent combination of SME coefficients. In the minimal SME, $\bar{s}^{(4)}(\hat{\boldsymbol{p}})$ is decomposed on spherical harmonics as

$$\bar{s}^{(4)}(\hat{\boldsymbol{p}}) = \sum_{jm} Y_{jm}(\hat{p}) \bar{s}_{jm}^{(SH)} \, , \tag{30}$$

where we explicitly indicated the (SH) to specify that these coefficients are spherical harmonic decomposition of the SME coefficients. The calculation of the dimensionless factor $F^w(d)$ for scalar particles, fermions and photons has been done in [56]. The integration of Equation (29) shows that if a cosmic ray of specie w is observed on Earth with an energy E_f after traveling a distance L along the direction $\hat{\boldsymbol{p}}$, this implies the following constraint on the SME coefficients

$$\bar{s}^{(d)}(\hat{\boldsymbol{p}}) < \sqrt{\frac{\mathcal{F}^w(d)}{G E_f^{2d-5} L}} \, , \tag{31}$$

where $\mathcal{F}^w(d) = (2d - 5)/F^w(d)$ is another dimensionless factor dependent on the matrix element of the tree-level process for graviton emission.

Using data for the energies and angular positions of 299 observed cosmic rays from different collaborations [188–195], Kostelecký and Tasson [56] derived lower and upper constraints on 80 SME coefficients, including the nine coefficients from the minimal SME whose constraints are given by the Table 7. In their analysis, they consider the coefficients from the different dimensions separately and did not fit all of them simultaneously. In addition, in the minimal SME, they did a fit for the $\bar{s}^{TT}$ parameter alone and another fit for the other 8 coefficients. The number of sources and their directional dependence across the sky allow one to disentangle the SME coefficients and to derive two-sided bounds from the Equation (31). The only coefficient that is one sided is $\bar{s}^{TT}$ because it produces isotropic effects. The bounds are severe for these coefficients, on the order of 10^{-13}. However, this analysis assumes that the matter sector coefficients vanish. Furthermore, several assumptions have been made in order to derive the bounds from Table 7. It is assumed that the cosmic ray primaries are nuclei of atomic weight $N = 56$ (iron), that the Čerenkov radiation is emitted by one of the fermionic partons in the nucleus that carries 10 % of the cosmic ray energy and that the travel distance of the cosmic ray is 10 Megaparsec (Mpc) [56]. Although only conservative assumptions are used for the astrophysical processes involved in the production of high-energy cosmic rays, the observations rely on the sources on the order of 10 Mpc distant, and thus the analysis is of a different nature than a controlled laboratory or even Solar-System test.

Table 7. Lower and upper limits on the SME coefficients decomposed in spherical harmonics derived from Čerenkov radiation [56].

Coefficient	Lower Bound	Upper Bound
$\bar{s}_{00}^{(SH)}$	-3×10^{-14}	
$\bar{s}_{10}^{(SH)}$	-1×10^{-13}	7×10^{-14}
Re $\bar{s}_{11}^{(SH)}$	-8×10^{-14}	8×10^{-14}
Im $\bar{s}_{11}^{(SH)}$	-7×10^{-14}	9×10^{-14}
$\bar{s}_{20}^{(SH)}$	-7×10^{-14}	1×10^{-13}
Re $\bar{s}_{21}^{(SH)}$	-7×10^{-14}	7×10^{-14}
Im $\bar{s}_{21}^{(SH)}$	-5×10^{-14}	8×10^{-14}
Re $\bar{s}_{22}^{(SH)}$	-6×10^{-14}	8×10^{-14}
Im $\bar{s}_{22}^{(SH)}$	-7×10^{-14}	7×10^{-14}

For the sake of completeness and to allow an easy comparison with the estimations of the other standard cartesian $\bar{s}^{\mu\nu}$ coefficients, the following relations give the links between the spherical harmonic decomposition and the standard cartesian decomposition of the SME coefficients:

$$\bar{s}_{00}^{(SH)} = \frac{4}{3}\sqrt{4\pi}\,\bar{s}^{TT}, \tag{32a}$$

$$\bar{s}_{10}^{(SH)} = -\sqrt{\frac{16\pi}{3}}\bar{s}^{TZ}, \qquad \text{Re}\,\bar{s}_{11}^{(SH)} = \sqrt{\frac{8\pi}{3}}\bar{s}^{TX}, \qquad \text{Im}\,\bar{s}_{11}^{(SH)} = -\sqrt{\frac{8\pi}{3}}\bar{s}^{TY}, \tag{32b}$$

$$\bar{s}_{20}^{(SH)} = -\sqrt{\frac{4\pi}{5}}\frac{1}{3}\left(\bar{s}^{XX} + \bar{s}^{YY} - 2\bar{s}^{ZZ}\right), \quad \text{Re}\,\bar{s}_{21}^{(SH)} = -\sqrt{\frac{8\pi}{15}}\bar{s}^{XZ}, \quad \text{Im}\,\bar{s}_{21}^{(SH)} = \sqrt{\frac{8\pi}{15}}\bar{s}^{YZ}, \tag{32c}$$

$$\text{Re}\,\bar{s}_{22}^{(SH)} = \sqrt{\frac{2\pi}{15}}\left(\bar{s}^{XX} - \bar{s}^{YY}\right), \qquad \text{Im}\,\bar{s}_{22}^{(SH)} = -2\sqrt{\frac{2\pi}{15}}\bar{s}^{XY}. \tag{32d}$$

In summary, observations of cosmic rays allow one to derive some stringent boundaries on the SME coefficients. The idea is that if Lorentz symmetry is broken, these high energy cosmic rays would have lost energy by emitting Čerenkov radiation that has not been observed. The boundaries on the spherical harmonic decomposition of the SME coefficients are given in the Table 7 (in order to compare these boundaries to other constraints, they have been transformed into boundaries on standard cartesian SME coefficients in Table 8). For the minimal SME, one can limit the isotropic $\bar{s}^{TT}$ (one sided bound) or the other eight other coefficients in $\bar{s}^{\mu\nu}$, but not all the nine simultaneously. These boundaries are currently the best available in the literature at the exception of $\bar{s}^{TT}$ whose constraint is only one sided. Nevertheless, several assumptions have been made in this analysis and the observations rely on sources located at very high distances. This analysis is therefore of a different nature than the other ones where more control on the measurements is possible.

Table 8. Summary of all estimations of the $\bar{s}^{\mu\nu}$ coefficients.

Coefficient	Atomic Grav. [48]	LLR [46]	Planetary Eph. [50]	Pulsars [53,54]	Čerenkov rad. [56] Lower Bound	Upper Bound
$\bar{s}^{TT}$				$< 2.8 \times 10^{-4}$	$-6 \times 10^{-15} <$	
$\bar{s}^{XX} - \bar{s}^{YY}$	$(4.4 \pm 11) \times 10^{-9}$	$(0.6 \pm 4.2) \times 10^{-11}$	$(-0.8 \pm 2.0) \times 10^{-10}$	$(0.2 \pm 9.9) \times 10^{-11}$	$-9 \times 10^{-14} <$	$< 1.2 \times 10^{-13}$
$\bar{s}^{XX} + \bar{s}^{YY} - 2\bar{s}^{ZZ}$			$(-0.8 \pm 2.7) \times 10^{-10}$	$(-0.05 \pm 12.25) \times 10^{-11}$	$-1.9 \times 10^{-13} <$	$< 1.3 \times 10^{-13}$
$\bar{s}^{XY}$	$(0.2 \pm 3.9) \times 10^{-9}$	$(-5.7 \pm 7.7) \times 10^{-12}$	$(-0.3 \pm 1.1) \times 10^{-10}$	$(0.05 \pm 3.55) \times 10^{-11}$	$-3.9 \times 10^{-14} <$	$< 6.2 \times 10^{-14}$
$\bar{s}^{XZ}$	$(-2.6 \pm 4.4) \times 10^{-9}$	$(-2.2 \pm 5.9) \times 10^{-12}$	$(-1.0 \pm 3.5) \times 10^{-11}$	$(0.0 \pm 2.0) \times 10^{-11}$	$-5.4 \times 10^{-14} <$	$< 5.4 \times 10^{-14}$
$\bar{s}^{YZ}$	$(-0.3 \pm 4.5) \times 10^{-9}$		$(5.5 \pm 5.2) \times 10^{-12}$	$(0.0 \pm 3.3) \times 10^{-11}$	$-3.9 \times 10^{-14} <$	$< 6.2 \times 10^{-14}$
$\bar{s}^{TX}$	$(-3.1 \pm 5.1) \times 10^{-5}$	$(-0.9 \pm 1.0) \times 10^{-8}$	$(-2.9 \pm 8.3) \times 10^{-9}$	$(0.05 \pm 5.25) \times 10^{-9}$	$2.8 \times 10^{-14} <$	$< 2.8 \times 10^{-14}$
$\bar{s}^{TY}$	$(0.1 \pm 5.4) \times 10^{-5}$		$(0.3 \pm 1.4) \times 10^{-8}$	$(0.5 \pm 8.0) \times 10^{-9}$	$3.1 \times 10^{-14} <$	$< 2.4 \times 10^{-14}$
$\bar{s}^{TZ}$	$(1.4 \pm 6.6) \times 10^{-5}$		$(-0.2 \pm 5.0) \times 10^{-8}$	$(-0.05 \pm 5.85) \times 10^{-9}$	$1.7 \times 10^{-14} <$	$< 2.4 \times 10^{-14}$
$\bar{s}^{TY} + 0.43\,\bar{s}^{TZ}$		$(6.2 \pm 7.9) \times 10^{-9}$				
$\bar{s}^{XX} + \bar{s}^{YY} - 2\bar{s}^{ZZ} - 4.5\,\bar{s}^{YZ}$		$(2.3 \pm 4.5) \times 10^{-11}$				

Coefficient	VLBI [55]	GPB [52]
$\bar{s}^{TT}$	$(-5 \pm 8) \times 10^{-5}$	
$\bar{s}^{TT} + 970\,(\bar{s}^{XX} - \bar{s}^{YY}) - 0.05\,(\bar{s}^{XX} + \bar{s}^{YY} - 2\bar{s}^{ZZ}) + 2895\,\bar{s}^{XY} - 3235\,\bar{s}^{XZ} - 11240\,\bar{s}^{YZ}$		$(0.7 \pm 3.1) \times 10^{-3}$
$\bar{s}^{XX} - \bar{s}^{YY} + 3.02\,\bar{s}^{XY} + 0.32\,\bar{s}^{XZ} + 1.05\,\bar{s}^{YZ}$		$(-1.1 \pm 3.8) \times 10^{-7}$

4.8. Summary and Combined Analysis

To summarize, several measurements have already successfully been used to constrain the minimal SME in the gravitational sector (i.e., the $\bar{s}^{\mu\nu}$ coefficients):

- Atom interferometry [47,48].
- Lunar Laser Ranging [45,46].
- Planetary ephemerides [49,50].
- Very Long Baseline Interferometry [55].
- Gravity Probe B [52].
- Pulsars timing [53,54].
- Čerenkov radiation [30,56].

A detailed description of all these analyses is provided in the previous subsections and the Table 8 summarizes the current estimates. It is also interesting to combine all these estimations together to provide the best estimates on the SME coefficients. In order to do this, we perform a large least-square fit including all the results from the Table 8 including the covariance matrices quoted in the previous subsections. The results from the Čerenkov radiation are not included since they rely on a very different type of observations. Two combined fits are presented: one without including the pulsars results and one including the pulsars results. This is due to the fact that pulsars are sensitive to a strong version of the SME coefficients that may include non perturbative strong field effects as described in Section 4.6. If this is the case, then the pulsars results cannot be directly combined with the weak gravitational field estimates on the SME coefficients. If no non perturbative strong field effect arises, then the right column from Table 9 presents a combined fit that includes these observations as well. The results from Table 9 include all the information currently available in the literature on the $\bar{s}^{\mu\nu}$ (estimations and correlation matrices). It can also be noted that the pulsars results improve significantly the marginalized estimations on $\bar{s}^{TY}$ and $\bar{s}^{TZ}$ by reducing strongly the correlation between these two coefficients.

In addition, several measurements have been used to constrain the non-minimal SME sectors:

- Short gravity experiment [57–59].
- Čerenkov radiation [56].
- Gravitational waves analysis [60].

A review of these measurements can be found in [30].

Table 9. Estimation of SME coefficients resulting from a fit combining results from: atomic gravimetry (see Table 1), VLBI (see Equation (13)), LLR (see Table 3), planetary ephemerides (see Table 4), Gravity Probe B (see Table 5). The correlation matrices from all these analyses have been used in the combined fit. The right column includes the pulsars results from Table 6 as well. The three estimates on $\bar{s}^{JJ}$ are obtained by using the traceless condition $\bar{s}^{TT} = \bar{s}^{XX} + \bar{s}^{YY} + \bar{s}^{ZZ}$.

Coefficient	Without Pulsars	With Pulsars
$\bar{s}^{TT}$	$(-5. \pm 8.) \times 10^{-5}$	$(-4.6 \pm 7.7) \times 10^{-5}$
$\bar{s}^{XX} - \bar{s}^{YY}$	$(-0.5 \pm 1.9) \times 10^{-11}$	$(-0.5 \pm 1.9) \times 10^{-11}$
$\bar{s}^{XX} + \bar{s}^{YY} - 2\bar{s}^{ZZ}$	$(1.6 \pm 3.1) \times 10^{-11}$	$(0.8 \pm 2.5) \times 10^{-11}$
$\bar{s}^{XY}$	$(-1.5 \pm 6.8) \times 10^{-12}$	$(-1.6 \pm 6.6) \times 10^{-12}$
$\bar{s}^{XZ}$	$(-1.0 \pm 4.1) \times 10^{-12}$	$(-0.8 \pm 3.9) \times 10^{-12}$
$\bar{s}^{YZ}$	$(2.6 \pm 4.7) \times 10^{-12}$	$(1.1 \pm 3.2) \times 10^{-12}$
$\bar{s}^{TX}$	$(-0.1 \pm 1.3) \times 10^{-9}$	$(-0.1 \pm 1.3) \times 10^{-9}$
$\bar{s}^{TY}$	$(0.5 \pm 1.1) \times 10^{-8}$	$(0.4 \pm 2.3) \times 10^{-9}$
$\bar{s}^{TZ}$	$(-1.2 \pm 2.7) \times 10^{-8}$	$(-0.6 \pm 5.5) \times 10^{-9}$
$\bar{s}^{XX}$	$(-1.7 \pm 2.7) \times 10^{-5}$	$(-1.5 \pm 2.6) \times 10^{-5}$
$\bar{s}^{YY}$	$(-1.7 \pm 2.7) \times 10^{-5}$	$(-1.5 \pm 2.6) \times 10^{-5}$
$\bar{s}^{ZZ}$	$(-1.7 \pm 2.7) \times 10^{-5}$	$(-1.5 \pm 2.6) \times 10^{-5}$

5. The future

In addition to all the improvements related to existing analysis suggested in the previous sections, there are a couple of sensitivity analyses that have been done within the SME framework. First of all, a thorough and detailed analysis of a lot of observables related to gravitation can be found in [39]. In addition, we will present in the next subsections a couple of analyses and ideas that may improve the SME coefficients estimates in the future.

5.1. The Gaia Mission

Launched in December 2013, the ESA Gaia mission [196] is scanning regularly the whole celestial sphere once every 6 months providing high precision astrometric data for a huge number ($\approx$1 billion) of celestial bodies. In addition to stars, it is also observing Solar System Objects (SSO), in particular asteroids. The high precision astrometry (at sub-mas level) will allow us to perform competitive tests of gravitation and to provide new constraints on alternative theories of gravitation.

First of all, the Gaia mission is expected to provide an estimate of the γ PPN parameter at the level of 10^{-6} [197] by measuring the deflection of the light on a 5 years timescale. Furthermore, in addition to this global determination of a global PPN parameter from observations of light deflection, it has been proposed to use Gaia observations to map the deflection angle in the sky and to look for a dependence of the γ PPN parameter with respect to the Sun impact parameter [198–202]. Such a dependence of the gravitational deflection with respect to the observation geometry is also a feature predicted by SME as shown in [82]. Therefore, the global mapping of the light deflection with Gaia can also be efficiently used to constrain some SME coefficients. A first sensitivity analysis can be found in [82] and is reported on Table 10. Note that proposals observations and missions like AGP [203] or LATOR [204] can in the long term improve these estimates further by improving the light deflection measurement.

Table 10. Sensitivity of the SME coefficients to the measurement of the light deflection by several space missions or proposals (these estimates are based on Table I from [82]).

Mission	$\bar{s}^{TT}$	$\bar{s}^{TJ}$	$\bar{s}^{IJ}$
Gaia [196]	10^{-6}	10^{-6}	10^{-5}
AGP [203]	10^{-7}	10^{-7}	10^{-6}
LATOR [204]	10^{-8}	10^{-8}	10^{-7}

In addition to gravitation tests performed by measuring the light deflection, Gaia also provides a unique opportunity to test gravitation by considering the orbital dynamics of SSO. One can estimate that about 360,000 asteroids will be regularly observed by Gaia at the sub-mas level, which will allow us to perform various valuable tests of gravitation [205,206]. In particular, realistic simulations of more than 250,000 asteroids have shown that Gaia will be able to constrain the β PPN parameter at the level of 10^{-3} [205]. The main advantage from Gaia is related to the huge number of bodies that will be observed with very different orbital parameters as illustrated on Figure 1. As a consequence, the huge correlations appearing in the planetary ephemerides analysis (see Section 4.4) will not appear when considering asteroids observations and the marginalized confidence intervals will be highly improved compared to planetary ephemerides analysis.

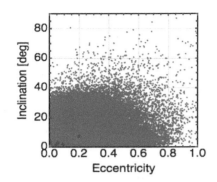 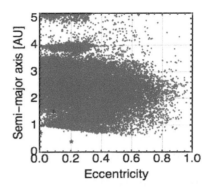

Figure 1. This figure represents the distribution of the orbital parameters for the Solar System Objects (SSOs) expected to be observed by the Gaia satellite. The red stars represent the innermost planets of the Solar System.

A realistic sensitivity analysis of Gaia SSOs observations within the SME framework has been performed (see also [206] for preliminary results). In this analysis, 360,000 asteroids have been considered over the nominal mission duration (i.e., five years) and a match between the SSO trajectories with the Gaia scanning law is performed to find the observation times for each SSO. Simultaneously with the equations of motion, we integrate the variational equations, the simulated SSO trajectories being transformed into astrometric observables as well as their partial derivatives with respect to the parameters considered in the covariance analysis. The covariance analysis leads to the estimated uncertainties presented in Table 11. These uncertainties are incredibly good, which is due to the variety of the asteroids orbital parameters as discussed above. Using our set of asteroids, the correlation matrix for the SME parameters is very reasonable: the most important correlation coefficients are 0.71, −0.68 and 0.46. All the other correlations are below 0.3. Therefore, Gaia offers a unique opportunity to constrain Lorentz violation through the SME formalism. Finally, the Gaia mission is likely to be extended to 10 years, therefore doubling the measurements baseline which will also impact significantly the expected uncertainties. Finally, it is worth mentioning that the Gaia dataset can be combined with radar observations [207] that are complementary in the time frame and orthogonal to astrometric telescopic observations.

Table 11. Sensitivity of the SME coefficients to the observations of 360,000 asteroids by the Gaia satellite during a period of 5 years.

SME Coefficients	Sensitivity $(1 - \sigma)$
$\bar{s}^{XX} - \bar{s}^{YY}$	3.7×10^{-12}
$\bar{s}^{XX} + \bar{s}^{YY} - 2\bar{s}^{ZZ}$	6.4×10^{-12}
$\bar{s}^{XY}$	1.6×10^{-12}
$\bar{s}^{XZ}$	9.2×10^{-13}
$\bar{s}^{YZ}$	1.7×10^{-12}
$\bar{s}^{TX}$	5.6×10^{-9}
$\bar{s}^{TY}$	8.8×10^{-9}
$\bar{s}^{TZ}$	1.6×10^{-8}

In summary, the Gaia space mission offers two opportunities to test Lorentz symmetry in the Solar System by looking at the deflection of light and by considering the orbital dynamics of SSO. The second type of observations is extremely interesting in the sense that the high number and the variety of orbital parameters of the observed SSO leads to decorrelate the SME coefficients.

5.2. Analysis of Cassini Conjunction Data

The space mission Cassini is exploring the Saturnian system since July 2004. During its cruising phase while the spacecraft was on its interplanetary journey between Jupiter and Saturn, a measurement of the gravitational time delay was performed [118]. This measurement occurred during a Solar conjunction in June 2002 and was made possible thanks to a multi-frequency radioscience link (at X and Ka-band) which allows a cancellation of the solar plasma noise [118]. The related data spans over 30 days and has been analyzed in the PPN framework leading to the best estimation of the γ PPN parameter so far given by $(2.1 \pm 2.3) \times 10^{-5}$ [118].

The exact same set of data can be reduced within the SME framework and is expected to improve our current $\bar{s}^{TT}$ estimation. The time delay within the SME framework has been derived in [80] and is given by Equation (9).

A simulation of the Cassini link during the 2002 conjunction within the full SME framework has been realized using the software presented in [148] (see also [208,209]). The signature produced by the $\bar{s}^{TT}$ coefficients on the 2-way Doppler link during the Solar conjunction is illustrated on Figure 2. In [80], a crude estimate of attainable sensitivities in estimate of the SME coefficients using the Cassini conjunction data is given (see Table I from [80]). It is shown that some combinations of the $\bar{s}^{IJ}$ coefficients can only be constrained at the level of 10^{-4}, which is 7 to 8 orders of magnitude worse than the current best constraints on these coefficients. It is therefore safe to neglect these and to concentrate only on the $\bar{s}^{TT}$ coefficient. A realistic covariance analysis performed over the 30 days of the Solar conjunction and assuming an uncertainty of the Cassini Doppler of 3 μm/s [210,211] shows that the $\bar{s}^{TT}$ parameter can be constrained at the level of 2×10^{-5} using the Cassini data allowing an improvement of a factor 4 with respect to the current best estimate coming from VLBI analysis (see Equation (13)). Therefore, a reanalysis of the 2002 Cassini data within the SME framework would be highly valuable.

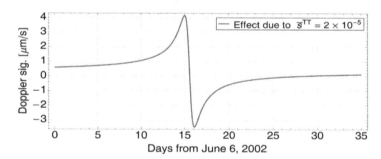

Figure 2. Doppler signature produced by $\bar{s}^{TT} = 2 \times 10^{-5}$ on the 2-way Doppler link Earth-Cassini-Earth during the 2002 Solar conjunction.

5.3. Satellite Laser Ranging (LAGEOS/LARES)

Searching for violations of Lorentz symmetry by using the orbital motion of planets (see Section 4.4), binary pulsars (see Section 4.6), the Moon (see Section 4.3) and asteroids (see Section 5.1) has turned out to be highly powerful. It is therefore logical to consider the motion of artificial satellite orbiting around Earth to search for Lorentz violations. In particular, laser ranging to the two LAGEOS and to the LARES satellites has successfully been used to test GR by measuring the impact of the Schwarzschild precession on the motion of the satellites [212–214]. It has also been claimed that the impact of the frame-dragging (or Lense-Thirring effect) due to the Earth's spin on the orbital motion of the satellites has been measured [215–220] although this claim remains controversial [221–226]. Similarly, the LAGEOS/LARES satellites can also be used to search for Lorentz violations. A sensitivity analysis has been done in [49] and it has been shown that the LAGEOS satellites are sensitive at the level of 10^{-4} to the $\bar{s}^{TJ}$ coefficients. Using LARES should improve significantly this value. Further numerical simulations are required in order to determine exactly the SME linear combinations to which the ranging to these satellites is sensitive to. A data analysis within the full

SME framework (i.e., including the integration of the SME equation of motion and including the SME coefficients with the other global parameters in the fit) would also be highly interesting. In addition, similar tests of Lorentz symmetry can also be included within the scientific goals of the LAser RAnged Satellites Experiment (LARASE) project [227] or within the OPTIS project [228].

5.4. Gravity-Matter Coefficients and Breaking of the Einstein Equivalence Principle

All the measurements mentioned in Section 4 can be analyzed by considering the gravity-matter coupling coefficients $\bar{a}_\mu$ and $\bar{c}_{\mu\nu}$ [29] that are breaking the EEP. Some atomic clocks measurements have already provided some constraints on the $\bar{a}_\mu$ coefficients [229–231]. In addition, in [50] the planetary ephemerides analysis is interpreted by considering the $\bar{a}_\mu$ coefficients and the atomic interferometry results from [48] and the LLR results from [45] are also reinterpreted by considering the gravity-coupling coefficients. Clearly this is a preliminary analysis that needs to be refined by more solid data reductions. Considering the increasing number of fitted parameters, it is of prime importance to increase the number of measurements used in the analysis and to produce combined analysis with as many types of observations as possible. The measurements developed in Section 4 are a first step in order to reach this goal. The gravity-coupling coefficients can also be constrained by more specific tests related to the EEP like for example tests of the Universality of Free Fall with MicroSCOPE [83,84], tests of the gravitational redshift with GNSS satellites [232], with the Atomic Clocks Ensemble in Space (ACES) project [233], or with the OPTIS project [228], . . .

6. Conclusions

Lorentz symmetry is at the heart of both GR and the Standard Model of particle physics. This symmetry is broken in various scenarios of unification, of quantum gravity and even in some models of Dark Matter and Dark Energy. Searching for violations of Lorentz symmetry is therefore a powerful tool to test fundamental physics. The last decades have seen the number of tests of Lorentz invariance arise dramatically in all sectors of physics [24]. In this review, we focused on searches for Lorentz symmetry breaking in the pure gravitational sector. Mainly two frameworks exist to parametrize violations of Lorentz invariance in the gravitation sector. First of all, the three $\alpha_{1,2,3}$ PPN parameters phenomenologically encode a violation of Lorentz symmetry at the level of the spacetime metric [2]. These parameters are constrained by LLR (see Section 4.3) and by pulsars timing measurements (see Section 4.6). In addition, it is interesting to notice that the corresponding PPN metric parametrizes also Einstein-Aether and Khronometric theories in the weak gravitational field limit [72] while these theories have a more complex strong field limit (and can show non perturbative effects) that have been constrained by pulsars observations (see Section 4.6 and [72,73]).

In addition to the PPN formalism, the SME formalism has been developed by including systematically all possible Lorentz violations terms that can be constructed at the level of the action. In the pure gravitational sector, the gravitational action within the SME formalism contains the usual Einstein-Hilbert action but also new Lorentz violating terms constructed by contracting new fields with some operators built from curvature tensors and covariant derivatives with increasing mass dimension [36]. The lower mass dimension term is known as the minimal SME. In the limit of linearized gravity, the observations within the minimal SME formalism depend on 9 coefficients, the $\bar{s}^{\mu\nu}$ symmetric traceless tensor. This formalism offers a new opportunity to search for deviations from GR in a framework different from the standard PPN formalism. We reviewed the different observations that have been used so far to constrain the SME coefficients. The main idea is to search for a signature (usually periodic) that arises from a dependence on the orientation of the system measured (the dependence on the orientation is typically due to the Earth's rotation, the orbital motion of the planets around the Sun, etc. . .) or from a dependence on the boost of the system observed (so far, only the binary pulsars $\bar{s}^{TT}$ constraint comes from this type of dependence [54]). Most of SME analyses are postfit analyses in the sense that analytical signatures due to SME are fitted in residuals noise obtained in a previous data reduction performed in pure GR. In Section 3, we showed that this approach can sometimes lead to overoptimistic constraint on the SME coefficients and that one should be careful in interpreting results obtained using such an approach.

In Section 4, we discussed in details the different measurements used so far to constrain the $\bar{s}^{\mu\nu}$ coefficients: atomic gravimetry (Section 4.1), VLBI (Section 4.2), LLR (Section 4.3), planetary ephemerides (Section 4.4), Gravity Probe B (Section 4.5), pulsars timing (Section 4.6) and Čerenkov radiation (Section 4.7). In each of these subsections, we describe the current analyses performed in order to constrain the SME coefficients and provide a critical discussion from each of them. We also provide a summary of these constraints on Table 8. In addition, we used all these results to produce a combined analysis of the SME coefficients. This fit is done by taking into account the correlation matrices for each individual analysis. The results of this combined fit are presented in Table 9 and are the current best estimates of the SME coefficients that are possible to derive with all available analyses. In addition to the minimal SME, there exists higher order Lorentz violating terms that have been considered and constrained by short-range gravity experiments [57–59], gravitational waves analysis [60] and Čerenkov radiation [30,56].

In Section 5, we discussed some opportunities to improve the current constraints on the SME coefficients. In particular, the European space mission Gaia offers an excellent opportunity to probe Lorentz symmetry through the measurement of light deflection and through the orbital motion of asteroids. The Cassini conjunction data also offers a way to constrain the $\bar{s}^{TT}$ coefficient that impacts severely the propagation of light. Finally, existing satellite laser ranging data can also be analyzed within the SME framework.

In addition, as mentioned in Section 5.4, all the analyses presented in this review can include gravity-matter coefficients [29]. While considering these, the number of coefficients fitted increase significantly and it becomes crucial to produce a fit combining several kinds of experiments. A preliminary analysis considering these coefficients for planetary ephemerides, LLR and atomic gravimetry has been performed in [50] but needs to be refined. In addition, some atomic clocks experiments have already been used to constrain matter-gravity coefficients [229–231].

In conclusion, though no violation of Lorentz symmetry has been observed so far, an incredible number of opportunities still exists for additional investigations. There remains a large area of unexplored coefficients space that can be explored by improved measurements or by new projects aiming at searching for breaking of Lorentz symmetry. In addition, the increasing number of parameters fitted (by including the gravity-matter coupling coefficients simultaneously with the pure gravity coefficients in the analyses) will deter the marginalized estimates of each coefficient. This verdict emphasizes the need to increase the types of measurements that can be combined together to explore the vast parameters space as efficiently as possible. The current theoretical questions related to the quest for a unifying theory or for a quantum theory of gravitation suggests that Lorentz symmetry will play an important role in the search for new physics. Hopefully, future searches for Lorentz symmetry breaking will help theoreticians to unveil some of the mysteries about Planck-scale physics [22].

Acknowledgments: A.H. is thankful to P. Wolf, S. Lambert, B. Lamine, A. Rivoldini, F. Meynadier, S. Bouquillon, G. Francou, M.-C. Angonin, D. Hestroffer, P. David and A. Fienga for interesting discussions about some part of this work. Q.G.B. was supported in part by the National Science Foundation under Grant No. PHY-1402890. A.B. and C.L.P.L. are grateful for the CNRS/GRAM and "Axe Gphys" of Paris Observatory Scientific Council. C.G. and Q.G.B. acknowledge support from Sorbonne Universités Emergence grant.

Author Contributions: All the authors wrote the paper.

Abbreviations

The following abbreviations are used in this manuscript:

CMB	Cosmic Microwave Background
ELPN	Éphéméride Lunaire Parisienne Numérique
GPB	Gravity Probe B
GR	General Relativity
GRAIL	Gravity Recovery And Interior Laboratory
INPOP	Intégrateur Numérique Planétaire de l'Observatoire de Paris
IVS	International VLBI Service for Geodesy and Astrometry
LARASE	LAser RAnged Satellites Experiment
LLR	Lunar Laser Ranging
LIV	Lorentz Invariance Violation
mas	milliarcsecond
Mpc	Megaparsec
NFT	Numerical Fourier Transform
PPN	Parametrized Post-Newtonian
SME	Standard-Model Extension
SSO	Solar System Object
TAI	International Atomic Time
TDB	Barycentric Dynamical Time
TT	Terrestrial Time
UTC	Universal Time Coordinate
VLBI	Very Long Baseline Interferometry
yr	year

References

1. Iorio, L. Editorial for the Special Issue 100 Years of Chronogeometrodynamics: The Status of the Einstein's Theory of Gravitation in Its Centennial Year. *Universe* **2015**, *1*, 38–81.

2. Will, C.M. The Confrontation between General Relativity and Experiment. *Living Rev. Relativ.* **2014**, *17*, 4.

3. Turyshev, S.G. REVIEWS OF TOPICAL PROBLEMS: Experimental tests of general relativity: Recent progress and future directions. *Phys. Uspekhi* **2009**, *52*, 1–27.

4. Abbott, B.P.; Abbott, R.; Abbott, T.D.; Abernathy, M.R.; Acernese, F.; Ackley, K.; Adams, C.; Adams, T.; Addesso, P.; Adhikari, R.X.; et al. Observation of Gravitational Waves from a Binary Black Hole Merger. *Phys. Rev. Lett.* **2016**, *116*, 061102.

5. Cervantes-Cota, J.; Galindo-Uribarri, S.; Smoot, G. A Brief History of Gravitational Waves. *Universe* **2016**, *2*, 22.

6. Debono, I.; Smoot, G.F. General Relativity and Cosmology: Unsolved Questions and Future Directions. *Universe* **2016**, *2*, 23.

7. Berti, E.; Barausse, E.; Cardoso, V.; Gualtieri, L.; Pani, P.; Sperhake, U.; Stein, L.C.; Wex, N.; Yagi, K.; Baker, T. et al. Testing general relativity with present and future astrophysical observations. *Class. Quantum Gravity* **2015**, *32*, 243001.

8. Colladay, D.; Kostelecký, V.A. CPT violation and the standard model. *Phys. Rev. D* **1997**, *55*, 6760–6774.

9. Kostelecký, V.A.; Samuel, S. Gravitational phenomenology in higher-dimensional theories and strings. *Phys. Rev. D* **1989**, *40*, 1886–1903.

10. Kostelecký, V.A.; Samuel, S. Spontaneous breaking of Lorentz symmetry in string theory. *Phys. Rev. D* **1989**, *39*, 683–685.

11. Kostelecký, V.A.; Potting, R. CPT and strings. *Nucl. Phys. B* **1991**, *359*, 545.

12. Gambini, R.; Pullin, J. Nonstandard optics from quantum space-time. *Phys. Rev. D* **1999**, *59*, 124021.

13. Amelino-Camelia, G. Quantum-Spacetime Phenomenology. *Living Rev. Relativ.* **2013**, *16*, 5.

14. Mavromatos, N.E. *CPT Violation and Decoherence in Quantum Gravity*; Lecture Notes in Physics; Kowalski-Glikman, J., Amelino-Camelia, G., Eds.; Springer: Berlin, Germany, 2005.

15. Myers, R.C.; Pospelov, M. Ultraviolet Modifications of Dispersion Relations in Effective Field Theory. *Phys. Rev. Lett.* **2003**, *90*, 211601.

16. Hayakawa, M. Perturbative analysis on infrared aspects of noncommutative QED on R^4. *Phys. Lett. B* **2000**, *478*, 394–400.

17. Carroll, S.M.; Harvey, J.A.; Kostelecký, V.A.; Lane, C.D.; Okamoto, T. Noncommutative Field Theory and Lorentz Violation. *Phys. Rev. Lett.* **2001**, *87*, 141601.

18. Bjorken, J.D. Cosmology and the standard model. *Phys. Rev. D* **2003**, *67*, 043508.

19. Burgess, C.P.; Martineau, P.; Quevedo, F.; Rajesh, G.; Zhang, R.J. Brane-antibrane inflation in orbifold and orientifold models. *J. High Energy Phys.* **2002**, *3*, 052.

20. Frey, A.R. String theoretic bounds on Lorentz-violating warped compactification. *J. High Energy Phys.* **2003**, *4*, 12.

21. Cline, J.M.; Valcárcel, L. Asymmetrically warped compactifications and gravitational Lorentz violation. *J. High Energy Phys.* **2004**, *3*, 032.

22. Tasson, J.D. What do we know about Lorentz invariance? *Rep. Prog. Phys.* **2014**, *77*, 062901.

23. Mattingly, D. Modern Tests of Lorentz Invariance. *Living Rev. Relativ.* **2005**, *8*, 5.

24. Kostelecký, V.A.; Russell, N. Data tables for Lorentz and CPT violation. *Rev. Mod. Phys.* **2011**, *83*, 11–32.

25. Colladay, D.; Kostelecký, V.A. Lorentz-violating extension of the standard model. *Phys. Rev. D* **1998**, *58*, 116002.

26. Thorne, K.S.; Will, C.M. Theoretical Frameworks for Testing Relativistic Gravity. I. Foundations. *Astrophys. J.* **1971**, *163*, 595.

27. Will, C.M. *Theory and Experiment in Gravitational Physics*; Cambridge University Press: Cambridge, UK, 1993.

28. Thorne, K.S.; Lee, D.L.; Lightman, A.P. Foundations for a Theory of Gravitation Theories. *Phys. Rev. D* **1973**, *7*, 3563–3578.

29. Kostelecký, V.A.; Tasson, J.D. Matter-gravity couplings and Lorentz violation. *Phys. Rev. D* **2011**, *83*, 016013.

30. Tasson, J.D. The Standard-Model Extension and Gravitational Tests. *Symmetry* **2016**, *8*, 111.

31. Fischbach, E.; Sudarsky, D.; Szafer, A.; Talmadge, C.; Aronson, S.H. Reanalysis of the Eotvos experiment. *Phys. Rev. Lett.* **1986**, *56*, 3–6.

32. Talmadge, C.; Berthias, J.P.; Hellings, R.W.; Standish, E.M. Model-independent constraints on possible modifications of Newtonian gravity. *Phys. Rev. Lett.* **1988**, *61*, 1159–1162.

33. Fischbach, E.; Talmadge, C.L. *The Search for Non-Newtonian Gravity*; Aip-Press Series; Springer: New York, NY, USA, 1999.

34. Adelberger, E.G.; Heckel, B.R.; Nelson, A.E. Tests of the Gravitational Inverse-Square Law. *Annu. Rev. Nucl. Part. Sci.* **2003**, *53*, 77–121.

35. Reynaud, S.; Jaekel, M.T. Testing the Newton Law at Long Distances. *Int. J. Mod. Phys. A* **2005**, *20*, 2294–2303.

36. Bailey, Q.G. Gravity Sector of the SME. In Proceedings of the Seventh Meeting on CPT and Lorentz Symmetry, Bloomington, IN, USA, 20–24 June 2016.

37. Kostelecký, V.A. Gravity, Lorentz violation, and the standard model. *Phys. Rev. D* **2004**, *69*, 105009.

38. Bluhm, R.; Kostelecký, V.A. Spontaneous Lorentz violation, Nambu-Goldstone modes, and gravity. *Phys. Rev. D* **2005**, *71*, 065008.

39. Bailey, Q.G.; Kostelecký, V.A. Signals for Lorentz violation in post-Newtonian gravity. *Phys. Rev. D* **2006**, *74*, 045001.

40. Bluhm, R. Nambu-Goldstone Modes in Gravitational Theories with Spontaneous Lorentz Breaking. *Int. J. Mod. Phys. D* **2007**, *16*, 2357–2363.

41. Bluhm, R.; Fung, S.H.; Kostelecký, V.A. Spontaneous Lorentz and diffeomorphism violation, massive modes, and gravity. *Phys. Rev. D* **2008**, *77*, 065020.

42. Bluhm, R. Explicit versus spontaneous diffeomorphism breaking in gravity. *Phys. Rev. D* **2015**, *91*, 065034.

43. Bailey, Q.G.; Kostelecký, V.A.; Xu, R. Short-range gravity and Lorentz violation. *Phys. Rev. D* **2015**, *91*, 022006.

44. Bonder, Y. Lorentz violation in the gravity sector: The t puzzle. *Phys. Rev. D* **2015**, *91*, 125002.

45. Battat, J.B.R.; Chandler, J.F.; Stubbs, C.W. Testing for Lorentz Violation: Constraints on Standard-Model-Extension Parameters via Lunar Laser Ranging. *Phys. Rev. Lett.* **2007**, *99*, 241103.

46. Bourgoin, A.; Hees, A.; Bouquillon, S.; Le Poncin-Lafitte, C.; Francou, G.; Angonin, M.C. Testing Lorentz symmetry with Lunar Laser Ranging. *arXiv* **2016**, arXiv:gr-qc/1607.00294

47. Müller, H.; Chiow, S.W.; Herrmann, S.; Chu, S.; Chung, K.Y. Atom-Interferometry Tests of the Isotropy of Post-Newtonian Gravity. *Phys. Rev. Lett.* **2008**, *100*, 031101.

48. Chung, K.Y.; Chiow, S.W.; Herrmann, S.; Chu, S.; Müller, H. Atom interferometry tests of local Lorentz invariance in gravity and electrodynamics. *Phys. Rev. D* **2009**, *80*, 016002.

49. Iorio, L. Orbital effects of Lorentz-violating standard model extension gravitomagnetism around a static body: A sensitivity analysis. *Class. Quantum Gravity* **2012**, *29*, 175007.

50. Hees, A.; Bailey, Q.G.; Le Poncin-Lafitte, C.; Bourgoin, A.; Rivoldini, A.; Lamine, B.; Meynadier, F.; Guerlin, C.; Wolf, P. Testing Lorentz symmetry with planetary orbital dynamics. *Phys. Rev. D* **2015**, *92*, 064049.

51. Bennett, D.; Skavysh, V.; Long, J. Search for Lorentz Violation in a Short-Range Gravity Experiment. In Proceedings of the Fifth Meeting on CPT and Lorentz Symmetry, Bloomington, IN, USA, 28 June–2 July 2010.

52. Bailey, Q.G.; Everett, R.D.; Overduin, J.M. Limits on violations of Lorentz symmetry from Gravity Probe B. *Phys. Rev. D* **2013**, *88*, 102001.

53. Shao, L. Tests of Local Lorentz Invariance Violation of Gravity in the Standard Model Extension with Pulsars. *Phys. Rev. Lett.* **2014**, *112*, 111103.

54. Shao, L. New pulsar limit on local Lorentz invariance violation of gravity in the standard-model extension. *Phys. Rev. D* **2014**, *90*, 122009.

55. Le Poncin-Lafitte, C.; Hees, A.; lambert, S. Lorentz symmetry and Very Long Baseline Interferometry. *arXiv* **2016**, arXiv:gr-qc/1604.01663.

56. Kostelecký, V.A.; Tasson, J.D. Constraints on Lorentz violation from gravitational Čerenkov radiation. *Phys. Lett. B* **2015**, *749*, 551–559.

57. Shao, C.G.; Tan, Y.J.; Tan, W.H.; Yang, S.Q.; Luo, J.; Tobar, M.E. Search for Lorentz invariance violation through tests of the gravitational inverse square law at short ranges. *Phys. Rev. D* **2015**, *91*, 102007.

58. Long, J.C.; Kostelecký, V.A. Search for Lorentz violation in short-range gravity. *Phys. Rev. D* **2015**, *91*, 092003.

59. Shao, C.G.; Tan, Y.J.; Tan, W.H.; Yang, S.Q.; Luo, J.; Tobar, M.E.; Bailey, Q.G.; Long, J.C.; Weisman, E.; Xu, R.; et al. Combined Search for Lorentz Violation in Short-Range Gravity. *Phys. Rev. Lett.* **2016**, *117*, 071102.

60. Kostelecký, V.A.; Mewes, M. Testing local Lorentz invariance with gravitational waves. *Phys. Lett. B* **2016**, *757*, 510–514.

61. Yunes, N.; Yagi, K.; Pretorius, F. Theoretical physics implications of the binary black-hole mergers GW150914 and GW151226. *Phys. Rev. D* **2016**, *94*, 084002.

62. Kostelecký, V.A.; Russell, N.; Tasson, J.D. Constraints on Torsion from Bounds on Lorentz Violation. *Phys. Rev. Lett.* **2008**, *100*, 111102.

63. Heckel, B.R.; Adelberger, E.G.; Cramer, C.E.; Cook, T.S.; Schlamminger, S.; Schmidt, U. Preferred-frame and CP-violation tests with polarized electrons. *Phys. Rev. D* **2008**, *78*, 092006.

64. Kostelecký, V.A.; Mewes, M. Signals for Lorentz violation in electrodynamics. *Phys. Rev. D* **2002**, *66*, 056005.

65. Kostelecký, V.A.; Mewes, M. Electrodynamics with Lorentz-violating operators of arbitrary dimension. *Phys. Rev. D* **2009**, *80*, 015020.

66. Kostelecký, V.A. Riemann-Finsler geometry and Lorentz-violating kinematics. *Phys. Lett. B* **2011**, *701*, 137–143.

67. Kostelecký, V.A.; Russell, N. Classical kinematics for Lorentz violation. *Phys. Lett. B* **2010**, *693*, 443–447.

68. Jacobson, T.; Mattingly, D. Gravity with a dynamical preferred frame. *Phys. Rev. D* **2001**, *64*, 024028.

69. Jackiw, R.; Pi, S.Y. Chern-Simons modification of general relativity. *Phys. Rev. D* **2003**, *68*, 104012.

70. Hernaski, C.A.; Belich, H. Lorentz violation and higher derivative gravity. *Phys. Rev. D* **2014**, *89*, 104027.

71. Balakin, A.B.; Lemos, J.P.S. Einstein-aether theory with a Maxwell field: General formalism. *Ann. Phys.* **2014**, *350*, 454–484.

72. Yagi, K.; Blas, D.; Yunes, N.; Barausse, E. Strong Binary Pulsar Constraints on Lorentz Violation in Gravity. *Phys. Rev. Lett.* **2014**, *112*, 161101.

73. Yagi, K.; Blas, D.; Barausse, E.; Yunes, N. Constraints on Einstein-AEther theory and Hořava gravity from binary pulsar observations. *Phys. Rev. D* **2014**, *89*, 084067.

74. Hernaski, C.A. Quantization and stability of bumblebee electrodynamics. *Phys. Rev. D* **2014**, *90*, 124036.

75. Seifert, M.D. Vector models of gravitational Lorentz symmetry breaking. *Phys. Rev. D* **2009**, *79*, 124012.

76. Kostelecký, V.A.; Potting, R. Gravity from local Lorentz violation. *Gen. Relativ. Gravit.* **2005**, *37*, 1675–1679.

77. Kostelecký, V.A.; Potting, R. Gravity from spontaneous Lorentz violation. *Phys. Rev. D* **2009**, *79*, 065018.

78. Altschul, B.; Bailey, Q.G.; Kostelecký, V.A. Lorentz violation with an antisymmetric tensor. *Phys. Rev. D* **2010**, *81*, 065028.

79. Kostelecký, V.A.; Vargas, A.J. Lorentz and C P T tests with hydrogen, antihydrogen, and related systems. *Phys. Rev. D* **2015**, *92*, 056002.

80. Bailey, Q.G. Time delay and Doppler tests of the Lorentz symmetry of gravity. *Phys. Rev. D* **2009**, *80*, 044004.

81. Bailey, Q.G. Lorentz-violating gravitoelectromagnetism. *Phys. Rev. D* **2010**, *82*, 065012.

82. Tso, R.; Bailey, Q.G. Light-bending tests of Lorentz invariance. *Phys. Rev. D* **2011**, *84*, 085025.

83. Touboul, P.; Rodrigues, M. The MICROSCOPE space mission. *Class. Quantum Gravity* **2001**, *18*, 2487–2498.

84. Touboul, P.; Métris, G.; Lebat, V.; Robert, A. The MICROSCOPE experiment, ready for the in-orbit test of the equivalence principle. *Class. Quantum Gravity* **2012**, *29*, 184010.

85. Kostelecký, V.A.; Tasson, J.D. Prospects for Large Relativity Violations in Matter-Gravity Couplings. *Phys. Rev. Lett.* **2009**, *102*, 010402.

86. Kostelecký, V.A.; Lehnert, R. Stability, causality, and Lorentz and CPT violation. *Phys. Rev. D* **2001**, *63*, 065008.

87. Nordtvedt, K., Jr.; Will, C.M. Conservation Laws and Preferred Frames in Relativistic Gravity. II. Experimental Evidence to Rule Out Preferred-Frame Theories of Gravity. *Astrophys. J.* **1972**, *177*, 775.

88. Warburton, R.J.; Goodkind, J.M. Search for evidence of a preferred reference frame. *Astrophys. J.* **1976**, *208*, 881–886.

89. Nordtvedt, K., Jr. Anisotropic parametrized post-Newtonian gravitational metric field. *Phys. Rev. D* **1976**, *14*, 1511–1517.

90. Bordeé, C.J. Atomic interferometry with internal state labelling. *Phys. Lett. A* **1989**, *140*, 10–12.

91. Farah, T.; Guerlin, C.; Landragin, A.; Bouyer, P.; Gaffet, S.; Pereira Dos Santos, F.; Merlet, S. Underground operation at best sensitivity of the mobile LNE-SYRTE cold atom gravimeter. *Gyroscopy Navig.* **2014**, *5*, 266–274.

92. Hauth, M.; Freier, C.; Schkolnik, V.; Senger, A.; Schmidt, M.; Peters, A. First gravity measurements using the mobile atom interferometer GAIN. *Appl. Phys. B Lasers Opt.* **2013**, *113*, 49–55.

93. Hu, Z.K.; Sun, B.L.; Duan, X.C.; Zhou, M.K.; Chen, L.L.; Zhan, S.; Zhang, Q.Z.; Luo, J. Demonstration of an ultrahigh-sensitivity atom-interferometry absolute gravimeter. *Phys. Rev. A* **2013**, *88*, 043610.

94. Tamura, Y. Bulletin d'Information Marées Terrestres; Royal Observatory of Belgium: Uccle, Belgium, 1987; Volume 99, p. 6813.

95. Dehant, V.; Defraigne, P.; Wahr, J.M. Tides for a convective Earth. *J. Geophys. Res.* **1999**, *104*, 1035–1058.

96. Egbert, G.D.; Bennett, A.F.; Foreman, M.G.G. TOPEX/POSEIDON tides estimated using a global inverse model. *J. Geophys. Res.* **1994**, *99*, 24821.

97. Peters, A.; Chung, K.Y.; Chu, S. High-precision gravity measurements using atom interferometry. *Metrologia* **2001**, *38*, 25–61.

98. Merlet, S.; Kopaev, A.; Diament, M.; Geneves, G.; Landragin, A.; Pereira Dos Santos, F. Micro-gravity investigations for the LNE watt balance project. *Metrologia* **2008**, *45*, 265–274.

99. Fey, A.L.; Gordon, D.; Jacobs, C.S.; Ma, C.; Gaume, R.A.; Arias, E.F.; Bianco, G.; Boboltz, D.A.; Böckmann, S.; Bolotin, S.; et al. The Second Realization of the International Celestial Reference Frame by Very Long Baseline Interferometry. *Astron. J.* **2015**, *150*, 58.

100. Soffel, M.; Klioner, S.A.; Petit, G.; Wolf, P.; Kopeikin, S.M.; Bretagnon, P.; Brumberg, V.A.; Capitaine, N.; Damour, T.; Fukushima, T.; et al. The IAU 2000 Resolutions for Astrometry, Celestial Mechanics, and Metrology in the Relativistic Framework: Explanatory Supplement. *Astron. J.* **2003**, *126*, 2687–2706.

101. Lambert, S.B.; Le Poncin-Lafitte, C. Determining the relativistic parameter γ using very long baseline interferometry. *Astron. Astrophys.* **2009**, *499*, 331–335.

102. Lambert, S.B.; Le Poncin-Lafitte, C. Improved determination of γ by VLBI. *Astron. Astrophys.* **2011**, *529*, A70.

103. Le Poncin-Lafitte, C.; Linet, B.; Teyssandier, P. World function and time transfer: General post-Minkowskian expansions. *Class. Quantum Gravity* **2004**, *21*, 4463–4483.

104. Teyssandier, P.; Le Poncin-Lafitte, C. General post-Minkowskian expansion of time transfer functions. *Class. Quantum Gravity* **2008**, *25*, 145020.

105. Le Poncin-Lafitte, C.; Teyssandier, P. Influence of mass multipole moments on the deflection of a light ray by an isolated axisymmetric body. *Phys. Rev. D* **2008**, *77*, 044029.

106. Hees, A.; Bertone, S.; Le Poncin-Lafitte, C. Relativistic formulation of coordinate light time, Doppler, and astrometric observables up to the second post-Minkowskian order. *Phys. Rev. D* **2014**, *89*, 064045.

107. Hees, A.; Bertone, S.; Le Poncin-Lafitte, C. Light propagation in the field of a moving axisymmetric body: Theory and applications to the Juno mission. *Phys. Rev. D* **2014**, *90*, 084020.

108. Finkelstein, A.M.; Kreinovich, V.I.; Pandey, S.N. Relativistic reductions for radiointerferometric observables. *Astrophys. Space Sci.* **1983**, *94*, 233–247.

109. Petit, G.; Luzum, B. *IERS Conventions (2010)*; Bundesamt für Kartographie und Geodäsie: Frankfurt am Main, Germany, 2010.

110. Chapront, J.; Chapront-Touzé, M.; Francou, G. Determination of the lunar orbital and rotational parameters and of the ecliptic reference system orientation from LLR measurements and IERS data. *Astron. Astrophys.* **1999**, *343*, 624–633.

111. Dickey, J.O.; Bender, P.L.; Faller, J.E.; Newhall, X.X.; Ricklefs, R.L.; Ries, J.G.; Shelus, P.J.; Veillet, C.; Whipple, A.L.; Wiant, J.R.; et al. Lunar Laser Ranging: A Continuing Legacy of the Apollo Program. *Science* **1994**, *265*, 482–490.

112. Nordtvedt, K. Equivalence Principle for Massive Bodies. I. Phenomenology. *Phys. Rev.* **1968**, *169*, 1014–1016.

113. Nordtvedt, K. Equivalence Principle for Massive Bodies. II. Theory. *Phys. Rev.* **1968**, *169*, 1017–1025.

114. Nordtvedt, K. Testing Relativity with Laser Ranging to the Moon. *Phys. Rev.* **1968**, *170*, 1186–1187.

115. Williams, J.G.; Turyshev, S.G.; Boggs, D.H. Progress in Lunar Laser Ranging Tests of Relativistic Gravity. *Phys. Rev. Lett.* **2004**, *93*, 261101.

116. Williams, J.G.; Turyshev, S.G.; Boggs, D.H. Lunar Laser Ranging Tests of the Equivalence Principle with the Earth and Moon. *Int. J. Mod. Phys. D* **2009**, *18*, 1129–1175.

117. Merkowitz, S.M. Tests of Gravity Using Lunar Laser Ranging. *Living Rev. Relativ.* **2010**, *13*, 7.

118. Bertotti, B.; Iess, L.; Tortora, P. A test of general relativity using radio links with the Cassini spacecraft. *Nature* **2003**, *425*, 374–376.

119. Müller, J.; Williams, J.G.; Turyshev, S.G. Lunar Laser Ranging Contributions to Relativity and Geodesy. In *Lasers, Clocks and Drag-Free Control: Exploration of Relativistic Gravity in Space*; Astrophysics and Space Science Library; Dittus, H., Lammerzahl, C., Turyshev, S.G., Eds.; Springer: Berlin/Heidelberg, Germany, 2008; Volume 349, pp. 457–472.

120. Folkner, W.M.; Williams, J.G.; Boggs, D.H.; Park, R.; Kuchynka, P. The Planetary and Lunar Ephemeris DE 430 and DE431. *IPN Prog. Report.* **2014**, *42*, 196.

121. Lupton, R. Statistics in theory and practice. *Econ. J.* **1993**, *43*, 688–690.

122. Gottlieb, A.D. Asymptotic equivalence of the jackknife and infinitesimal jackknife variance estimators for some smooth statistics. *Ann. Inst. Stat. Math.* **2003**, *55*, 555–561.

123. Konopliv, A.S.; Park, R.S.; Yuan, D.N.; Asmar, S.W.; Watkins, M.M.; Williams, J.G.; Fahnestock, E.; Kruizinga, G.; Paik, M.; Strekalov, D.; et al. High-resolution lunar gravity fields from the GRAIL Primary and Extended Missions. *Gepphys. Res. Lett.* **2014**, *41*, 1452–1458.

124. Lemoine, F.G.; Goossens, S.; Sabaka, T.J.; Nicholas, J.B.; Mazarico, E.; Rowlands, D.D.; Loomis, B.D.; Chinn, D.S.; Neumann, G.A.; Smith, D.E.; et al. GRGM900C: A degree 900 lunar gravity model from GRAIL primary and extended mission data. *Gepphys. Res. Lett.* **2014**, *41*, 3382–3389.

125. Arnold, D.; Bertone, S.; Jäggi, A.; Beutler, G.; Mervart, L. GRAIL gravity field determination using the Celestial Mechanics Approach. *Icarus* **2015**, *261*, 182–192.

126. Standish, E.M. The JPL planetary ephemerides. *Celest. Mech.* **1982**, *26*, 181–186.

127. Newhall, X.X.; Standish, E.M.; Williams, J.G. DE 102—A numerically integrated ephemeris of the moon and planets spanning forty-four centuries. *Astron. Astrophys.* **1983**, *125*, 150–167.

128. Standish, E.M., Jr. The observational basis for JPL's DE 200, the planetary ephemerides of the Astronomical Almanac. *Astron. Astrophys.* **1990**, *233*, 252–271.

129. Standish, E.M. Testing alternate gravitational theories. *IAU Symp.* **2010**, *261*, 179–182.

130. Standish, E.M.; Williams, J.G. Orbital Ephemerides of the Sun, Moon, and Planets. In *Explanatory Supplement to the Astronomical Almanac*, 3rd ed.; Urban, S.E., Seidelmann, P.K., Eds.; Univeristy Science Books: Herndon, VA, USA, 2012; Chapter 8, pp. 305–346.

131. Hees, A.; Folkner, W.M.; Jacobson, R.A.; Park, R.S. Constraints on modified Newtonian dynamics theories from radio tracking data of the Cassini spacecraft. *Phys. Rev. D* **2014**, *89*, 102002.

132. Fienga, A.; Manche, H.; Laskar, J.; Gastineau, M. INPOP06: A new numerical planetary ephemeris. *Astron. Astrophys.* **2008**, *477*, 315–327.

133. Fienga, A.; Laskar, J.; Morley, T.; Manche, H.; Kuchynka, P.; Le Poncin-Lafitte, C.; Budnik, F.; Gastineau, M.; Somenzi, L. INPOP08, a 4-D planetary ephemeris: From asteroid and time-scale computations to ESA Mars Express and Venus Express contributions. *Astron. Astrophys.* **2009**, *507*, 1675–1686.

134. Fienga, A.; Laskar, J.; Kuchynka, P.; Le Poncin-Lafitte, C.; Manche, H.; Gastineau, M. Gravity tests with INPOP planetary ephemerides. *IAU Symp.* **2010**, *261*, 159–169.

135. Fienga, A.; Laskar, J.; Kuchynka, P.; Manche, H.; Desvignes, G.; Gastineau, M.; Cognard, I.; Theureau, G. The INPOP10a planetary ephemeris and its applications in fundamental physics. *Celest. Mech. Dyn. Astron.* **2011**, *111*, 363–385.

136. Verma, A.K.; Fienga, A.; Laskar, J.; Manche, H.; Gastineau, M. Use of MESSENGER radioscience data to improve planetary ephemeris and to test general relativity. *Astron. Astrophys.* **2014**, *561*, A115.

137. Fienga, A.; Laskar, J.; Exertier, P.; Manche, H.; Gastineau, M. Numerical estimation of the sensitivity of INPOP planetary ephemerides to general relativity parameters. *Celest. Mech. Dyn. Astron.* **2015**, *123*, 325–349.

138. Pitjeva, E.V. High-Precision Ephemerides of Planets EPM and Determination of Some Astronomical Constants. *Sol. Syst. Res.* **2005**, *39*, 176–186.

139. Pitjeva, E.V. EPM ephemerides and relativity. *IAU Symp.* **2010**, *261*, 170–178.

140. Pitjeva, E.V.; Pitjev, N.P. Relativistic effects and dark matter in the Solar system from observations of planets and spacecraft. *Mon. Not. R. Astrono. Soc.* **2013**, *432*, 3431–3437.

141. Pitjeva, E.V. Updated IAA RAS planetary ephemerides-EPM2011 and their use in scientific research. *Sol. Syst. Res.* **2013**, *47*, 386–402.

142. Pitjeva, E.V.; Pitjev, N.P. Development of planetary ephemerides EPM and their applications. *Celest. Mech. Dyn. Astron.* **2014**, *119*, 237–256.

143. Konopliv, A.S.; Asmar, S.W.; Folkner, W.M.; Karatekin, O.; Nunes, D.C.; Smrekar, S.E.; Yoder, C.F.; Zuber, M.T. Mars high resolution gravity fields from MRO, Mars seasonal gravity, and other dynamical parameters. *Icarus* **2011**, *211*, 401–428.

144. Pitjev, N.P.; Pitjeva, E.V. Constraints on dark matter in the solar system. *Astron. Lett.* **2013**, *39*, 141–149.

145. Milgrom, M. MOND effects in the inner Solar system. *Mon. Not. R. Astrono. Soc.* **2009**, *399*, 474–486.

146. Blanchet, L.; Novak, J. External field effect of modified Newtonian dynamics in the Solar system. *Mon. Not. R. Astrono. Soc.* **2011**, *412*, 2530–2542.

147. Hees, A.; Famaey, B.; Angus, G.W.; Gentile, G. Combined Solar system and rotation curve constraints on MOND. *Mon. Not. R. Astrono. Soc.* **2016**, *455*, 449–461.

148. Hees, A.; Lamine, B.; Reynaud, S.; Jaekel, M.T.; Le Poncin-Lafitte, C.; Lainey, V.; Füzfa, A.; Courty, J.M.; Dehant, V.; Wolf, P. Radioscience simulations in General Relativity and in alternative theories of gravity. *Class. Quantum Gravity* **2012**, *29*, 235027.

149. de Sitter, W. Einstein's theory of gravitation and its astronomical consequences. *Mon. Not. R. Astrono. Soc.* **1916**, *76*, 699–728.

150. Lense, J.; Thirring, H. Über den Einfluß der Eigenrotation der Zentralkörper auf die Bewegung der Planeten und Monde nach der Einsteinschen Gravitationstheorie. *Phys. Z.* **1918**, *19*, 156.

151. Schiff, L.I. Possible New Experimental Test of General Relativity Theory. *Phys. Rev. Lett.* **1960**, *4*, 215–217.

152. Pugh, G.E. Proposal for a Satellite Test of the Coriolis Prediction of General Relativity. In *Nonlinear Gravitodynamics: The Lense-Thirring Effect*; Word Scientific Publishing: Singapore, 1959.

153. Everitt, C.W.F.; Debra, D.B.; Parkinson, B.W.; Turneaure, J.P.; Conklin, J.W.; Heifetz, M.I.; Keiser, G.M.; Silbergleit, A.S.; Holmes, T.; Kolodziejczak, J.; et al. Gravity Probe B: Final Results of a Space Experiment to Test General Relativity. *Phys. Rev. Lett.* **2011**, *106*, 221101.

154. Stella, L.; Vietri, M. kHz Quasiperiodic Oscillations in Low-Mass X-Ray Binaries as Probes of General Relativity in the Strong-Field Regime. *Phys. Rev. Lett.* **1999**, *82*, 17–20.

155. Hulse, R.A.; Taylor, J.H. Discovery of a pulsar in a binary system. *Astrophys. J.* **1975**, *195*, L51–L53.

156. Taylor, J.H.; Hulse, R.A.; Fowler, L.A.; Gullahorn, G.E.; Rankin, J.M. Further observations of the binary pulsar PSR 1913+16. *Astrophys. J.* **1976**, *206*, L53–L58.

157. Taylor, J.H.; Fowler, L.A.; McCulloch, P.M. Measurements of general relativistic effects in the binary pulsar PSR 1913+16. *Nature* **1979**, *277*, 437–440.

158. Damour, T.; Deruelle, N. General relativistic celestial mechanics of binary systems. II. The post-Newtonian timing formula. *Ann. Inst. Henri Poincaré Phys. Théor.* **1986**, *44*, 263–292.

159. Stairs, I.H. Testing General Relativity with Pulsar Timing. *Living Rev. Relativ.* **2003**, *6*, 5.

160. Lorimer, D.R. Binary and Millisecond Pulsars. *Living Rev. Relativ.* **2008**, *11*, 8.

161. Damour, T.; Deruelle, N. General relativistic celestial mechanics of binary systems. I. The post-Newtonian motion. *Ann. Inst. Henri Poincaré Phys. Théor.* **1985**, *43*, 107–132.

162. Wex, N. The second post-Newtonian motion of compact binary-star systems with spin. *Class. Quantum Gravity* **1995**, *12*, 983–1005.

163. Edwards, R.T.; Hobbs, G.B.; Manchester, R.N. TEMPO2, a new pulsar timing package - II. The timing model and precision estimates. *Mon. Not. R. Astrono. Soc.* **2006**, *372*, 1549–1574.

164. Wex, N. Testing Relativistic Gravity with Radio Pulsars. In *Frontiers in Relativistic Celestial Mechanics*; Applications and Experiments; Kopeikin, S., Ed.; De Gruyter: Berlin, Germany, 2014; Volume 2.

165. Kramer, M. Pulsars as probes of gravity and fundamental physics. *Int. J. Mod. Phys. D* **2016**, *25*, 14.

166. Kramer, M.; Stairs, I.H.; Manchester, R.N.; McLaughlin, M.A.; Lyne, A.G.; Ferdman, R.D.; Burgay, M.; Lorimer, D.R.; Possenti, A.; D'Amico, N.; et al. Tests of General Relativity from Timing the Double Pulsar. *Science* **2006**, *314*, 97–102.

167. Damour, T.; Esposito-Farese, G. Tensor-multi-scalar theories of gravitation. *Class. Quantum Gravity* **1992**, *9*, 2093–2176.

168. Damour, T.; Taylor, J.H. Strong-field tests of relativistic gravity and binary pulsars. *Phys. Rev. D* **1992**, *45*, 1840–1868.

169. Damour, T.; Esposito-Farèse, G. Tensor-scalar gravity and binary-pulsar experiments. *Phys. Rev. D* **1996**, *54*, 1474–1491.

170. Freire, P.C.C.; Wex, N.; Esposito-Farèse, G.; Verbiest, J.P.W.; Bailes, M.; Jacoby, B.A.; Kramer, M.; Stairs, I.H.; Antoniadis, J.; Janssen, G.H. The relativistic pulsar-white dwarf binary PSR J1738+0333 - II. The most stringent test of scalar-tensor gravity. *Mon. Not. R. Astrono. Soc.* **2012**, *423*, 3328–3343.

171. Ransom, S.M.; Stairs, I.H.; Archibald, A.M.; Hessels, J.W.T.; Kaplan, D.L.; van Kerkwijk, M.H.; Boyles, J.; Deller, A.T.; Chatterjee, S.; Schechtman-Rook, A.; et al. A millisecond pulsar in a stellar triple system. *Nature* **2014**, *505*, 520–524.

172. Damour, T.; Esposito-Farese, G. Nonperturbative strong-field effects in tensor-scalar theories of gravitation. *Phys. Rev. Lett.* **1993**, *70*, 2220–2223.

173. Foster, B.Z. Strong field effects on binary systems in Einstein-aether theory. *Phys. Rev. D* **2007**, *76*, 084033.

174. Wex, N.; Kramer, M. A characteristic observable signature of preferred-frame effects in relativistic binary pulsars. *Mon. Not. R. Astrono. Soc.* **2007**, *380*, 455–465.

175. Shao, L.; Wex, N. New tests of local Lorentz invariance of gravity with small-eccentricity binary pulsars. *Class. Quantum Gravity* **2012**, *29*, 215018.

176. Nordtvedt, K. Probing gravity to the second post-Newtonian order and to one part in 10 to the 7th using the spin axis of the sun. *Astrophys. J.* **1987**, *320*, 871–874.

177. Shao, L.; Caballero, R.N.; Kramer, M.; Wex, N.; Champion, D.J.; Jessner, A. A new limit on local Lorentz invariance violation of gravity from solitary pulsars. *Class. Quantum Gravity* **2013**, *30*, 165019.

178. Bell, J.F.; Damour, T. A new test of conservation laws and Lorentz invariance in relativistic gravity. *Class. Quantum Gravity* **1996**, *13*, 3121–3127.

179. Gonzalez, M.E.; Stairs, I.H.; Ferdman, R.D.; Freire, P.C.C.; Nice, D.J.; Demorest, P.B.; Ransom, S.M.; Kramer, M.; Camilo, F.; Hobbs, G.; et al. High-precision Timing of Five Millisecond Pulsars: Space Velocities, Binary Evolution, and Equivalence Principles. *Astrophys. J.* **2011**, *743*, 102.

180. Lorimer, D.R.; Kramer, M. *Handbook of Pulsar Astronomy*; Cambridge University Press: Cambridge, UK, 2004.

181. Jennings, R.J.; Tasson, J.D.; Yang, S. Matter-sector Lorentz violation in binary pulsars. *Phys. Rev. D* **2015**, *92*, 125028.

182. Moore, G.D.; Nelson, A.E. Lower bound on the propagation speed of gravity from gravitational Cherenkov radiation. *J. High Energy Phys.* **2001**, *9*, 023.

183. Kiyota, S.; Yamamoto, K. Constraint on modified dispersion relations for gravitational waves from gravitational Cherenkov radiation. *Phys. Rev. D* **2015**, *92*, 104036.

184. Elliott, J.W.; Moore, G.D.; Stoica, H. Constraining the New Aether: Gravitational Cherenkov radiation. *J. High Energy Phys.* **2005**, *8*, 066.

185. Kimura, R.; Yamamoto, K. Constraints on general second-order scalar-tensor models from gravitational Cherenkov radiation. *J. Cosmol. Astropart. Phys.* **2012**, *7*, 050

186. De Laurentis, M.; Capozziello, S.; Basini, G. Gravitational Cherenkov Radiation from Extended Theories of Gravity. *Mod. Phys. Lett. A* **2012**, *27*, 1250136.

187. Kimura, R.; Tanaka, T.; Yamamoto, K.; Yamashita, Y. Constraint on ghost-free bigravity from gravitational Cherenkov radiation. *Phys. Rev. D* **2016**, *94*, 064059.

188. Takeda, M.; Hayashida, N.; Honda, K.; Inoue, N.; Kadota, K.; Kakimoto, F.; Kamata, K.; Kawaguchi, S.; Kawasaki, Y.; Kawasumi, N.; et al. Small-Scale Anisotropy of Cosmic Rays above 10^{19} eV Observed with the Akeno Giant Air Shower Array. *Astrophys. J.* **1999**, *522*, 225–237.

189. Bird, D.J.; Corbato, S.C.; Dai, H.Y.; Elbert, J.W.; Green, K.D.; Huang, M.A.; Kieda, D.B.; Ko, S.; Larsen, C.G.; Loh, E.C.; et al. Detection of a cosmic ray with measured energy well beyond the expected spectral cutoff due to cosmic microwave radiation. *Astrophys. J.* **1995**, *441*, 144–150.

190. Wada, M. *Catalogue of Highest Energy Cosmic Rays. Giant Extensive Air Showers. No._1. Volcano Ranch, Haverah Park*; Institute of Physical and Chemical Research: Tokyo, Japan, 1980.

191. High Resolution Fly'S Eye Collaboration.; Abbasi, R.U.; Abu-Zayyad, T.; Allen, M.; Amman, J.F.; Archbold, G.; Belov, K.; Belz, J.W.; BenZvi, S.Y.; Bergman, D.R.; et al. Search for correlations between HiRes stereo events and active galactic nuclei. *Astropart. Phys.* **2008**, *30*, 175–179.

192. Aab, A.; Abreu, P.; Aglietta, M.; Ahn, E.J.; Al Samarai, I.; Albuquerque, I.F.M.; Allekotte, I.; Allen, J.; Allison, P.; Almela, A.; et al. Searches for Anisotropies in the Arrival Directions of the Highest Energy Cosmic Rays Detected by the Pierre Auger Observatory. *Astrophys. J.* **2015**, *804*, 15.

193. Winn, M.M.; Ulrichs, J.; Peak, L.S.; McCusker, C.B.A.; Horton, L. The cosmic-ray energy spectrum above 10^{17} eV. *J. Phys. G Nucl. Phys.* **1986**, *12*, 653–674.

194. Abbasi, R.U.; Abe, M.; Abu-Zayyad, T.; Allen, M.; Anderson, R.; Azuma, R.; Barcikowski, E.; Belz, J.W.; Bergman, D.R.; Blake, S.A.; et al. Indications of Intermediate-scale Anisotropy of Cosmic Rays with Energy Greater Than 57 EeV in the Northern Sky Measured with the Surface Detector of the Telescope Array Experiment. *Astrophys. J.* **2014**, *790*, L21.

195. Pravdin, M.I.; Glushkov, A.V.; Ivanov, A.A.; Knurenko, S.P.; Kolosov, V.A.; Makarov, I.T.; Sabourov, A.V.; Sleptsov, I.Y.; Struchkov, G.G. Estimation of the giant shower energy at the Yakutsk EAS Array. *Int. Cosm. Ray Conf.* **2005**, *7*, 243.

196. De Bruijne, J.H.J. Science performance of Gaia, ESA's space-astrometry mission. *Astrophys. Space Sci.* **2012**, *341*, 31–41.

197. Mignard, F.; Klioner, S.A. Gaia: Relativistic modelling and testing. *IAU Symp.* **2010**, *261*, 306–314.

198. Jaekel, M.T.; Reynaud, S. Gravity Tests in the Solar System and the Pioneer Anomaly. *Mod. Phys. Lett. A* **2005**, *20*, 1047–1055.

199. Jaekel, M.T.; Reynaud, S. Post-Einsteinian tests of linearized gravitation. *Class. Quantum Gravity* **2005**, *22*, 2135–2157.

200. Jaekel, M.T.; Reynaud, S. Post-Einsteinian tests of gravitation. *Class. Quantum Gravity* **2006**, *23*, 777–798.

201. Reynaud, S.; Jaekel, M.T. Long Range Gravity Tests and the Pioneer Anomaly. *Int. J. Mod. Phys. D* **2007**, *16*, 2091–2105.

202. Reynaud, S.; Jaekel, M.T. Tests of general relativity in the Solar System. *Atom Opt. Space Phys.* **2009**, *168*, 203–207.

203. Gai, M.; Vecchiato, A.; Ligori, S.; Sozzetti, A.; Lattanzi, M.G. Gravitation astrometric measurement experiment. *Exp. Astron.* **2012**, *34*, 165–180.

204. Turyshev, S.G.; Shao, M. Laser Astrometric Test of Relativity: Science, Technology and Mission Design. *Int. J. Mod. Phys. D* **2007**, *16*, 2191–2203.

205. Mouret, S. Tests of fundamental physics with the Gaia mission through the dynamics of minor planets. *Phys. Rev. D* **2011**, *84*, 122001.

206. Hees, A.; Hestroffer, D.; Le Poncin-Lafitte, C.; David, P. Tests of gravitation with GAIA observations of Solar System Objects. In Proceedings of the Annual meeting of the French Society of Astronomy and Astrophysics (SF2A-2015), Toulouse, France, 2–5 June 2015; Martins, F., Boissier, S., Buat, V., Cambrésy, L., Petit, P., Eds.; pp. 125–131.

207. Margot, J.L.; Giorgini, J.D. Probing general relativity with radar astrometry in the inner solar system. *IAU Symp.* **2010**, *261*, 183–188.

208. Hees, A.; Lamine, B.; Poncin-Lafitte, C.L.; Wolf, P. How to Test the SME with Space Missions? In Proceedings of the Sixth Meeting CPT and Lorentz Symmetry, Bloomington, IN, USA, 17–21 June 2013; Kostelecky, A., Ed.; pp. 107–110.

209. Hees, A.; Lamine, B.; Reynaud, S.; Jaekel, M.T.; Le Poncin-Lafitte, C.; Lainey, V.; Füzfa, A.; Courty, J.M.; Dehant, V.; Wolf, P. Simulations of Solar System Observations in Alternative Theories of Gravity. In Proceedings of the Thirteenth Marcel Grossmann Meeting: On Recent Developments in Theoretical and Experimental General Relativity, Astrophysics and Relativistic Field Theories, Stockholm, Sweden, 1–7 July 2012; Rosquist, K., Ed.; pp. 2357–2359.

210. Iess, L.; Asmar, S. Probing Space-Time in the Solar System: From Cassini to Bepicolombo. *Int. J. Mod. Phys. D* **2007**, *16*, 2117–2126.

211. Kliore, A.J.; Anderson, J.D.; Armstrong, J.W.; Asmar, S.W.; Hamilton, C.L.; Rappaport, N.J.; Wahlquist, H.D.; Ambrosini, R.; Flasar, F.M.; French, R.G.; et al. Cassini Radio Science. *Space Sci. Rev.* **2004**, *115*, 1–70.

212. Iorio, L.; Ciufolini, I.; Pavlis, E.C. Measuring the relativistic perigee advance with satellite laser ranging. *Class. Quantum Gravity* **2002**, *19*, 4301–4309.

213. Lucchesi, D.M.; Peron, R. Accurate Measurement in the Field of the Earth of the General-Relativistic Precession of the LAGEOS II Pericenter and New Constraints on Non-Newtonian Gravity. *Phys. Rev. Lett.* **2010**, *105*, 231103.

214. Lucchesi, D.M.; Peron, R. LAGEOS II pericenter general relativistic precession (1993–2005): Error budget and constraints in gravitational physics. *Phys. Rev. D* **2014**, *89*, 082002.

215. Ciufolini, I.; Pavlis, E.C. A confirmation of the general relativistic prediction of the Lense-Thirring effect. *Nature* **2004**, *431*, 958–960.

216. Ciufolini, I.; Paolozzi, A.; Pavlis, E.C.; Koenig, R.; Ries, J.; Gurzadyan, V.; Matzner, R.; Penrose, R.; Sindoni, G.; Paris, C.; et al. A test of general relativity using the LARES and LAGEOS satellites and a GRACE Earth gravity model. Measurement of Earth's dragging of inertial frames. *Eur. Phys. J. C* **2016**, *76*, 120.

217. Ciufolini, I.; Paolozzi, A.; Pavlis, E.C.; Ries, J.C.; Koenig, R.; Matzner, R.A.; Sindoni, G.; Neumayer, H. Towards a One Percent Measurement of Frame Dragging by Spin with Satellite Laser Ranging to LAGEOS, LAGEOS 2 and LARES and GRACE Gravity Models. *Space Sci. Rev.* **2009**, *148*, 71–104.

218. Ciufolini, I.; Paolozzi, A.; Pavlis, E.; Ries, J.; Gurzadyan, V.; Koenig, R.; Matzner, R.; Penrose, R.; Sindoni, G. Testing General Relativity and gravitational physics using the LARES satellite. *Eur. Phys. J. Plus* **2012**, *127*, 133.

219. Ciufolini, I.; Pavlis, E.C.; Paolozzi, A.; Ries, J.; Koenig, R.; Matzner, R.; Sindoni, G.; Neumayer, K.H. Phenomenology of the Lense-Thirring effect in the Solar System: Measurement of frame-dragging with laser ranged satellites. *New Astron.* **2012**, *17*, 341–346.

220. Paolozzi, A.; Ciufolini, I.; Vendittozzi, C. Engineering and scientific aspects of LARES satellite. *Acta Astronaut.* **2011**, *69*, 127–134.

221. Iorio, L. Towards a 1% measurement of the Lense-Thirring effect with LARES? *Adv. Space Res.* **2009**, *43*, 1148–1157.

222. Iorio, L. Will the recently approved LARES mission be able to measure the Lense-Thirring effect at 1%? *Gen. Relativ. Gravit.* **2009**, *41*, 1717–1724.

223. Iorio, L. An Assessment of the Systematic Uncertainty in Present and Future Tests of the Lense-Thirring Effect with Satellite Laser Ranging. *Space Sci. Rev.* **2009**, *148*, 363–381.

224. Iorio, L.; Lichtenegger, H.I.M.; Ruggiero, M.L.; Corda, C. Phenomenology of the Lense-Thirring effect in the solar system. *Astrophys. Space Sci.* **2011**, *331*, 351–395.

225. Renzetti, G. Are higher degree even zonals really harmful for the LARES/LAGEOS frame-dragging experiment? *Can. J. Phys.* **2012**, *90*, 883–888.

226. Renzetti, G. First results from LARES: An analysis. *New Astron.* **2013**, *23*, 63–66.

227. Lucchesi, D.M.; Anselmo, L.; Bassan, M.; Pardini, C.; Peron, R.; Pucacco, G.; Visco, M. Testing the gravitational interaction in the field of the Earth via satellite laser ranging and the Laser Ranged Satellites Experiment (LARASE). *Class. Quantum Gravity* **2015**, *32*, 155012.

228. Lämmerzahl, C.; Ciufolini, I.; Dittus, H.; Iorio, L.; Müller, H.; Peters, A.; Samain, E.; Scheithauer, S.; Schiller, S. OPTIS–An Einstein Mission for Improved Tests of Special and General Relativity. *Gen. Relativ. Gravit.* **2004**, *36*, 2373–2416.

229. Hohensee, M.A.; Chu, S.; Peters, A.; Müller, H.   Equivalence Principle and Gravitational Redshift. *Phys. Rev. Lett.* **2011**, *106*, 151102.

230. Hohensee, M.A.; Leefer, N.; Budker, D.; Harabati, C.; Dzuba, V.A.; Flambaum, V.V. Limits on Violations of Lorentz Symmetry and the Einstein Equivalence Principle using Radio-Frequency Spectroscopy of Atomic Dysprosium. *Phys. Rev. Lett.* **2013**, *111*, 050401.

231. Hohensee, M.A.; Müller, H.; Wiringa, R.B. Equivalence Principle and Bound Kinetic Energy. *Phys. Rev. Lett.* **2013**, *111*, 151102.

232. Delva, P.; Hees, A.; Bertone, S.; Richard, E.; Wolf, P. Test of the gravitational redshift with stable clocks in eccentric orbits: Application to Galileo satellites 5 and 6. *Class. Quantum Gravity* **2015**, *32*, 232003.

233. Cacciapuoti, L.; Salomon, C. Atomic clock ensemble in space. *J. Phys. Conf. Ser.* **2011**, *327*, 012049.

4

Einstein and Beyond: A Critical Perspective on General Relativity

Ram Gopal Vishwakarma

Unidad Académica de Matemáticas, Universidad Autónoma de Zacatecas, Zacatecas, ZAC C.P. 98000, Mexico; vishwa@uaz.edu.mx

Academic Editors: Lorenzo Iorio and Elias C. Vagenas

Abstract: An alternative approach to Einstein's theory of General Relativity (GR) is reviewed, which is motivated by a range of serious theoretical issues inflicting the theory, such as the cosmological constant problem, presence of non-Machian solutions, problems related with the energy-stress tensor T^{ik} and unphysical solutions. The new approach emanates from a critical analysis of these problems, providing a novel insight that the matter fields, together with the ensuing gravitational field, are already present inherently in the spacetime without taking recourse to T^{ik}. Supported by lots of evidence, the new insight revolutionizes our views on the representation of the source of gravitation and establishes the spacetime itself as the source, which becomes crucial for understanding the unresolved issues in a unified manner. This leads to a new paradigm in GR by establishing equation $R^{ik} = 0$ as the field equation of gravitation plus inertia in the very presence of matter.

Keywords: gravitation; general relativity; fundamental problems and general formalism; Mach's principle

1. Introduction

The year 2015 marks the centenary of the advent of Albert Einstein's theory of General Relativity (GR), which constitutes the current description of gravitation in modern physics. It is undoubtedly one of the towering theoretical achievements of 20th-century physics, which is recognized as an intellectual achievement par excellence.

Einstein first revolutionized, in 1905, the concepts of absolute space and absolute time by superseding them with a single four-dimensional spacetime fabric, which only had an absolute meaning. He discovered this in his theory of Special Relativity (SR), which he formulated by postulating that the laws of physics are the same in all non-accelerating reference frames and the speed of light in vacuum never changes. He then made a great leap from SR to GR through his penetrating insight that the gravitational field in a small neighborhood of spacetime is indistinguishable from an appropriate acceleration of the reference frame (principle of equivalence), and hence gravitation can be added to SR (which is valid only in the absence of gravitation) by generalizing it for the accelerating observers. This leads to a curved spacetime.

This dramatically revolutionized the Newtonian notion of gravitation as a force by heralding that gravitation is a manifestation of the dynamically curved spacetime created by the presence of matter. The principle of general covariance (the laws of physics should be the same in all coordinate systems, including the accelerating ones) then suggests that the theory must be formulated by using the language of tensors. This leads to the famous Einstein equation:

$$G^{ik} \equiv R^{ik} - \frac{1}{2}g^{ik}R = -\frac{8\pi G}{c^4}T^{ik} \qquad (1)$$

which represents how geometry, encoded in the left-hand side (which is a function of the spacetime curvature), behaves in response to matter encoded in the energy-momentum-stress tensor T^{ik}. [Here, as usual, g^{ik} is the contravariant form of the metric tensor g_{ik} representing the spacetime geometry, which is defined by $ds^2 = g_{ik}dx^i dx^k$. R^{ik} is the Ricci tensor defined by $R^{ik} = g_{hj}R^{hijk}$ in terms of the Riemann tensor R^{hijk}. $R = g_{ik}R^{ik}$ is the Ricci scalar and G^{ik} the Einstein tensor. T^{ik} is the energy-stress tensor of matter (which can very well absorb the cosmological constant or any other candidate of dark energy). G is the Newtonian constant of gravitation and c the speed of light in vacuum. The Latin indices range and sum over the values 0, 1, 2, 3 unless stated otherwise.] This, in a sense, completes the identification of gravitation with geometry. It turns out that the spacetime geometry is no longer a fixed inert background, rather it is a key player in physics, which acts on matter and can be acted upon. This constitutes a profound paradigm shift.

The theory has made remarkable progress on both theoretical and observational fronts [1–5]. It is remarkable that, born a century ago out of almost pure thought, the theory has managed to survive extensive experimental/observational scrutiny and describes accurately all gravitational phenomena ranging from the solar system to the largest scale—the Universe itself. Nevertheless, a number of questions remain open. On the one hand, the theory requires the dark matter and dark energy—two of the largest contributions to T^{ik}—which have entirely mysterious physical origins and do not have any non-gravitational or laboratory evidence. On the other hand, the theory suffers from profound theoretical difficulties, some of which are reviewed in the following. Nonetheless, if a theory requires more than 95% of "dark entities" in order to describe the observations, it is an alarming signal for us to turn back to the very foundations of the theory, rather than just keep adding epicycles to it.

Although Einstein, and then others, were mesmerized by the "inner consistency" and elegance of the theory, many theoretical issues were discovered even during the lifetime of Einstein which were not consistent with the founding principles of GR. In the following, we provide a critical review of the historical development of GR and some ensuing problems, most of which are generally ignored or not given the proper attention they deserve. This review will differ from the conventional reviews in the sense that, unlike most of the traditional reviews, it will not recount a well-documented story of the discovery of GR, rather it will focus on some key problems which insinuate an underlying new insight on a geometric theory of gravitation, thereby providing a possible way out in the framework of GR itself.

2. Issues Warranting Attention: Mysteries of the Present with Roots in the Past

Mach's Principle: Mach's principle, akin to the equivalence principle, was the primary motivation and guiding principle for Einstein in the formulation of GR. (The name "Mach's principle" was coined by Einstein for the general inspiration that he found in Mach's works on mechanics [6], even though the principle itself was never formulated succinctly by Mach himself.) Though in the absence of a clear statement from Ernst Mach, there exist a number of formulations of Mach's principle, in essence the principle advocates to shun all vestiges of the unobservable absolute space and time of Newton in favor of the directly observable background matter in the Universe, which determines its geometry and the inertia of an object.

As the principle of general covariance (non-existence of a privileged reference frame) emerges as a consequence of Mach's denial of absolute space, Einstein expected that his theory would automatically obey Mach's principle. However, it turned out not to be so, as there appear several anti-Machian features in GR. According to Mach's principle, the presence of a material background is essential for defining motion and a meaningful spacetime geometry. This means that an isolated object in an otherwise empty Universe should not possess any inertial properties. However, this is clearly violated by the Minkowski solution, which possesses timelike geodesics and a well-defined notion of inertia in the total absence of T^{ik}. Similarly, the cosmological constant also violates Mach's principle (if it does not represent the vacuum energy, but just a constant of nature—as is believed by some authors) in the sense in which the geometry should be determined completely by the mass distribution. In the same vein, there exists a class of singularity-free curved solutions, which admit Einstein's equations in the absence of T^{ik}. Furthermore, a global rotation, which is not allowed by Mach's principle (in the absence of an absolute frame of reference), is revealed in the Gödel solution [7], which describes a Universe with a uniform rotation in the whole spacetime.

After failing to formulate GR in a fully Machian sense, Einstein himself moved away from Mach's principle in his later years. Nevertheless, the principle continued to attract a lot of sympathy due to its aesthetic appeal and enormous impact, and it is widely believed that a viable theory of gravitation must be Machian. Moreover, the consistency of GR with SR, which abolishes the absolute space akin to Mach's principle, also persuades us that GR must be Machian. This characterization has however remained just wishful thinking.

Equivalence Principle: The equivalence principle—the physical foundation of any metric theory of gravitation—first expressed by Galileo and later reformulated by Newton, was assumed by Einstein as one of the defining principles of GR. According to the principle, one can choose a locally inertial coordinate system (LICS) (*i.e.*, a freely-falling one) at any spacetime point in an arbitrary gravitational field such that within a sufficiently small region of the point in question, the laws of nature take the same form as in unaccelerated Cartesian coordinate systems in the absence of gravitation [8]. As has been mentioned earlier, this equivalence of gravitation and accelerated reference frames paved the way for the formulation of GR. Since the principle rests on the conviction that the equality of the gravitational and inertial mass is exact [8,9], one expects the same to hold in GR solutions. However, the inertial and the (active) gravitational mass have remained unequal in general. For instance, for the case of T^{ik} representing a perfect fluid:

$$T^{ik} = (\rho + p)u^i u^k - pg^{ik} \tag{2}$$

various solutions of Equation (1) indicate that the inertial mass density (=passive gravitational mass density) $= (\rho + p)/c^2$, while the active gravitational mass density $= (\rho + 3p)/c^2$, where ρ is the energy density of the fluid (which includes all the sources of energy of the fluid except the gravitational field energy) and p is its pressure. The binding energy of the gravitational field is believed to be responsible for this discrepancy. However, why the contributions from the gravitational energy to the different masses are not equal, has remained a mystery.

T^{ik} **and Gravitational Energy:** Appearing as the source term in Equation (1), T^{ik} is expected to include all the inertial and gravitational aspects of matter, *i.e.*, all the possible sources of gravitation. However, this requirement does not seem to be met on at least two counts. Firstly, T^{ik} fails to support, in a general spacetime with no symmetries, an unambiguous definition of angular momentum, which is a fundamental and unavoidable characteristic of matter, as is witnessed from the subatomic to the galactic scales. While a meaningful notion of the angular momentum in GR always needs the introduction of some additional structure in the form of symmetries, quasi-symmetries, or some other background structure, it can be unambiguously defined only for isolated systems [10,11].

Secondly, T^{ik} fails to include the energy of the gravitational field, which also gravitates. Einstein and Grossmann emphasized that, *akin to all other fields, the gravitational field must also have an energy-momentum tensor which should be included in the "source term"* [9]. However, after failing to find a tensor representation of the gravitational field, Einstein then commented that *"there may very well be gravitational fields without stress and energy density"* [12] and finally admitted that *"the energy tensor can be regarded only as a provisional means of representing matter"* [13]. Alas, a century-long dedicated effort to discover a unanimous formulation of the energy- stress tensor of the gravitational field, has failed concluding that a proper energy-stress tensor of the gravitational field does not exist. [It can be safely said that despite the century-long dedicated efforts of many luminaries, like Einstein, Tolman, Papapetrou, Landau-Lifshitz, Möller and Weinberg, the attempts to discover a unanimous formulation of the gravitational field energy has failed due to the following three reasons: (i) the non-tensorial character of the energy-stress 'complexes' (pseudo tensors) of the gravitational field; (ii) the lack of a unique agreed-upon formula for the gravitational field pseudo tensor in view of various formulations thereof, which may lead to different distributions even in the same spacetime background. Moreover, a pseudo tensor, unlike a true tensor, can be made to vanish at any pre-assigned point by an appropriate transformation of coordinates, rendering its status rather nebulous; (iii) according to the equivalence principle, the gravitational energy cannot be localized.] Since then, neither Einstein nor anyone else has been able to discover the true form of T^{ik}, although it is at the heart of the current efforts to reconcile GR with quantum mechanics.

It is an undeniable fact that the standards of T^{ik}, in terms of elegance, consistency and mathematical completeness, do not match the vibrant geometrical side of Equation (1), which is determined almost uniquely by pure mathematical requirements. Einstein himself conceded this fact when he famously remarked: *"GR is similar to a building, one wing of which is made of fine marble, but the other wing of which is built of low grade wood"*. It was his obsession that attempts should be directed to convert the "wood" into "marble".

The doubt envisioned by Einstein about representing matter by T^{ik}, is further strengthened by a recent study which discovers some surprising inconsistencies and paradoxes in the formulation of the energy-stress tensor of the matter fields, concluding that the formulation of T^{ik} does not seem consistent with the geometric description of gravitation [14]. This is reminiscent of the view expressed about four decades ago by J. L. Synge, one of the most distinguished mathematical physicists of the 20th Century: *"the concept of energy-momentum* (tensor) *is simply incompatible with general relativity"* [15] (which may seem radical from today's mainstream perspective).

Unphysical Solutions: Since its very inception, GR started having observational support which substantiated the theory. Its predictions have been well-tested in the limit of the weak gravitational field in the solar system, and in the stronger fields present in the systems of binary pulsars. This has been done through two solutions—the Schwarzschild and Kerr solutions.

However, there exist many other 'vacuum' solutions of Equation (1) which are considered *unphysical*, since they represent curvature in the absence of any conventional source. The solutions falling in this category are the de Sitter solution, Taub-NUT Solution, Ozsváth–Schücking solution and two newly discovered [16,17] solutions (given by Equations (6) and (7) in the following). (Another solution, which falls in this category, is the Gödel solution which admits closed timelike-curves and hence permits a possibility to travel in the past, violating the concepts of causality and creating paradoxes: "what happens if you go back in the past and kill your father when he was a baby!") Hence the theory has been supplemented by additional "physical grounds" that are used to exclude otherwise exact solutions of Einstein's equation.

This situation is very reminiscent of what Kinnersley wrote about the GR solutions, *"most of the known exact solutions describe situations which are frankly unphysical"* [18].

This is however misleading because not only does it reject *a priori* the majority of the exact solutions claiming "unphysical" and "extraneous", but also mars the general validity of the theory and introduces an element of subjectivity in it. Perhaps we fail to interpret a solution correctly and pronounce it unphysical because the interpretation is done in the framework of the conventional wisdom, which may not be correct [14,19].

Interior Solutions: As mentioned earlier, GR successfully describes the gravitational field outside the Sun in terms of the Schwarzschild (exterior) and Kerr solutions. Nevertheless, the theory has not been that successful in describing the interior of a massive body.

Soon after discovering his famous and successful (exterior) solution (with $T^{ik} = 0$), Schwarzschild discovered another solution of Equation (1) (with a non-zero T^{ik}) representing the interior of a static, spherically symmetric non-rotating massive body, generally called the Schwarzschild interior solution. SInce then, many other, similar interior solutions have been discovered with different matter distributions. It appears, however, that the picture the conventional interiors provide is not conceptually satisfying. For example, the Schwarzschild-interior solution assumes a static sphere of matter consisting of an incompressible perfect fluid of constant density (in order to obtain a mathematically simple solution). Hence, the solution turns out to be unphysical, since the speed of sound $= c\sqrt{dp/d\rho}$ becomes infinite in the fluid with a constant density ρ and a variable pressure p.

The Kerr solution, representing the exterior of a rotating mass, has remained unmatched to any known non-vacuum solution that could represent the interior of a rotating mass. It seems that we have been searching for the interior solutions in the wrong place [17].

Dark Matter and Dark Energy: Soon after formulating GR, Einstein applied his theory to model the Universe. At that time, Einstein believed in a static Universe, perhaps guided by his religious conviction that the Universe must be eternal and unchanging. As Equation (1) in its original form does not permit a static Universe, he inserted a term—the famous 'cosmological constant Λ' to force the equation to predict a static Universe. However, it was realized later that this gave an unstable Universe. It was then realized that a naive prediction of Equation (1) was an expanding Universe, which was subsequently found consistent with the observations. Realizing this, Einstein retracted the introduction of Λ terming it his *"biggest blunder"*.

The cosmological constant has however reentered the theory in the guise of dark energy. As has been mentioned earlier, in order to explain various observations, the theory requires two mysterious, invisible, and as yet unidentified ingredients—dark matter and dark energy—and Λ is the principal candidate of dark energy.

One the one hand, the theory predicts that about 27% of the total content of the Universe is made of non-baryonic dark matter particles, which should certainly be predicted by some extension of the Standard Model of particles physics. However, there is no indication of any new physics beyond the Standard Model which has been successfully verified at the Large Hadron Collider. Curious discrepancies also appear to exist between the predicted clustering properties of dark matter on small scales and observations. Obviously, the dark matter has eluded our every effort to bring it out of the shadows.

On the other hand, the dark energy is believed to constitute about 68% of the total content of the Universe. The biggest mystery is not that the majority of the content of T^{ik} cannot be seen, but that it cannot be comprehended. Moreover, the most favored candidate of dark energy—the cosmological constant Λ—poses serious conceptual issues, including the cosmological constant problem—"why does Λ appear to take such an unnatural value?" That is, "why is the observed value of the energy associated with Λ so small (by a factor of $\approx 10^{-120}$!) compared to its value (Planck mass) predicted by the quantum field theory?" and the coincidence problem—"why is this observed value so close to the present matter density?".

The cosmological constant problem in fact arises from a structural defect of the field Equation (1). While in all non-gravitational physics, the dynamical equations describing a system do not change if we shift the "zero point" of energy, this symmetry is not respected by Equation (1) wherein all sources of energy and stress appear through T^{ik} and hence gravitate (*i.e.*, affect the curvature). As the Λ-term can very well be assimilated in T^{ik}, adding this constant to Equation (1) changes the solution. It may be noted that no dynamical solution of the cosmological constant problem is possible within the existing framework of GR [20].

Horizon Problem: Why does the cosmic microwave background (CMB) radiation look the same in all directions despite being emitted from regions of space failing to be causally connected? The size of the largest coherent region on the last scattering surface, in which the homogenizing signals passed at sound speed, can be measured in terms of the sound horizon. In the standard cosmology, this implies, however, that the CMB ought to exhibit large anisotropies (*not isotropy*) for angular scales of theorder of $1°$ or larger—a result contrary to what is observed [8]. Hence, it seems that the isotropy of the CMB cannot be explained in terms of some physical process operating under the principle of causality in the standard paradigm.

Inflation comes to the rescue. It is generally believed that inflation made the Universe smooth and left the seeds of structures, on the surface of the last scatter, of the order of the Hubble distance at that time. However, inflation has its own problems either unsolved or fundamentally unresolvable. There is no consensus on which (if any) the inflation model is correct, given that there are many different inflation models. A physical mechanism that could cause inflation is not known, though there are many speculations. There are also difficulties on how to turn off the inflation once it starts—the "graceful exit" problem.

Flatness Problem: In the standard cosmology, the total energy density ρ in the early Universe appears to be extremely fine-tuned to its critical value $\rho_c = 3H^2 / (8\pi G)$, which corresponds to a flat spatial geometry of the Universe, where H is the Hubble parameter. Since ρ departs rapidly from ρ_c over cosmic time, even a small deviation from this value would have had massive effects on the nature of the present Universe. For instance, the theory requires ρ at the Planck time to be within one part in 10^{57} of ρ_c in order to meet the observed uncertainties in ρ at present! That is, the Universe was almost flat just after the Big Bang—but how?

If a theory predicts a fine-tuned value for some parameter, there should be some underlying physical symmetry in the theory. In the present case however, this appears just an unnatural and *ad hoc* assumption in order to reproduce observation. Inflation comes to the rescue again. Irregularities in the geometry were evened out by inflation's rapid accelerated expansion causing space to become flatter and hence forcing ρ toward its critical value, no matter what its initial value was.

However, it should also be mentioned that flatness and horizon problems are not problems of GR. Rather, they are problems concerned with the cosmologist's conception of the Universe, very much in the same vein as was Einstein's conception of a static Universe.

Scale Invariance: It is well-known that GR, unlike the rest of physics, is not scale invariant in the field Equation (1) [21]. As scale invariance is one of the most fundamental symmetries of physics, any physical theory, including GR, is desired to be scale invariant.

3. A New Perspective on Gravity

Hence, with a substantial amount of anomalies, paradoxes and unexplained phenomena, one would question whether the pursued approach to GR is correct. It appears that we have misunderstood the true nature of a geometric theory of gravitation because of the way the theory has evolved. Taken at face value, these problems insinuate that our understanding of gravitation in terms of the conventional GR is grossly incomplete (if not incorrect) and we need yet another paradigm shift.

Science advances more from what we do not understand than by what we do understand. From a careful re-examination of the above-mentioned problems, a new insight with deeper vision of a geometric theory of gravitation emerges, which appears as the missing piece of the theory. It may appear surprising at first sight though that these seemingly disconnected problems can lead to any coherent, meaningful solution. Nevertheless, as we shall see in the following, the analysis develops drastic revolutionary changes in our conventional views of GR and offers an enlightened view wherein all the above-mentioned difficulties disappear.

3.1. Revisiting Mach's Principle

Guided by the principle of covariance, GR has been formulated in the language of tensors. As the principle of covariance results as a consequence from Mach's principle, one naturally expects the theory to be perfectly Machian, as Einstein did. Then, why do some of the solutions of GR contradict Mach's philosophy? Perhaps we have missed the real message these solutions want to convey. Particularly, the curious presence of the timelike geodesics and a well-defined notion of inertia in the solutions of Equation (1) obtained in the absence of T^{ik} must not be just coincidental and there must be some source.

In order to witness this, let us try to impose the philosophy of Mach on the existing framework of GR by quantifying Mach's principle with a precise formulation in which matter and geometry appear to be in one-to-one correspondence. The key insight is the observation that not only inertia, but also space and time emerge from the interaction of matter. As space is an abstraction from the totality of distance-relations between matter, it follows that the existence of matter (fields) is a necessary and sufficient condition for the existence of spacetime. This idea can be formulated in terms of the following postulate:

Postulate: *Spacetime cannot exist in the absence of fields.*

The postulate posits that spacetime is not something to which one can ascribe a separate existence, independently of the matter fields, and the very existence of spacetime signifies the presence of the matter (fields). This is very much in the spirit of Mach's principle which implies that the existence of a spacetime structure has any meaning only in the presence of matter, which is bound so tightly to the former that one can not exist without the other.

Inspired by this, Einstein had envisioned that *"space as opposed to 'what fills space', has no separate existence"* [22] thought he could not implement it in his field Equation (1), wherein the "space" (represented in the left-hand side of the equation) and "what fills space" (represented by its right-hand side) do have separate existence: as has been mentioned earlier, there exist various meaningful spacetime solutions of Equation (1) in the total absence of T^{ik}. The adopted postulate, on the other hand, emphasizes that spacetime has no independent existence without a material background, which is present universally regardless of the geometry of the spacetime.

As the matter field is always accompanied by the ensuing gravitational field and since the latter also gravitates, an important consequence of the adopted postulate is that the geometry of the resulting spacetime should be determined by the net contribution from the two fields. Thus, the metric field is entirely governed by considered matter fields, as one should expect from a Machian theory.

3.2. Fields without T^{ik}: An Inescapable Consequence of Mach's Principle

The theoretical appeal of the above-described hypothesis is that it is naive, self evident and plausible. However, more than that, it has potential to shape a theory and gives rise to a new vision of GR with novel, dramatic implications. For instance, it makes a powerful prediction that the resulting theory should not have any bearing on the energy-stress tensor T^{ik} in order to represent the source fields. [The source of curvature in a solution of Einstein's field Equation (1), in the absence of T^{ik}, is conventionally attributed to a singularity. This prescription is however rendered nebulous by the presence of various singularity-free curved solutions of Equation (1) in the absence of T^{ik}]. Let us recall that Equation (1) does admit various meaningful spacetime solutions in the absence of the "source" term T^{ik}.

According to the postulate, as fields are present universally in all spacetimes irrespective of their geometry, the flat Minkowskian spacetime should not be an exception, and it must also be endowed with the matter fields and the ensuing gravitational field. Now let us recall that the Minkowski spacetime appears as a solution of Einstein's field Equation (1) only in the absence of T^{ik}, in which case the effective field equation yields

$$R^{ik} = 0 \tag{3}$$

However, if the fields can exist in the Minkowski spacetime (as asserted by the founding postulate) in the absence of T^{ik}, they can also exist in other spacetimes in the absence of T^{ik}. Hence, the requirement of uniqueness of the field equation of a viable theory dictates that T^{ik} must not be the carrier of the source fields in a theory resulting from the adopted postulate, and, thus, the canonical Equation (3) emerges as the field equation of the resulting theory in the very presence of matter. In fact, this is what happens if we accept, at their face value, the implications of Mach's principle applied to GR.

This novel feature that GR would acquire—that the spacetime solutions of Equation (3), including the Minkowskian one, are not devoid of fields—provides an appealing first principle approach and a linchpin to understand various unsolved issues in a unified scheme. It becomes remarkably decisive for the theory on Machianity. It was the earlier-mentioned characteristic of the Minkowski and other solutions of Equation (3) to possess timelike geodesics and a well-defined notion of inertia, that pronounced these solutions non-Machian, as they are conventionally regarded to represent empty spacetimes. The new insight, however, renders them perfectly Machian and physically meaningful by bestowing a matter-full dignity on them. Moreover, this novel feature of the Minkowski solution also explains another so-far unexplained issue: It has been noticed that the Noether current associated with an arbitrary vector field in the Minkowski solution is non-zero in general [23], which remains unexplained in the conventional 'empty' Minkowskian spacetime.

Though the proposed scheme of having matter fields in the absence of T^{ik} may sound surprising and orthogonal to the prevailing perspective, it seems to have many advantages over the conventional approach, as we shall see in the following. The issue is whether it can be made realistic. That is, if Equation (3) is claimed to constitute the field equation of a viable theory of gravitation in the very presence of matter, its solutions must possess some imprint of this matter. Thus, do we have any evidence of such imprints in the solutions of Equation (3)? The answer is, yes.

3.3. Evidence of the Presence of Fields in the Absence of T^{ik}

As Mach's principle denies unobservable absolute spacetime in favor of the observable quantities (the background matter) which determine its geometry, the principle would expect the source of curvature in a solution to be attributable entirely to some directly observable quantity, such as mass-energy, momentum, and angular momentum or their densities. Thus, if GR is correct and it must be Machian, these quantities are expected to be supported by some dimension-full parameters appearing in the curved spacetime solutions in such a way that the parameters vanish as the observable quantities vanish, reducing the solutions to the Minkowskian form.

Interestingly, it has been shown recently [16,17] that it is always possible to write a curved solution of Equation (3) in a form containing some dimension-full parameters, which appear in the Riemann tensor generatively and can be attributed to the source of curvature. The study further shows that these parameters can support physical observable quantities such as the mass-energy, momentum or angular momentum or their densities. For instance, the source of curvature in the Schwarzschild solution

$$ds^2 = \left(1 + \frac{K}{r}\right) c^2 dt^2 - \frac{dr^2}{(1 + K/r)} - r^2 d\theta^2 - r^2 \sin^2\theta \, d\phi^2 \tag{4}$$

can be attributed to the mass m (of the isotropic matter situated at $r = 0$) through the parameter $K = -2Gm/c^2$. Similarly, the dimension-full parameters present in the Kerr solution can be attributed to the mass and the angular momentum of the source mass; those in the Taub-NUT solution to the mass and the momentum of the source; and the parameters in the Kerr-NUT solution to the mass, momentum and angular momentum [16,17].

A remarkable piece of evidence of the presence of fields in the absence of T^{ik} is provided by the Kasner solution, which exemplifies that even in the standard paradigm, all the well-known curved solutions of Equation (3) do not represent space outside a gravitating mass in an empty space. [It is conventionally believed that only those curved solutions of Equation (3) are meaningful which represent space outside some source matter, otherwise the solutions represent an empty spacetime. However, Equation (3) cannot decipher just from the symmetry of a solution that it necessarily belongs to a spacetime structure in an empty space outside a mass, since the same symmetry can also be shared by a spacetime structure inside a matter distribution.] Although the Kasner solution in its standard form does not contain any dimension-full parameter that can be attributed to its curvature, the solution can however be transformed to the form

$$ds^2 = c^2 dt^2 - (1 + nt)^{2p_1} dx^2 - (1 + nt)^{2p_2} dy^2 - (1 + nt)^{2p_3} dz^2 \tag{5}$$

where n is an arbitrary constant parameter (which is dimension-full) and the dimensionless parameters p_1, p_2, p_3 satisfy $p_1 + p_2 + p_3 = 1 = p_1^2 + p_2^2 + p_3^2$.

A dimensional analysis suggests that, in order to meet its natural dimension (which is of the dimension of the inverse of time), the parameter n can support only the densities of the observables energy, momentum or angular momentum and *not* the energy, momentum or angular momentum themselves [such that Equation (5) becomes Minkowskian when the observables vanish]. However, the energy density and the angular momentum density vanish here: while the symmetries of Equation (5) discard any possibility for the angular momentum density, the energy density disappears as it is canceled by the negative gravitational energy [17,24]. That is, the parameter n in Equation (5) can be expressed in terms of the momentum density $\mathcal{P}$ as $n = \gamma\sqrt{G\mathcal{P}/c}$, where γ is a dimensionless constant. This indicates that Equation (5) results from a (uniform) matter distribution (throughout space) and not from a spacetime outside a point mass as in the cases of the Schwarzschild and Kerr solutions. Thus, the Kasner solution represents a homogeneous distribution of matter expanding and contracting anisotropically (at different rates in different directions), which can give rise to a net non-zero momentum density represented through the parameter n serving as the source of curvature, thus demystifying the solution.

This new insight on the source of curvature is authenticated by two new solutions of Equation (3) discovered in [16,17] whose discovery is facilitated by the new insight. The first solution, whose source of curvature cannot be explained with the conventional wisdom (as it is singularity-free), provides a powerful support to the Machian strategy of representing the source in terms of the dimension-full source-carrier parameters (here ℓ). The solution is given by

$$ds^2 = \left(1 - \frac{\ell^2 x^2}{8}\right)c^2dt^2 - dx^2 - dy^2 - \left(1 + \frac{\ell^2 x^2}{8}\right)dz^2 + \ell x(cdt - dz)dy + \frac{\ell^2 x^2}{4}cdt\,dz \quad (6)$$

which has been derived by defining the parameter ℓ in terms of the angular momentum density $\mathcal{J}$ via $\ell = G\mathcal{J}/c^3$ [16]. The fact that the parameter ℓ can support only the density of angular momentum and *not* the angular momentum itself asserts that Equation (6) results from a rotating *matter distribution* (confined to $-\frac{2\sqrt{2}}{|\ell|} < x < \frac{2\sqrt{2}}{|\ell|}$) and not from a spacetime outside a point mass as are the cases of the Schwarzschild and Kerr solutions. This is in perfect agreement with the founding postulate that the fields are not different from the spacetime.

Equation (6) as a new solution of field Equation (3) is important in its own right. Moreover, it illuminates the so far obscure source of curvature in the well-known Ozsváth–Schücking solution, which would otherwise be in stark contrast with the new strategy in the absence of any free parameter. It has been shown in [16] that the Ozsváth-Schücking solution results from Equation (6) by assigning a particular value to the parameter ℓ.

Following the new insight, another new solution of Equation (3) has been discovered recently in [17], whose curvature is supported by the energy density (the author recently came to know that solution Equation (7) has also been reported in [25]). The solution is given by:

$$ds^2 = \frac{(1 + 4\mu z^2)}{(1 + \mu r^2)^2}c^2dt^2 - \frac{dr^2}{(1 + \mu r^2)^4} - r^2d\phi^2 - \frac{dz^2}{(1 + 4\mu z^2)(1 + \mu r^2)^2} \quad (7)$$

which represents a inhomogeneous axisymmetric distribution of matter, with the parameter μ given in terms of the energy density $\mathcal{E}$ as $\mu = G\mathcal{E}/c^4$. As Equation (7) is curved but singularity-free for all finite values of the coordinates, it provides, in the absence of any conventional source there, a strong support to the new strategy of source representation.

3.4. A New Vision of Gravity in the Framework of GR: Spacetime Becomes a Physical Entity

What does the presence of these dimension-full parameters we witness in the solutions of the field Equation (3) signify? As the physical observable quantities sustained by the parameters—*i.e.*, energy, momentum, angular momentum and their densities have any meaning only in the presence of matter, the presence of such parameters in the solutions of Equation (3) must not be just a big coincidence, and, at face value, their ubiquitous presence in the solutions of Equation (3) insinuates that fields are universally present in the spacetime in Equation (3).

Not only does this provide a strong support to the founding postulate establishing GR as a Machian theory, but also establishes, on firm grounds, Equation (3) as the field equation of a feasible theory of gravitation in the very presence of fields. More than that, there emerges a radically new vision of a geometric theory of gravitation through drastic revolutionary changes in our views on the representation of the source of gravitation, which must be through the geometry and *not* through T^{ik}. By reconceptualizing our previous notions of spacetime, this constitutes a paradigm shift in GR wherein the spacetime itself becomes a physical entity, we may call it the "emergent matter" in a relativistic/geometric theory of gravitation. From the ubiquitous presence of fields in all geometries, it becomes clear that there is no empty space solution in the new paradigm, as one should expect from a Machian theory. The same was also envisioned by Einstein (though could not be achieved).

One may wonder how the properties of matter can be incorporated into the dynamical equations of the new theory without taking recourse to T^{ik}. This can be achieved by applying the conservation laws and symmetry principles to the new conviction that all spacetimes harbor fields, inertial and gravitational, whose net contribution determines their geometry. For instance, by assuming that the sum of the gravitational and inertial energies in a uniform matter distribution should be vanishing [17,24], it has been shown recently that the homogeneous, isotropic Universe in the new paradigm leads to the Friedmann equation of the standard "concordance" cosmology [17]. This should not be a surprise, as the Friedmann equation for dust can also be derived in Newtonian cosmology or in a kinematic theory (like the Milne model) by using the continuity equation and the Navier–Stokes equation of fluid dynamics [26,27].

3.5. Equivalence Principle in the New Perspective

The perfect equivalence between gravitational and inertial masses, first noted by Galileo and Newton, was more or less accidental. For Einstein, however, this served as a key to a deeper understanding of inertia and gravitation. From his valuable insight that the kinematic acceleration and the acceleration due to gravity are intrinsically identical, he was able to unearth a hitherto unknown mystery of nature—that gravitation is a geometric phenomenon.

It however seems that the full implications of the equivalence principle have not yet been appreciated. If gravitation is a geometric phenomenon, then through the (local) equivalence of gravitation and inertia, the inertia of matter should also be considered geometrical in nature, at least when it appears in a geometric theory of gravitation. A purely geometrical interpretation of gravitation would be impossible unless the gravitational as well as the inertial properties of matter are intrinsically geometrical. This would, however, have revolutionary implications. Considering T^{ik} (which represents the inertial fields) of a purely geometric origin, Equation (1) would imply

$$G^{ik} + \frac{8\pi G}{c^4} T^{ik} \equiv \chi^{ik} = 0 \qquad (8)$$

where χ^{ik} appears a tensor of purely geometric origin. This would however be nothing else but the Ricci tensor R^{ik} (with a suitable g_{ik}), since the only tensor of rank two having a purely geometric origin (emerging from the Riemann tensor), is the Ricci tensor. That is, Equation (8) would reduce to the field Equation (3)! In this way, the consequences of the equivalence principle would be in perfect agreement with the adopted Machian postulate—that spacetime has no separate existence from matter, *i.e.*, the parameters of the spacetime geometry determine entirely the combined effects of gravitation and inertia.

Therefore, the consequence of the equivalence principle—that the gravitational and inertial fields are entirely geometrical by nature—takes GR to its logical extreme in that the spacetime emerges from the interaction of matter. This reconceptualizes the previous notion of spacetime by establishing it as the very source of gravitation. The matter is in fact more intrinsically related to the geometry than is believed in the conventional GR and all the aspects of matter fields (including the ensuing gravitational field) are already present inherently in the spacetime geometry. This establishes Equation (3) as a competent field equation of gravitation plus inertia. This is well-supported by our observation that while the gravitational field is present in the Schwarzschild, Kerr and Taub-NUT solutions (as these represent the spacetimes outside the source mass), the inertial as well as the gravitational fields are present in Equations (5)–(7) including the Minkowskian one, which represent matter distribution.

A precise specification of the fields, which are being claimed to be present in the spacetime, is possible only when a precise formulation thereof is available. Nevertheless, in view of the newly gained insight, at least this much can be declared that the matter fields present in the geometry of Equation (3) are those which are attempted to be introduced in Equation (1) or (8) via T^{ik} (which has now been absorbed in Equation (3)).

4. A Closer Look at the Conventional Four-Dimensional Formulation of Matter

Modeling matter by T^{ik} in Equation (1) has modified at the deepest level the way we used to think about the source of gravitation. As mass density is the source of gravitation in Newtonian theory, the energy density was expected to take over this role in the relativistic generalization of Poisson's equation. To our surprise, however, all ten (independent) components of T^{ik} become contributing sources of gravitation. We need not doubt this novelty, as new theories originated from innovative ideas are expected to have innovative features. However, the way the non-conventional sources appear in the dynamical equations, appears to create inconsistencies and paradoxes, which warrants a second look at the relativist formulation of matter given by T^{ik}.

Everyone will agree that, like the conservation of momentum, the conservation of energy of an isolated system is an absolute symmetry of nature and this fundamental principle is expected to be respected by any physical theory. Nonetheless, the principle is violated in GR in many different situations including the cosmological scenarios (see, for example, [28]). The blame rests with the energy of the gravitational field, which has been of an obscure nature and a controversial history, as has been mentioned earlier. We shall, however, see that the gravitational energy is not to be blamed for the trouble. This is ascertained beyond a doubt in the following analysis by filtering out the gravitational energy from the equations.

4.1. Problems with T^{ik}

As is well-known, the formulation of the energy-stress tensor T^{ik} given by Equation (2) is obtained by first deriving it in the absence of gravity in SR, by considering a fluid element in a small neighborhood of an LICS, which exists admittedly at all points of spacetime (by courtesy of the principle of equivalence). Then, the expression for the tensor in the presence of gravity is imported, from SR to GR, through a coordinate transformation. It would be insightful to reconsider the same LICS to understand the mysterious implications of T^{ik}, since the subtleties of gravitation and the gravitational energy disappear locally in this coordinate system. Let us then study the divergence of T^{ik} in the considered LICS, which is known for describing the mechanical behavior of the fluid. Through the vanishing divergence of G^{ik}, Equation (1) implies that $T^{ij}{}_{;j} = 0$, which, in the chosen coordinates, reduces to

$$\frac{\partial T^{ij}}{\partial x^j} = 0 \tag{9}$$

For the case of a perfect fluid given by Equation (2), it is easy to show that Equation (9), in the chosen LICS, yields [29]

$$\frac{\partial p}{\partial x} + \frac{(\rho + p)}{c^2} \frac{du_x}{dt} = 0 \tag{10}$$

for the case $i = 1$, where du_x/dt is the acceleration of the considered fluid element in the x-direction. As any role of gravity and gravitational energy is absent in this equation, it can be interpreted as the relativistic analogue of the Newtonian law of motion: the fluid element of unit volume, which moves under the action of the force applied by the pressure gradient $\partial p/\partial x$, has got the inertial mass $(\rho + p)/c^2$. Let us, however, recall that the term ρ in Equation (2) includes in it, by definition, not only the rest mass of the individual particles of the fluid but also their kinetic energy, internal energy (for example, the energy of compression, energy of nuclear binding, *etc.*) and *all other sources of mass-energy* [11]. Therefore, the additional contribution to the inertial mass entering through the term p, appears to violate the celebrated law of the conservation of energy. Though Equation (10) is usually interpreted as a momentum conservation equation, an alternative (but viable) interpretation is not expected to defy the energy conservation.

Similar problems seem to afflict the temporal component of Equation (9) for $i = 0$, which can be written as the following [29]:

$$\frac{d}{dt}(\rho \delta v) + p \frac{d}{dt}(\delta v) = 0 \tag{11}$$

where δv is the proper volume of the fluid element. The usual interpretation to this equation says: the rate of change in the energy of the fluid element is given in terms of the work done against the external pressure. This seems reasonable at first sight, but cracks seem to appear in it after a little reflection. The concern, as also noticed by Tolman [29], is that the fluid of a finite size can be divided into similar fluid elements and the same Equation (11) can be applied to each of these elements, meaning that the proper energy $(\rho \delta v)$ of every element is decreasing when the fluid is expanding or increasing when the fluid is contracting. This leads to a paradoxical situation that the sum of the proper energies of the fluid elements which make up an isolated system, is not constant. Tolman overlooked this problem by assuming a possible role of the gravitational energy in it. We note, however, that no such possibility exists as Equation (11) has been derived in an LICS.

The total energy E, including the gravitational energy, of an isolated time-independent fluid sphere comprised of perfect fluid given by Equation (2) and occupying volume V of the three-space $x^0 = $ constant, is given by the Tolman formula [29]:

$$E = \int_V (\rho + 3p) \sqrt{|g_{00}|}\, dV \tag{12}$$

which measures the strength of the gravitational field produced by the fluid sphere. The formula is believed to be consistent, for the case of the disordered radiation ($p = \rho/3$), with the observed deflection of starlight (twice as much as predicted by a heuristic argument made in Newtonian gravity), when it passes the Sun. Ironically, this expectation is contradicted by the weak-field approximation of the same Equation (12). In a weak field, like that of the Sun, where Newtonian gravitation can be regarded as a satisfactory approximation, Equation (12) can be written, following Tolman (see page 250 of [29]), as $E = \int \rho dV + (1/2c^2) \int \rho\psi dV$, where ψ is the Newtonian gravitational potential. As ψ is negative, we note that the general relativistic active gravitational mass E/c^2 of the gravitating body, here the Sun, is obviously less than its Newtonian value $(1/c^2) \int \rho dV$ and is expected to give a lower value for the gravitational deflection of light than the corresponding Newtonian value. (Let us recall that the correct interpretation of the observations of the bending of starlight, when it passes past the Sun, comes from the correct geometry around Sun resulting from the Schwarzschild solution.)

As has been mentioned earlier, the (active) gravitational and inertial mass are in general unequal in GR solutions (the discrepancy thereof is supposed to be accounted by the gravitational energy). Thus, in an LICS, which nullifies gravitation and hence gravitational mass locally, we expect a unique value for the mass in the equations. To check this, let us calculate the Tolman integral Equation (12) (density) in the considered LICS wherein it reduces to

$$E = \int (\rho + 3p)\, dV \tag{13}$$

which may now be valid for a sufficiently small volume of the fluid. Surprisingly, we still encounter different unequal values of mass (density) in Equations (10), (11) and (13). [Equation (11) can be written alternatively as $\delta v\, d\rho/dt + (\rho + p)d(\delta v)/dt = 0$.] While Equations (10) and (11) give this value as $(\rho + p)/c^2$, Equation (13) provides a different value $(\rho + 3p)/c^2$. Perhaps the origin of the problem is not in the gravitational energy but in T^{ik} itself.

Given this backdrop, it thus appears that the relativistic formulation of matter given by T^{ik} suffers from some subtle inherent problems. The point to note is that there is no role of the notorious (pseudo) energy of the gravitational field in these problems. It would not be correct to conclude that the above-analysis advocates denial of fluid pressure in GR (as the problems are evaded in the absence of pressure). Rather it insinuates that the four-dimensional description of matter in terms of T^{ik} is not compatible with the geometric description of gravitation. It is perhaps not correct to patchwork a four-dimensional tensor from two basically distinct kinds of three-dimensional quantities—(i) the energy density, a non-directional quantity and (ii) the momenta and stresses, directional quantities. The tensor, however, treats them on equal footing by recognizing a component T^{ik} as a scalar (irrespective of the values of i and k) linked with the surface specified by i and k in the *hypothetical* four-dimensional fluid, in the same way as the component G^{ik} is linked with the curvature of the same surface. This leads to sound mathematics, and we do not notice any inconsistency until we relate the tensor T^{ik} with the real fluid, which is three-dimensional and not four-dimensional.

Does it then mean that Einstein's "wood" is not only low grade compared to the standards of his "marble" but it is also infested? It should be noted that the relativistic formulation of the matter, in terms of the tensor T^{ik}, has never been tested in any direct experiment. It may be recalled that the crucial tests of GR, which have substantiated the theory beyond doubt, are based on the solutions of Equation (3) only, *viz.* the Schwarzschild and Kerr solutions.

It thus becomes increasingly clear that the development of GR was led astray by formulating matter in terms of T^{ik}. This is corroborated by the fact that whenever the theory takes recourse to T^{ik} in Equation (1), trouble shows up in the form of either the dark energy or the inviabilities of Godel's solution and Schwarzschild's interior solution, *etc.* In view of the new finding, this assertion acquires a new meaning—we have been searching for the matter in the wrong place. The correct place to search for it is the geometry. We have seen in innumerable examples that matter is already present in the geometry of Equation (3) without taking recourse to T^{ik}. That is, the "wood" is already included into the 'marble', dramatically fulfilling Einstein's obsession.

5. Successes of the Novel Gravity Formulation

5.1. Observational Support for the New Paradigm

The last words on a putative theory have to be spoken by observations and experiments. The consistency of the field Equation (3) with the local observations in the solar system and binary pulsars, has already been established in the standard tests of GR—the only satisfactory testimonial of the theory among the conventional tests, which do not require any epicycle of the dark sectors.

Interestingly, as has been shown recently [27], all the cosmological observations can also be explained successfully in terms of a homogeneous, isotropic solution of Equation (3). This solution can be obtained by solving Equation (3) for the Robertson–Walker metric, yielding

$$ds^2 = c^2 dt^2 - c^2 t^2 \left(\frac{dr^2}{1 + r^2} + r^2 d\theta^2 + r^2 \sin^2\theta\, d\phi^2 \right) \tag{14}$$

which represents the homogeneous, isotropic Universe in the new paradigm. It may be mentioned that solution Equation (14) (which is generally recognized as the Milne model), wherein the Universe appears dynamic in terms of the comoving coordinates and the cosmic time t, can be reduced to the Minkowskian form by using the locally defined measures of space and time [27].

The observational tests considered in [27] include the observations of the high-redshift supernovae (SNe) Ia, the observations of high-redshift radio sources, observations of starburst galaxies, the CMB observations and compatibility of the age of the Universe with the oldest objects in it (for instance, the globular clusters) for the currently measured values of the Hubble parameter. It may also be mentioned that, by preforming a rigorous statistical test on a much bigger sample of SNe Ia, (by taking account of the empirical procedure by which corrections are made to their absolute magnitudes), a recent study has found only marginal evidence for an accelerated expansion, and the data are quite consistent with the Milne model [30].

One may wonder how the new model, which does not possess dark energy (and hence is not an accelerated expansion), manages to reconcile with the observations. The mystery lies in the special expansion dynamics of the model at a constant rate throughout the evolution, as is clear from Equation (14), wherein the Robertson–Walker scale factor $S = ct$. We note that, unlike the standard cosmology, Equation (14) provides efficiently different measures of distances without requiring any input from the matter fields. For instance, the luminosity distance d_L of a source of redshift z, in the present case, is given by

$$d_L = cH_0^{-1}(1+z)\sinh[\ln(1+z)] \tag{15}$$

where H_0 represents the present value of the Hubble parameter $H = S/S$. As has been shown in Figure 1, the luminosity distance of an object of redshift z in the new cosmology is almost the same as that in the standard cosmology for $z \lesssim 1.3$. This explains why both models are equally consistent with the SNe Ia data wherein the majority of the SNe belong to this range of redshift. However, for $z > 1.3$, the new model departs significantly from the standard cosmology, as is clear from the figure. Hence, observations of more SNe Ia at higher redshifts will be decisive for both paradigms.

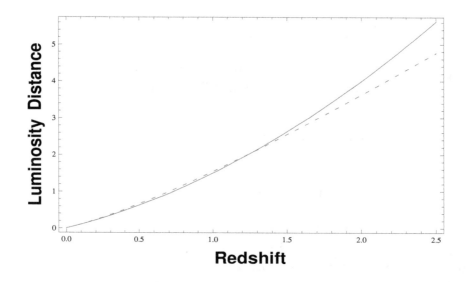

Figure 1. Luminosity distance in the new model (continuous curve) is compared with that in the Λ CDM concordance model $\Omega_m = 1 - \Omega_\Lambda = 0.3$ (broken curve). Distances shown on the vertical axis are measured in units of cH_0^{-1}. The two models significantly depart for $z \gtrsim 1.3$.

5.2. Different Pieces Fit Together

As the dark energy can be assimilated in the energy-stress tensor, and since the latter is absent from the dynamical equations in the new paradigm (wherein the fields appear through the geometry), the dark energy and its associated problems, for instance the cosmological constant problem (which appears due to a conflict between the energy-stress tensor T^{ik} in Equation (1) and the energy density of vacuum in the quantum field theory) and the coincidence problem, are evaded in the new paradigm.

For the same reason, the flatness problem is circumvented due to the absence of the energy-stress tensor in the new paradigm.

As has been mentioned earlier, the observed isotropy of CMB cannot be explained in the standard paradigm in terms of some homogenization process that has taken place in the baryon-photon plasma operating under the principle of causality, since a finite value for the particle horizon $d_{\mathrm{PH}}(t) = cS(t) \int_0^t dt'/S(t')$ (the largest distance from which light could have reached the present observer) exists in the theory. As $d_{\mathrm{PH}} = \infty$ always for $S = ct$, no horizon exists in the new paradigm, and the whole Universe is always causally connected, which explains the observed overall uniformity of CMB without invoking inflation [19].

As the Big Bang singularity is a breakdown of the laws of physics and the geometrical structure of spacetime, there have been attempts to discover singularity-free cosmological solutions of Einstein equations, which are usually achieved by violating the energy conditions.

Although Equation (14), which represents the cosmological model in the new paradigm, has well-behaved metric potentials at $t = 0$, the volume of the spatial slices vanishes there, resulting in a blowup in the accompanied matter density. However, this is just a coordinate effect which can be removed in the Minkowskian form of solution Equation (14) by considering the locally defined coordinates of space and time.

Moreover, as the locally defined time scale τ is related with the cosmic time t through the transformation $\tau = t_0 \ln(t/t_0)$ [27], the epoch corresponding to the Big Bang, is pushed back to the infinite past giving an infinite age to the Universe which can accommodate even older objects than the standard cosmology can. Interestingly, even in terms of the cosmic time t, wherein the Universe appears dynamic, the age of the Universe appears higher than that in the standard paradigm [27].

As has been mentioned earlier, the conventional 'source' term T^{ik} in Equation (1) fails to include the energy, momentum or angular momentum of the gravitational field. Remarkably, these quantities, akin to the matter fields, are inherently present in the geometry of Equation (3), substantiating the new strategy of the new paradigm to represent the source through geometry. For instance, the term $K/r = -2Gm/(c^2r)$ in the Schwarzschild solution Equation (4) contains the gravitational energy at the point r. It perfectly agrees with the Newtonian estimate of the gravitational energy given by $-Gm/r$, indicating that the term $-2Gm/(c^2r)$ is just its relativistic analogue. Assigning the gravitational energy to K/r is also supported by the locality of GR, which becomes an intrinsic characteristic of the theory as soon as the Newtonian concept of gravitation as a force (action-at-a-distance) is superseded by the curvature. Being a local theory, GR then assigns the curvature present at a particular point, to the source present at that very point. Thus, the agent responsible for the curvature in Equation (4) must be the gravitational energy, since matter exists only at $r = 0$, whereas Equation (4) is curved at all finite values of r. Hence, the presence of curvature in the Schwarzschild solution implies that the gravitational energy does gravitate just as does every other form of energy, and the gravitational field is obviously present in the geometry of Equation (3).

Similarly, the angular momentum of the gravitational field, arising from the rotation of the mass m, is revealed through the geometry of the Kerr solution, and its momentum in the Taub-NUT solution. Thus, the long-sought-after gravitational field energy-momentum-angular momentum of GR is already present in the geometry.

It may be interesting to note that new interior solutions, based on the solutions of Equation (3), have been formulated in the new paradigm that forms the Schwarzschild interior and the Kerr interior [17]. The new interiors are conceptually satisfying and free from the earlier mentioned problems.

As the Newtonian theory of gravitation provides excellent approximations under a wide range of astrophysical cases, the first crucial test of any theory of gravitation is that it reduces to the Newtonian gravitation in the limit of a weak gravitational field. In this context, it has been shown recently [17] that the new paradigm consistently admits the Poisson equation in the case of a slowly varying weak gravitational field when the concerned velocities are considered much less than c, provided we take into account the inertial as well as the gravitational properties of matter, as should correctly be expected in a true Machian theory. The standard paradigm on the other hand fails to fulfill this requirement as Equation (1), in the limit of the weak field, does not reduce to the Poisson equation in the presence of a non-zero Λ (or any other candidate of dark energy), which becomes unavoidable in the standard paradigm. In addition, it would not be correct to argue that a Λ as small as $\approx 10^{-56}$ cm^{-2} (as inferred from the cosmological observations) cannot contribute to the physics appreciably in the local problems. It has been shown recently that even this value of Λ does indeed contribute to the bending of light and to the advance of the perihelion of planets [31].

Interestingly, the new paradigm becomes scale invariant, since the new field Equation (3) is manifestly scale invariant. This becomes a remarkable achievement in the sense that one of the most common ways for a theory with continuous field to be renormalizable is for it to be scale-invariant.

Since the Universe in the new paradigm is flat, the symmetries of its Minkowskian form make it possible to validate the conservation of energy, solving the long-standing problems associated with the conservation of energy. As has been shown by Noether, it is the symmetry of the Minkowskian space that is the cause of the conservation of the energy momentum of a physical field [32,33].

5.3. Geometrization of Electromagnetism in the New Paradigm

How can the electromagnetic field be added to the new paradigm? While the equivalence principle renders the gravitational and inertial fields essentially geometrical (owing to the fact that the ratio of the gravitational and inertial mass is strictly unity for all matter), this is not so in the case of the electromagnetic field (since the ratio of electric charge to mass varies from particle to particle). Hence, the addition of the electromagnetic energy tensor E^{ik} to Equation (1), results in

$$R^{ik} = -\frac{8\pi G}{c^4} E^{ik} \tag{16}$$

since T^{ik} is absorbed in the geometry (as we have noted earlier), and $g_{ik}E^{ik} = 0$ reduces $R = 0$ identically. The tensor E_k^i is given, in terms of the skew-symmetric electromagnetic field tensor F_{ik}, as usual:

$$E^{ik} = \nu\left[-g^{k\ell}F^{ij}F_{\ell j} + \frac{1}{4}g^{ik}F_{\ell j}F^{\ell j}\right] \tag{17}$$

where ν is a constant. It has already been shown that Equation (16), taken together with the 'source-free' Maxwell equations,

$$\left.\begin{aligned}\frac{\partial F_{ik}}{\partial x^\ell} + \frac{\partial F_{k\ell}}{\partial x^i} + \frac{\partial F_{\ell i}}{\partial x^k} &= 0\\ \frac{\partial}{\partial x^k}\left(\sqrt{-g}F^{ik}\right) &= 0\end{aligned}\right\}, \quad g = \det((g_{ik})) \tag{18}$$

consistently represents the electromagnetic field in the presence of gravitation [17]. As the existence of charge is intimately related with the existence of the charge-carrier matter, and since the new paradigm claims the inherent presence of matter in the geometry, it is reasonable to expect the charge also to appear through the geometry. This view is indeed supported not only by the Reissner–Nordstrom and Kerr–Newman solutions, but also by the cosmological solutions—the so-called "electrovac universes" [17], wherein the charge does appear through the geometry. [Let us note that unlike the Reissner–Nordstrom and Kerr–Newman solutions (which represent the field outside the charged matter), the electrovac solutions are not expected to contain any "outside" where the charge-carrier matter can exist.]

Thus, Equations (16)–(18) of restricted validity in the standard paradigm [wherein they are believed to represent the electromagnetic field in vacuum, very much in the same vein as Equation (3) is believed to represent the gravitational field in vacuum] get full validity and represent a unified theory of gravitation, inertia and electromagnetism.

Interestingly, Misner and Wheeler also expressed similar views long ago and advocated to represent *"gravitation, electromagnetism, unquantized charge and unquantized mass as properties of curved empty space"* [34]. Although they failed to realize the presence of fields in the flat spacetimes; nonetheless, they also realized that Equations (16)–(18) provide a unified theory of electromagnetism and gravitation. [The removal of charge (by switching off E^{ik} from Equation (16), in which case the "electrovac universes" become flat) does not mean that mass (which was carrying charge) must necessarily disappear from these solutions.]

6. Summary and Conclusions: What Next?

GR is undoubtedly a theory of unrivaled elegance. The theory indoctrinates that gravitation is a manifestation of the spacetime geometry—one of the most precious insights in the history of science. It has emerged as a highly successful theory of gravitation and cosmology, predicting several new phenomena, most of them have already been confirmed by observations. The theory has passed every observational test ranging from the solar system to the largest scale, the Universe itself.

Nevertheless, GR ceases to be the ultimate description of gravitation, an epitome of a perfect theory, despite all these feathers in its cap. Besides its much-talked-about incompatibility with quantum mechanics, the theory suffers from many other conceptual problems, most of which are generally ignored. If in a Universe where, according to the standard paradigm, some 95% of the total content is still missing, it is an alarming signal for us to turn back to the very foundations of the theory. In view of these problems (discussed in the paper), we are led to believe that the historical development of GR was indeed on the wrong track, and the theory requires modification or at least reformulation.

By a critical analysis of Mach's principle and the equivalence principle, a new insight with a deeper vision of a geometric theory of gravitation emerges: matter, in its entirety of gravitational, inertial and electromagnetic properties, can be fashioned out of spacetime itself. This revolutionizes our views on the representation of the source of curvature/gravitation by dismissing the conventional source representation through T^{ik} and establishing spacetime itself as the source.

This appears as the missing link of the theory and posits that spacetime does not exist without matter, the former is just an offshoot of the latter. The conventional assumption that matter only fills the already existing spacetime, does not seem correct. This establishes the canonical equation $R^{ik} = 0$ as the field equation of gravitation plus inertia in the very presence of matter, giving rise to a new paradigm in the framework of GR. Though there seems to exist some emotional resistance in the community to tinkering with the elegance of GR, the new paradigm dramatically enhances the beauty of the theory in terms of the deceptively simple new field equation $R^{ik} = 0$. Remarkably, the new paradigm explains the observations at all scales without requiring the epicycle of dark energy.

This review provides an increasingly clear picture that the new paradigm is a viable possibility in the framework of GR, which is valid at all scales, avoids the fallacies, dilemmas and paradoxes, and answers the questions that the old framework could not address.

Though we have witnessed numerous evidences of the presence of fields in the solutions of the field Equation (3), however, the challenge to discover, from more fundamental considerations, a concrete mathematical formulation of the fields in purely geometric terms is still to be met. This formulation is expected to use the gravito-electromagnetic features of GR in the new paradigm and is expected to achieve the following:

1. It should explain the observed flat rotation curves of galaxies without requiring the ad-hoc dark matter.
2. The net field in a homogeneous and isotropic background must be vanishing.

Acknowledgments: The author thanks the IUCAA for hospitality where part of this work was done during a visit. Thanks are also due to two anonymous referees: to one for making critical, constructive comments which helped improving the manuscript; and to other for pointing out an old work of Misner and Wheeler [34], which is deeply connected with and strongly supporting the present work.

References

1. Ni, W.-T. Empirical Foundations of the Relativistic Gravity. *Int. J. Mod. Phys. D* **2005**, *14*, 901–921.
2. Will, C.M. The confrontation between general relativity and experiment. *Living Rev. Relativ.* **2006**, *9*, 3.
3. Turyshev, S.G. Experimental tests of general relativity: Recent progress and future directions. *Phys. Usp.* **2009**, *52*, 1–27.
4. Ashtekar, A. (Ed.) *100 Years of Relativity. Space-Time Structure: Einstein and Beyond*; World Scientific: Singapore, 2005.
5. Pdmanabhan, T. One hundred years of General Relativity: Summary, status and prospects. *Curr. Sci.* **2015**, *109*, 1215–1219.
6. Mach, E. *The Science of Mechanics: A Critical and Historical Account of Its Development*; Open Court Publishing: London, UK, 1919.
7. Gödel, K. An example of a new type of cosmological solution of Einstein's field equations of gravitation. *Rev. Mod. Phys.* **1949**, *21*, 447–450.
8. Weinberg, S. *Gravitation and Cosmology: Principles and Applications of the General Theory of Relativity*; John Wiley & Sons: New York, NY, USA, 1972.
9. Einstein, A.; Grossmann, M. Outline of a generalized theory of relativity and of a theory of gravitation. *Z. Math. Phys.* **1913**, *62*, 225–261.
10. Jaramillo, J.L.; Gourgoulhon, E. Mass and angular momentum in general relativity. In *Mass and Motion in General Relativity*; Springer: Dordrecht, The Netherlands, 2011.
11. Misner, C.W., Thorn, K.S.; Wheeler, J.A. *Gravitation*; W. H. Freeman and Company: New York, NY, USA, 1970.
12. Einstein, A. Note on E. Schrödinger's Paper: The energy components of the gravitational field. *Phys. Z.* **1918**, *19*, 115–116.
13. Einstein, A. *The Meaning of Relativity*; Princeton University Press: Princeton, NJ, USA, 1922.
14. Vishwakarma, R.G. On the relativistic formulation of matter. *Astrophys. Space Sci.* **2012**, *340*, 373–379.
15. Cooperstock, F.I.; Dupre, M.J. Covariant energy-momentum and an uncertainty principle for general relativity. *Ann. Phys.* **2013**, *339*, 531–541.
16. Vishwakarma, R.G. A new solution of Einstein's vacuum field equations. *Pramana J. Phys.* **2015**, *85*, 1101–1110.
17. Vishwakarma, R.G. A Machian approach to General Relativity. *Int. J. Geom. Methods Mod. Phys.* **2015**, *12*, 1550116.
18. Kinnersley, W. Recent progress in exact solutions. In Proceedings of the 7th International Conference on General Relativity and Gravitation (GR7), Tel-Aviv, Israel, 23–28 June 1974.
19. Vishwakarma, R.G. Mysteries of $R^{ik} = 0$: A novel paradigm in Einstein's theory of gravitation. *Front. Phys.* **2014**, *9*, 98–112.
20. Weinberg, S. The cosmological constant problem. *Rev. Mod. Phys.* **1989**, *61*, 1–23.
21. Hoyle, F.; Burbidge, G.; Narlikar, J.V. The basic theory underlying the Quasi-Steady-State cosmology. *Proc. R. Soc. Lond. A* **1995**, *448*, 191–212.
22. Einstein, A. *Relativity: The Special and the General Theory*; Create Space Independent Publishing Platform: London, UK, 2015.
23. Padmanabhan, T. Momentum density of spacetime and the gravitational dynamics. *Gen. Relativ. Grav.* **2016**, *48*, 4.
24. Hawking, S.; Milodinow, L. *The Grand Design*; Bantam Books: New York, NY, USA, 2010.
25. Giardino, S. Axisymmmetric empty space: Light propagation, orbits and dark matter. *J. Mod. Phys.* **2014**, *5*, 1402–1411.
26. Narlikar, J.V. *An Introduction to Cosmology*; Cambridge University Press: Cambridge, UK, 2002.
27. Vishwakarma, R.G. A curious explanation of some cosmological phenomena. *Phys. Scripta* **2013**, *87*, 5.
28. Harrison, E.R. Mining Energy in an Expanding Universe. *Astrophys. J.* **1995**, *446*, 63.
29. Tolman, R. C. *Relativity, Thermodynamics and Cosmology*; Oxford University Press: Oxford, UK, 1934.
30. Nielsen, J.T.; Guffanti, A.; Sarkar, S. Marginal evidence for cosmic acceleration from Type Ia supernovae. 2015, arXiv:1506.01354.
31. Ishak, M.; Rindler, W.; Dossett, J.; Moldenhauer, J.; Allison, C. A new independent limit on the cosmological constant/dark energy from the relativistic bending of light by galaxies and clusters of galaxies. *Mom. Not. R. Astron. Soc.* **2008**, *388*, 1279–1283.

32. Noether, E. Invariant variation problems. *Transp. Theory Stat. Phys.* **1971**, *1*, 186–207;

33. Baryshev, Y.V. Energy-momentum of the gravitational field: Crucial point for gravitation physics and cosmology. *Pract. Cosmol.* **2008**, *1*, 276–286.

34. Misner, C.W.; Wheeler, J.A. Classical physics as geometry: Gravitation, electromagnetism, unquantized charge, and mass as properties of empty space. *Ann. Phys.* **1957**, *2*, 525–603.

5

Testing General Relativity with the Radio Science Experiment of the BepiColombo Mission to Mercury

Giulia Schettino [1,*] and Giacomo Tommei [2]

[1]   IFAC-CNR, Via Madonna del Piano 10, 50019 Sesto Fiorentino (FI), Italy
[2]   Department of Mathematics, University of Pisa, Largo Bruno Pontecorvo 5, 56127 Pisa, Italy; giacomo.tommei@unipi.it
*   Correspondence: g.schettino@ifac.cnr.it.

Academic Editors: Lorenzo Iorio and Elias C. Vagenas

Abstract: The relativity experiment is part of the Mercury Orbiter Radio science Experiment (MORE) on-board the ESA/JAXA BepiColombo mission to Mercury. Thanks to very precise radio tracking from the Earth and accelerometer, it will be possible to perform an accurate test of General Relativity, by constraining a number of post-Newtonian and related parameters with an unprecedented level of accuracy. The Celestial Mechanics Group of the University of Pisa developed a new dedicated software, ORBIT14, to perform the simulations and to determine simultaneously all the parameters of interest within a global least squares fit. After highlighting some critical issues, we report on the results of a full set of simulations, carried out in the most up-to-date mission scenario. For each parameter we discuss the achievable accuracy, in terms of a formal analysis through the covariance matrix and, furthermore, by the introduction of an alternative, more representative, estimation of the errors. We show that, for example, an accuracy of some parts in 10^{-6} for the Eddington parameter β and of 10^{-5} for the Nordtvedt parameter η can be attained, while accuracies at the level of 5×10^{-7} and 1×10^{-7} can be achieved for the preferred frames parameters α_1 and α_2, respectively.

Keywords: general relativity and gravitation; experimental studies of gravity; Mercury; BepiColombo mission

1. Introduction

BepiColombo is a mission for the exploration of the planet Mercury, jointly developed by the European Space Agency (ESA) and the Japan Aerospace eXploration Agency (JAXA). The mission is scheduled for launch in April 2018 and for orbit insertion around Mercury at the end of 2024. The science mission consists of two separated spacecraft, which will be inserted in two different orbits around the planet: the Mercury Planetary Orbiter (MPO), devoted to the study of the surface and internal composition [1], and the Mercury Magnetospheric Orbiter (MMO), designed for the study of the planet's magnetosphere [2]. In particular, the Mercury Orbiter Radio science Experiment (MORE) is one of the experiments on-board the MPO spacecraft.

Thanks to the state-of-the-art on-board and on-ground instrumentation [3], MORE will enable a better understanding of both Mercury geophysics and fundamental physics. The main goals of the MORE radio science experiment are concerned with the gravity of Mercury [4–8], the rotation of Mercury [9–11] and General Relativity (GR) tests [12–16]. The global experiment consists in determining the value and the formal uncertainty (as defined in Section 2.2) of a number of parameters of general interest, with the addition of further parameters characterizing each specific goal of the experiment. The quantities to be determined can be partitioned as follows:

(a) spacecraft state vector (position and velocity) at given times (*Mercurycentric orbit determination*);

(b) spherical harmonics of the gravity field of Mercury [17] and tidal Love number k_2 [18], in order to constrain physical models of the interior of Mercury (*gravimetry experiment*);

(c) parameters defining the model of the Mercury's rotation (*rotation experiment*);

(d) digital calibrations for the Italian Spring Accelerometer (ISA) [19,20];

(e) state vector of Mercury and Earth-Moon Barycenter (EMB) orbits at some reference epoch, in order to improve the ephemerides (*Mercury and EMB orbit determination*);

(f) post-Newtonian (PN) parameters [12,13,21,22], together with some related parameters, like the solar oblateness factor $J_{2\odot}$, the solar gravitational factor $\mu_\odot = GM_\odot$, where G is the gravitational constant and $M_\odot$ the Sun's mass, and its time variation, $\zeta = (1/\mu_\odot)\,d\mu_\odot/dt$, in order to test gravitational theories (*relativity experiment*).

A good initial guess for each of the above parameters will also be necessary: for example, for the PN parameters we use the GR values, while for gravimetry we refer to the most recent MESSENGER results as nominal values in simulations [23].

To cope with the extreme complexity of MORE and its challenging goals, the Celestial Mechanics Group of the University of Pisa developed (under an Italian Space Agency commission) a dedicated software, ORBIT14, which is now ready for use. The software enables the generation of the simulated observables and the determination of the solve-for parameters by means of a global least squares fit.

In this paper, the results of a full set of simulations are discussed, carried out in the most up-to-date mission scenario [24], focusing on the parameters of interest for the relativity experiment. The paper is organized as follows: in Section 2 we describe in details the structure of the ORBIT14 software, while in Section 3 we introduce the mathematical models adopted to perform a full relativistic and coherent analysis of the observations. The simulation scenario and the adopted assumptions are detailed in Section 4, while the results are presented in Section 5, together with a comprehensive discussion on the achievable accuracies. Finally, conclusions are drawn in Section 6.

2. The ORBIT14 Software

The ORBIT14 software system has been developed by the Celestial Mechanics Group of the University of Pisa starting since 2007 as a new dedicated software for the MORE experiment. In Section 2.1 we describe the global structure of the software, while in Section 2.2 we briefly recall the non-linear least squares method. Details on the adopted multi-arc strategy are given in Section 2.3.

2.1. Global Structure

The global structure of the code is outlined in Figure 1. The main programs belong to two categories: **data simulator** (short: simulator) and **differential corrector** (short: corrector). Code is written in Fortran90 language. The simulator is needed to predict possible scientific results of the experiment. It generates simulated observables (range and range-rate, accelerometer readings) and the nominal value for orbital elements of the Mercurycentric orbit of the spacecraft, of Mercury and of the EMB orbits. The program structure of the simulator is quite simple if compared with the differential corrector, the most demanding part being the implementation of the dynamical, observational and error models.

The actual core of the code is the corrector, which solves for all the parameters **u** which can be determined by a least squares fit (possibly constrained and/or decomposed in a multi-arc structure). The corrector structure has been designed in order to exploit parallel computing, especially for the most computationally expensive portion of the processing. An outline of the steps involved in a single corrector's iteration is shown in Figure 2.

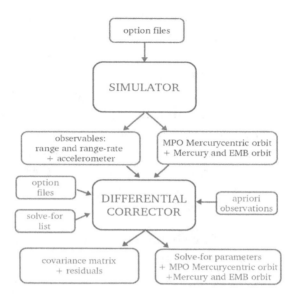

Figure 1. Block diagram of the ORBIT14 code: simulation and differential corrections stages. Green arrows refer to simulator inputs/outputs and orange arrows to corrector inputs/outputs. The input option files for simulator and corrector are similar and include, for example, the state vector of the spacecraft at the initial epoch, the number of considered arcs, the time steps for the orbit propagation of the spacecraft, Mercury and EMB, the time sampling for range, range-rate and accelerometer data.

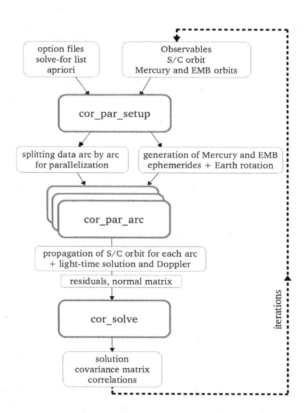

Figure 2. Block diagram of a differential corrector decomposed in three steps: (1) in "cor_par_setup" all the input options are read, data are split for the following parallel computation and the orbits of Mercury and EMB are propagated; (2) "cor_par_arc" contains most of the computationally expensive processing and is parallelized, by executing multiple copies of the same code, without need for interprocess communication; at this stage, the orbit of the spacecraft is propagated at each arc, the light-time computation is performed and residuals and normal matrix are given as output for the next step; (3) in "cor_solve" the covariance matrix and the LS solution are computed.

2.2. Non-Linear Least Squares Fit

Following a classical approach (see, for instance, [25]—Chapter 5), the non-linear least squares (LS) fit leads to compute a set of parameters $\mathbf{u}$ which minimizes the following target function:

$$Q(\mathbf{u}) = \frac{1}{m}\boldsymbol{\xi}^T(\mathbf{u})W\boldsymbol{\xi}(\mathbf{u}) = \frac{1}{m}\sum_{i=1}^{m} w_i \xi_i^2(\mathbf{u}), \tag{1}$$

where m is the number of observations and $\boldsymbol{\xi}(\mathbf{u}) = \mathcal{O} - \mathcal{C}(\mathbf{u})$ is the vector of *residuals*, i.e., the difference between the observed $\mathcal{O}$ and the predicted quantities $\mathcal{C}(\mathbf{u})$, computed following suitable mathematical models and assumptions. In our case, $\mathcal{O}$ are tracking data (range, range-rate and non-gravitational accelerations from the accelerometer), while $\mathcal{C}(\mathbf{u})$ are the results of the light-time computation (see [22] for details) as a function of all the parameters $\mathbf{u}$. Finally, w_i is the weight associated to the i-th observation. Among the parameters $\mathbf{u}$, the ones introduced in Section 1 in (a), (b), (c) and (d) occur in the equation of motion of the Mercurycentric orbit of the spacecraft, while those in e) and f) occur in the equations of the orbits of Mercury and the Earth-Moon barycenter with respect to the Solar System Barycenter (SSB). Other information required for such orbit propagations are supposed to be known: positions and velocities of the other planets of the Solar System are obtained from the JPL ephemerides DE421 [26], while the rotation of the Earth is provided by the interpolation table made public by the International Earth Rotation Service (IERS: http://www.iers.org.) and the coordinates associated with the ground stations are expected to be available.

The procedure to compute $\mathbf{u}^*$, the set of parameters which minimizes $Q(\mathbf{u})$, is based on a modified Newton's method known in the literature as *differential corrections method*. All the details can be found in [25]—Chapter 5. Let us define:

$$B = \frac{\partial \boldsymbol{\xi}}{\partial \mathbf{u}}(\mathbf{u}), \qquad C = B^T W B,$$

which are called *design matrix* and *normal matrix*, respectively. Then, the correction

$$\Delta \mathbf{u} = C^{-1}D \quad \text{with } D = -B^T W \boldsymbol{\xi}$$

is applied iteratively until either Q does not change meaningfully from one iteration to the other or $\Delta \mathbf{u}$ becomes smaller than a given tolerance. Introducing the inverse of the normal matrix, $\Gamma = C^{-1}$, we always adopt the probabilistic interpretation of Γ as the *covariance matrix* of the vector $\mathbf{u}$, considered as a multivariate Gaussian distribution with mean $\mathbf{u}^*$ in the space of parameters.

2.3. Pure and Constrained Multi-Arc Strategy

The tracking measurements from the Earth to the spacecraft are not continuous because of the mutual geometric configuration between the observing station and the antenna on the probe. The following visibility conditions are defined to account for: (i) the occultation of the spacecraft behind Mercury as seen from the Earth; (ii) the elevation of Mercury above the horizon at the observing station (the data received when Mercury is below a minimum elevation of 15° from the horizon are discarded because they are too noisy and could degrade the results); (iii) the angle between Mercury and the Sun as seen from the Earth. As a result, the observations are split in arcs, with a duration of ~24 h. Considering two observing stations (see Section 4), the adopted visibility conditions provide tracking sessions with an average duration of 15–16 h, called *observed arcs*, followed by a period of some hours without observations.

In orbit determination, the estimation approach consists of a combined solution called **multi-arc strategy** (see, e.g., [25]). According to this method, every single arc of observations has its own set of initial conditions (position and velocity at the reference central epoch of the considered time interval), as it belongs to a different object. In this way, due to lack of knowledge in the dynamical models, the actual errors in the orbit propagation can be reduced by an over-parameterization of the initial conditions. A different choice has been made in ORBIT14, implementing the so called **constrained multi-arc strategy** [10,14,27]. The method is based on the idea that each observed arc belongs to the same object (the spacecraft). First of all, an *extended arc* is defined as the observed arc broadened to half the preceding and to half the following periods without tracking, as shown in Figure 3. The orbits of two consecutive extended arcs should coincide at the connection time in the middle of the non-observed interval. We refer to [10,27] for a complete description of the constrained multi-arc strategy.

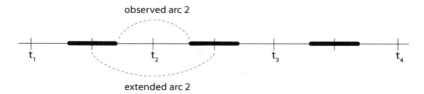

Figure 3. Schematic representation of observed and extended arc. The times t_i ($i = 1, .., 4$) are the central epoch of each arc; the black bars correspond to dark intervals, without tracking from Earth. See the text for more explanation.

In the constrained multi-arc approach, the parameters **u** can be classified, depending on the arc they refer to, as:

- **Global Parameters (g):** parameters that affect the dynamical equations of every observed (and extended) arc. The PN parameters and the spherical harmonic coefficients of Mercury are an example.
- **Local Parameters (l^k):** parameters that affect the dynamical equations of a single observed arc k. The state vector of the Mercurycentric orbit associated with the arc and the desaturation manoeuvres applied during the tracking are few examples.
- **Local External Parameters ($le^{k,k+1}$):** parameters that affect only the dynamical equations in the period without tracking between two subsequent observed arcs k and $k + 1$. These are the desaturation maneuvres taking place out of the observed arcs.

To implement the constrained multi-arc strategy in the framework of the LS fit described in Section 2.2, we define the *discrepancy vector* between the k and $k + 1$ arcs, $\mathbf{d}^{k,k+1}$, as:

$$\mathbf{d}^{k,k+1} = \Phi(t_c^k; t_0^{k+1}, \mathbf{X}_0^{k+1}) - \Phi(t_c^k; t_0^k, \mathbf{X}_0^k),$$

where t_0^k and t_0^{k+1} denote the central time of the k and $k + 1$ arc, respectively, t_c^k is the connection time between the two extended k and $k + 1$ arcs, $\mathbf{X}_0^k$ and $\mathbf{X}_0^{k+1}$ are the state vector at t_0^k and t_0^{k+1}, respectively, and $\Phi(t_c^k; t_0^{k+1}, \mathbf{X}_0^{k+1})$ is the image of $(t_0^{k+1}, \mathbf{X}_0^{k+1})$ under the flow of the vector field associated with the Mercurycentric orbit at time $t = t_c^k$ (analogously for $\Phi(t_c^k; t_0^k, \mathbf{X}_0^k)$). Thus, we have:

$$\mathbf{d}^{k,k+1} = \mathbf{d}^{k,k+1}(g, l^k, le^{k,k+1}, l^{k+1}),$$

where g, l^k, $le^{k,k+1}$ are the parameters **u** included in the LS fit, classified, respectively, as global (g), local (l^k) for the k-th arc and local external ($le^{k,k+1}$) between the k-th and $(k + 1)$-th arcs. The constrained multi-arc strategy consists in minimizing the target function:

$$Q(\mathbf{u}) = \frac{1}{m + 6(n - 1)} \sum_{i=1}^{m} w_i \xi_i^2 +$$

$$+ \frac{1}{\mu} \frac{1}{m + 6(n - 1)} \sum_{k=1}^{n-1} \mathbf{d}^{k,k+1} \cdot C^{k,k+1} \mathbf{d}^{k,k+1},$$

where n is the total number of extended arcs, μ is a *penalty parameter* and $C^{k,k+1}$ is a weight matrix for the discrepancy vectors. Two possible approaches can be followed. The first is called **internally constrained multi-arc strategy**. In this case, we consider the confidence ellipsoids associated with $\mathbf{X}_0^k$ and $\mathbf{X}_0^{k+1}$ at t_0^k and t_0^{k+1}, respectively, and we propagate them to t_c^k through the corresponding state transition matrices. This means that we expect $\mathbf{d}^{k,k+1}$ to be normally distributed with mean $\Phi(t_c^k; t_0^{k+1}, \mathbf{X}_0^{k+1}) - \Phi(t_c^k; t_0^k, \mathbf{X}_0^k)$ and covariance

$$\Gamma_c := (C^k)^{-1} + (C^{k+1})^{-1},$$

where C^k and C^{k+1} are the 6×6 normal matrices associated with $\mathbf{X}_0^k$ and $\mathbf{X}_0^{k+1}$ at t_0^k and t_0^{k+1} and propagated to t_c^k, respectively. It follows that

$$C^{k,k+1} = \Gamma_c^{-1} \qquad \text{and} \qquad \mu = 1.$$

The second approach is called **apriori constrained multi-arc strategy** and it takes care of the degeneracy in orbit determination due to the orbit geometry (details can be found in [28]). In particular, we deal with an approximated version of the exact symmetry described in [28], where the small parameter of the perturbation is the angle of displacement of the Earth-Mercury vector in an inertial frame. In this case, the normal matrix has one eigenvalue significantly smaller than the others. As a consequence of this weakness, the confidence ellipsoid associated with the discrepancy and defined by $C^{k,k+1}$ could be very elongated. The basic idea of this approach is to constrain the discrepancy $\mathbf{d}^{k,k+1}$ inside a sphere of given radius, that can be suitably shrunk by varying μ. This can be interpreted as adding apriori observations. On the contrary, in the internally constrained multi-arc strategy, the discrepancy is constrained inside the intersection of the two ellipsoids propagated from t_0^k and t_0^{k+1}. All the details are extensively explained in [27]. For the results presented in this review, we will always adopt an apriori constrained multi-arc strategy. Finally, it can be noted that in the multi-arc method the residuals ζ depend only on global and local parameters, and this applies to the target function Q defined in Equation (1) as well.

3. Mathematical Models

The purpose of the MORE relativity experiment is to perform a test of General Relativity comparing theory with experiment. The majority of the Solar System tests of gravitation can be set in the context of the slow-motion, weak field limit [29], usually known as the *post-Newtonian* approximation. In this limit, the space-time metric can be written as an expansion about the Minkowski metric in terms of dimensionless gravitational potentials. In the *parametrized* PN formalism, each potential term in the metric is expressed by a specific parameter, which measures a general property of the metric. The basic idea of the MORE relativity experiment is to investigate the dependence of the equation of motion from the PN parameters. By isolating the effects of each parameter on the motion, it is possible to constrain the parameters values within some accuracy threshold, testing the validity of GR predictions.

The PN parameters of interest for our analysis are the following:

- the *Eddington parameters* β and γ. β accounts for the modification of the non-linear three-body general relativistic interaction and γ parametrizes the velocity-dependent modification of the two-body interaction and accounts also for the space-time curvature through the Shapiro effect [30]. These are the only non-zero PN parameters in GR (they are both equal to unity);
- the *Nordtvedt parameter* η. The effect of η in the equations of motion is to produce a polarization of the Mercury and Earth orbits in the gravitational field of the other planets and it is related to possible violations of the Strong Equivalence Principle (see, e.g., the discussion in [31]);
- the *preferred frame effects parameters* α_1 and α_2. They phenomenologically describe the effects due to the presence of a gravitationally preferred frame; we follow the standard assumption to identify the preferred frame with the rest frame of the cosmic microwave background [32].

Beside the five PN parameters mentioned above, we consider a few additional parameters. These do not properly produce relativistic effects, but the uncertainty in their knowledge generates orbital effects at least comparable with the perturbations expected from the tested relativistic parameters. We consider two kinds of "Newtonian" orbital effects: a small change in the Sun's gravity oblateness $J_{2\odot}$ and a small change in the Sun's gravitational factor $\mu_\odot = GM_\odot$, where G is the gravitational constant and $M_\odot$ is the Sun's mass. Regarding the second quantity, since many alternative theories of gravitation allow for a possible time variation of the gravitational constant, we included in the solve-for parameters the time derivative of $\mu_\odot$ described by the parameter $\zeta = \frac{1}{\mu_\odot}\frac{d\mu_\odot}{dt}$. Indeed, within the MORE experiment we cannot discriminate between a time variation of $M_\odot$ or G, but assuming an independent estimate for the rate of $M_\odot$, it could be possible to draw information on the time rate of G.

The effects linked to each PN parameter, corresponding to modifications of the space-time metric, affect both the propagation of the tracking signal (range and range-rate) and the equations of motion. In the following we describe the mathematical models adopted in our analysis: in Section 3.1 we define the computed tracking observables in a coherent relativistic background and in Section 3.2 we present the Langrangian formulation of the planetary dynamics in the context of the first-order PN approximation. Finally, in Section 3.3 we define the Mercurycentric dynamical model.

3.1. Computation of Observables

In a radio science experiment, the observational technique is complicated by many factors (for example plasma reduction) but in simulations it can be merely considered as a tracking from an Earth-based station, giving range and range-rate information (see, e.g., [3]). In order to compute the range distance from the ground station on the Earth to the spacecraft around Mercury (or in an interplanetary trajectory), and the corresponding range-rate, we introduce the following state vectors, each one evolving according to a specific dynamical model (see Figure 4):

- the Mercurycentric position of the spacecraft, $\mathbf{x}_{sat}$;
- the SSB positions of Mercury and of the EMB, $\mathbf{x}_M$ and $\mathbf{x}_{EM}$;
- the geocentric position of the ground antenna, $\mathbf{x}_{ant}$;
- the position of the Earth barycenter with respect to the EMB, $\mathbf{x}_E$.

They can be combined to define the range distance using the following formula, as a first approximation:

$$r = |\mathbf{r}| = |(\mathbf{x}_{sat} + \mathbf{x}_M) - (\mathbf{x}_{EM} + \mathbf{x}_E + \mathbf{x}_{ant})|. \tag{2}$$

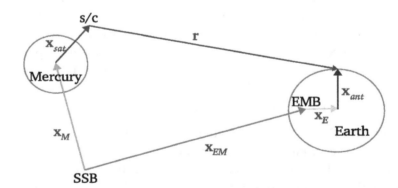

Figure 4. Vectors involved in the multiple dynamics for the tracking of the spacecraft from the Earth.

As explained in [22], Equation (2) corresponds to model the space as a flat arena (r is the Euclidean distance) and the time as an absolute parameter. Obviously, this is not a suitable assumption, since it is clear that beyond some threshold of accuracy, as expected for the BepiColombo radio science experiment, space and time must be formulated in the framework of General Relativity. Moreover, we have to take into account the different times at which the events have to be computed: the transmission of the signal at the transmitting time (t_t), the signal at the Mercury orbiter at the time of bounce (t_b) and the reception of the signal at the receiving time (t_r). Equation (2) can be used as a starting point to construct a correct relativistic formulation, containing not all the possible relativistic effects, but the ones which are measurable in the experiment.

The five vectors in Equation (2) have to be computed at their own time, which corresponds to the epoch of different events: e.g., x_{ant}, x_{EM} and x_E are computed at both the antenna transmission time t_t and receiving time t_r of the signal, while x_M and x_{sat} are computed at the bounce time t_b (when the signal has reached the orbiter and is sent back, with correction for the delay of the transponder). To perform the vectorial sums and differences, these vectors must be converted to a common space-time reference system, the only possible choice being some realization of the BCRS (Barycentric Celestial Reference System). We adopt a realization of the BCRS that we call SSB (Solar System Barycentric) reference frame in which the time is a re-definition of the TDB (Barycentric Dynamic Time), according to the IAU 2006 Resolution B3 (https://www.iau.org/static/resolutions/IAU2006_Resol3.pdf). Other possible choices, such as TCB (Barycentric Coordinate Time), can only differ by linear scaling. The TDB choice of the SSB time scale entails also the appropriate linear scaling of space-coordinates and planetary masses as described, for instance, in [33,34].

The vectors x_M, x_E, and x_{EM} are already in the SSB reference frame as provided by numerical integration and external ephemerides, while the vectors x_{ant} and x_{sat} must be converted to SSB from the geocentric and Mercurycentric systems, respectively. Of course, the conversion of reference systems implies also the conversion of the time coordinate. There are three different time coordinates to be considered. The currently published planetary ephemerides are provided in TDB. The observations from the Earth are based on averages of clock and frequency measurements on the Earth surface: this defines another time coordinate called TT (Terrestrial Time). Thus for each observation the times of transmission t_t and reception t_r need to be converted from TT to TDB to find the corresponding positions of the planets. This time conversion step is necessary for the accurate processing of each set of interplanetary tracking data. The main term in the difference TT-TDB is periodic, with period 1 year and amplitude $\simeq 1.6 \times 10^{-3}$ s, while there is essentially no linear trend, as a result of a suitable definition of the TDB. Finally, the equation of motion of the spacecraft orbiting Mercury has been approximated, to the required level of accuracy, following what done in [35] for the case of a near-Earth spacecraft in a geocentric frame of reference. We consider the Newtonian dynamics in a local Mercurycentric frame assuming as independent variable a suitably defined time coordinate. Moreover, we add the relativistic perturbative acceleration from the one-body Schwarzschild isotropic metric for Mercury and the acceleration due to the geodesic precession, as explained in [10]. Thus, for MORE we have defined a new time coordinate TDM (Mercury Dynamic Time), as described in [13], containing terms of 1-PN order depending mostly upon the distance from the Sun and velocity of Mercury.

In general, the differential equation giving the local time T (in our case TT or TDM) as a function of the SSB time t, which we are currently assuming to be TDB, is the following:

$$\frac{dT}{dt} = 1 - \frac{1}{c^2}\left[U + \frac{v^2}{2} - L\right], \tag{3}$$

where U is the gravitational potential (the list of contributing bodies depends upon the required accuracy: in our implementation we use Sun, Mercury to Neptune, Moon) at the planet center and v is the velocity with respect to the SSB of the same planet. The constant term L is used to perform the conventional rescaling motivated by removal of secular terms [36].

The space-time transformations needed to coherently compute the vector $\mathbf{x}_{ant}$ involve essentially the position of the antenna and of the orbiter. The geocentric coordinates of the antenna should be transformed into TDB-compatible coordinates [34]. The transformation is expressed by:

$$\mathbf{x}_{ant}^{TDB} = \mathbf{x}_{ant}^{TT}\left(1 - \frac{U}{c^2} - L_C\right) - \frac{1}{2}\left(\frac{\mathbf{v}_E^{TDB} \cdot \mathbf{x}_{ant}^{TT}}{c^2}\right)\mathbf{v}_E^{TDB},$$

where U is the gravitational potential at the geocenter (excluding the Earth mass), $L_C = 1.48082686741 \times 10^{-8}$ is a scaling factor given as definition [37], supposed to be a good approximation for removing secular terms from the transformation and $\mathbf{v}_E^{TDB}$ is the barycentric velocity of the Earth. The following equation contains the effect on the velocities of the time coordinate change, which should be consistently used together with the coordinate change:

$$\mathbf{v}_{ant}^{TDB} = \left[\mathbf{v}_{ant}^{TT}\left(1 - \frac{U}{c^2} - L_C\right) - \frac{1}{2}\left(\frac{\mathbf{v}_E^{TDB} \cdot \mathbf{v}_{ant}^{TT}}{c^2}\right)\mathbf{v}_E^{TDB}\right]\left[\frac{dT}{dt}\right].$$

Note that the previous formula contains the factor dT/dt (expressed by Equation (3)) that deals with a time transformation: T is the local time for Earth, that is TT, and t is the corresponding TDB time. To compute the coordinates of the orbiter (vector $\mathbf{x}_{sat}$) we adopt similar equations, as discussed in [22], where we neglected the terms of the SSB acceleration of the planet center [38], because they contain, beside $1/c^2$, the additional small parameter *distance from planet center divided by planet distance to the Sun*, which is of the order of 10^{-4} even for a Mercury orbiter.

In Figure 5 the behaviour of the range and range-rate observable, computed with and without the just explained relativistic corrections, is shown. It is quite evident that the differences are significant, at a signal-to-noise ratio $S/N \simeq 1$ for range, much larger for range-rate, with an especially strong signature from the orbital velocity of the Mercurycentric orbit (with $S/N > 50$).

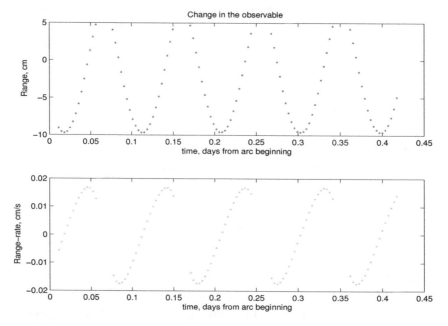

Figure 5. Cont.

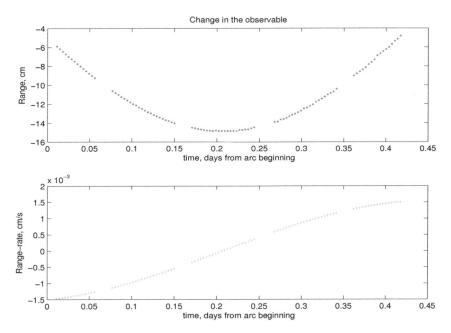

Figure 5. The difference in the observables range and range-rate for one pass of Mercury above the horizon for a ground station, by using an hybrid model in which the position and velocity of the orbiter have not been transformed to TDB-compatible quantities and a correct model in which all quantities are TDB-compatible. Gaps of the signal are due to spacecraft passage behind Mercury as seen for the Earth station. (**Top**): for a hybrid model with the satellite position and velocity not transformed to TDB-compatible; (**Bottom**): for a hybrid model with the position and velocity of the antenna not transformed to TDB-compatible.

3.2. Dynamical Relativistic Models

To constrain the PN and related parameters, we need to determine the orbit of Mercury. A relativistic model for the motion of the planet is necessary. We choose to start from a Lagrangian formulation (see, e.g., [21]) in order to compute the terms of accelerations to be included in the right-hand side of the differential equations for Mercury and EMB (the dynamics of the other planets, as well as the relative EMB-Earth position, are given by the JPL ephemerides). In particular, let us assume that the motion of the considered planets is described by the sum of different Lagrangians:

$$L = L_{New} + L_{GR} + (\gamma - 1)L_\gamma + (\beta - 1)L_\beta + \zeta L_\zeta + J_{2\odot}L_{J2\odot} + \alpha_1 L_{\alpha_1} + \alpha_2 L_{\alpha_2} + \eta L_\eta , \qquad (4)$$

where L_{New} is the Lagrangian of the Newtonian N-body problem, L_{GR} is the corrective term taking into account General Relativity in the post-Newtonian approximation, L_γ and L_β are the terms taking into account the PN parameters γ and β, respectively, L_ζ is the Lagrangian for the time variation of the gravitational parameter of the Sun $\mu_\odot = GM_\odot$, $L_{J2\odot}$ takes into account the effect of the oblateness of the Sun, L_{α_1}, L_{α_2} describe the preferred-frame effects through the parameters α_1, α_2, and, finally, L_η checks for possible violations of the strong equivalence principle (see [12,13,31]).

To express the Lagrangian terms, we follow the notation of [35] and we introduce two parameters: the total mass of EMB, μ_{EMB}, which is the sum of the gravitational parameter of the Earth and the Moon, and the mass ratio, $\bar{\mu}$, defined as the ratio of the gravitational parameter of the Earth over that of the Moon; both quantities μ_{EMB} and $\bar{\mu}$ are assumed to be fixed, hence we do not solve for them. Defining $\mu_E = \bar{\mu}/(1 + \bar{\mu})$ and $\mu_{Moon} = 1/(1 + \bar{\mu})$, we have:

$$\mathbf{r}_{EM} = \mu_E \, \mathbf{r}_{Earth} + \mu_{Moon} \, \mathbf{r}_{Moon}$$

with analogous expressions for velocity and acceleration.

Notice that the usual Lagrangians are multiplied by G, so that only $\mu_i = G M_i$ appear in the overall Lagrangian. Indeed, the gravitational constant cannot be determined from any form of orbit determination (apart artificial systems). In the following we give the explicit expressions for each Lagrangian term in Equation (4).

- **N-body Newtonian Lagrangian:**

$$L_{New} = \frac{1}{2} \sum_i \mu_i v_i^2 + \frac{1}{2} \sum_i \sum_{j \neq i} \frac{\mu_i \mu_j}{r_{ij}} .$$

- **Post-Newtonian General Relativistic Lagrangian:**

$$
\begin{aligned}
L_{GR} &= \frac{1}{8 c^2} \sum_i \mu_i v_i^4 - \frac{1}{2 c^2} \sum_i \sum_{j \neq i} \sum_{k \neq i} \frac{\mu_i \mu_j \mu_k}{r_{ij} r_{ik}} + \\
&+ \frac{1}{2} \sum_i \sum_{j \neq i} \frac{\mu_i \mu_j}{r_{ij}} \left[\frac{3}{2 c^2} (v_i^2 + v_j^2) - \frac{7}{2 c^2} (\mathbf{v_i} \cdot \mathbf{v_j}) - \frac{1}{2 c^2} (\mathbf{n_{ij}} \cdot \mathbf{v_i})(\mathbf{n_{ij}} \cdot \mathbf{v_j}) \right] .
\end{aligned}
$$

- **Lagrangian for PN parameter γ:**

$$L_\gamma = \frac{1}{2 c^2} \sum_i \sum_{j \neq i} \frac{\mu_i \mu_j}{r_{ij}} (\mathbf{v_i} - \mathbf{v_j})^2 .$$

- **Lagrangian for PN parameter β:**

$$L_\beta = -\frac{1}{c^2} \sum_i \sum_{j \neq i} \sum_{k \neq i} \frac{\mu_i \mu_j \mu_k}{r_{ij} r_{ik}} .$$

- **Lagrangian for parameter ζ:**
  L_ζ describes the effect of a time variation of the gravitational parameter of the Sun, $\mu_\odot$:

$$\mu_\odot = \mu_\odot(t_0) + \dot{\mu}_\odot(t_0)(t - t_0) + \dots ;$$

defining

$$\zeta = \frac{\dot{\mu}_\odot(t_0)}{\mu_\odot(t_0)} = \frac{d}{dt} \ln \mu_\odot(t_0) ,$$

we have:

$$L_\zeta = (t - t_0) \sum_{i \neq 0} \frac{\mu_\odot \mu_i}{r_{0i}} .$$

- **Lagrangian for $J_{2\odot}$ effect:**

$$L_{J_{2\odot}} = -\frac{1}{2} \sum_{i \neq 0} \frac{\mu_0 \mu_i}{r_{0i}} \left(\frac{R_\odot}{r_{0i}} \right)^2 [3(\mathbf{n}_{0i} \cdot \mathbf{e}_0)^2 - 1] ,$$

where $R_\odot$ is the radius of the Sun, $\mathbf{n}_{0i} = \mathbf{r}_{0i}/r_{0i}$ is the heliocentric position of body i and $\mathbf{e}_0$ is the unit vector along the rotation axis of the Sun. The unit vector $\mathbf{e}_0$ is given in standard equatorial coordinates with equinox J2000 at epoch J2000.0 (JD 2451545.0 TCB): $\alpha_0 = 286.13°$, $\delta_0 = 63.87°$ [39].

- **Lagrangian for preferred frame effects, PN α_1 and α_2:**

$$L_{\alpha_1} = -\frac{1}{4 c^2} \sum_j \sum_{i \neq j} \frac{\mu_i \mu_j}{r_{ij}} (\mathbf{z}_i \cdot \mathbf{z}_j) ,$$

$$L_{\alpha_2} = \frac{1}{4\,c^2} \sum_j \sum_{i\neq j} \frac{\mu_i\,\mu_j}{r_{ij}} \left[(\mathbf{z}_i \cdot \mathbf{z}_j) - (\mathbf{n_{ij}} \cdot \mathbf{z}_i)(\mathbf{n_{ij}} \cdot \mathbf{z}_j) \right],$$

where $\mathbf{z}_i = \mathbf{w} + \mathbf{v_i}$ and $\mathbf{w}$ is the velocity of the considered reference system with respect to the PN preferred reference frame, which is a reference frame whose outer regions are at rest with respect to the universe rest frame (see [21]). In the case of the SSB reference frame, that could be the one of cosmic microwave background, $|\mathbf{w}| = 370 \pm 10$ km/s, in the direction $(\alpha, \delta) = (168°, 7°)$ in the Equatorial J2000 reference frame (see [12]). Notice that we can combine the two previous Lagrangians and the parameters α_1 and α_2 obtaining an unique Lagrangian for the preferred frame effects:

$$L_\alpha = \alpha_1\,L_{\alpha_1} + \alpha_2\,L_{\alpha_2} = \frac{\alpha_2 - \alpha_1}{4\,c^2} \sum_j \sum_{i\neq j} \frac{\mu_i\,\mu_j}{r_{ij}}\,(\mathbf{v_i} + \mathbf{w}) \cdot (\mathbf{v_j} + \mathbf{w}) +$$

$$- \frac{\alpha_2}{4\,c^2} \sum_j \sum_{i\neq j} (\mathbf{r_{ji}} \cdot (\mathbf{v_j} + \mathbf{w}))(\mathbf{r_{ji}} \cdot (\mathbf{v_i} + \mathbf{w})) \frac{\mu_i\,\mu_j}{r_{ij}^3}.$$

- **Lagrangian for possible violation of the equivalence principle, PN η:**
  With the Lagrangian multiplied by G, the Newtonian kinetic energy is:

$$T = \frac{1}{2} \sum_i \mu_i\,v_i^2,$$

where we assume that the inertial mass and the gravitational mass are the same (or, at least, exactly proportional). If some form of mass has a different gravitational coupling, there are, for each body i, two quantities μ_i and μ_i^I, one appearing in the gravitational potential (including the relativistic part) and the other appearing in the kinetic energy. If there is a violation of the strong equivalence principle involving body i, with a fraction Ω_i of its mass due to gravitational self-energy (for the moment we are using the approximation of constant density: $\Omega_i = -3\mu_i/5Rc^2$; notice that Ω_i is $\mathcal{O}(c^{-2})$):

$$\mu_i = (1 + \eta\Omega_i)\,\mu_i^I \iff \mu_i^I = (1 - \eta\Omega_i)\,\mu_i + \mathcal{O}(\eta^2)$$

with η the PN parameter for this violation. Neglecting $\mathcal{O}(\eta^2)$ terms (and also $\mathcal{O}[\eta\,(\gamma - 1)], ..$) this is expressed by a Lagrangian term $\eta\,L_\eta$, where:

$$L_\eta = -\frac{1}{2} \sum_i \Omega_i\,\mu_i\,v_i^2.$$

Considering an inertial reference system, the equations of motion for the i-th body are described by the Lagrangian equations:

$$\frac{d}{dt} \frac{\partial L}{\partial \mathbf{v}_i} = \frac{\partial L}{\partial \mathbf{r}_i},$$

which in general give an implicit expression for the acceleration of the form $f(\mathbf{a}_i) = g(\mathbf{r}_i, \mathbf{v}_i)$. However, since the main term is the N-body Newtonian acceleration $\mathbf{a}_i^{New}$ and the other terms are small perturbations, we can use the following approximation for the total acceleration of the i-th body:

$$\mu_i\mathbf{a}_i = \mu_i\mathbf{a}_i^{New} + \frac{\partial(L - L_{New})}{\partial \mathbf{r}_i} - \left[\frac{d}{dt} \frac{\partial(L - L_{New})}{\partial \mathbf{v}_i} \right]\Bigg|_{\mathbf{a}_i=\mathbf{a}_i^{New}}.$$

If we call $Y = [\mathbf{r}_1, \mathbf{r}_{EM}, \mathbf{v}_1, \mathbf{v}_{EM}]^T$ the 12-dimensional state vector (for Mercury and EMB) we want to propagate, we can write the equations of motion in the more complete form:

$$\frac{d}{dt}Y = \begin{pmatrix} \mathbf{v}_1 \\ \mathbf{v}_{EM} \\ \mathbf{a}_1 \\ \mathbf{a}_{EM} \end{pmatrix} = F(\mathbf{r}_1, \mathbf{r}_{EM}, \mathbf{v}_1, \mathbf{v}_{EM}, \ldots).$$

The reference system for these dynamics is centered in the SSB, and it is inertial in the PN approximation. On the other hand, if we consider possible violations, as it happens in the case of parameterized PN formalism, we need to reassess the total linear momentum conservation theorem. Using Noether's theorem we can compute the integral of the total linear momentum of the system:

$$\frac{d}{dt}\mathbf{P} = 0, \qquad \text{where} \qquad \mathbf{P} = \sum_i \frac{\partial L}{\partial \mathbf{v_i}}.$$

Since L_β, L_ζ, $L_{J2\odot}$ do not depend on velocities and, because of the antisimmetry, we have that $\sum_i \frac{\partial L_\gamma}{\partial \mathbf{v_i}} = 0$ and L_γ does not contribute; thus, the total linear momentum of the system reads:

$$\mathbf{P} = \sum_i \frac{\partial(L_{New} + L_{GR} + L_\alpha + \eta L_\eta)}{\partial \mathbf{v_i}}.$$

In the PN approximation the total linear momentum is simply:

$$\mathbf{P} = \sum_i \frac{\partial(L_{New} + L_{GR})}{\partial \mathbf{v_i}} =$$

$$= \sum_i \mu_i \mathbf{v_i}\left[1 + \frac{1}{2c^2}v_i^2 - \frac{1}{2c^2}\sum_{k\neq i}\frac{\mu_k}{r_{ik}}\right] - \frac{1}{2c^2}\sum_i\sum_{k\neq i}\frac{\mu_i\mu_k}{r_{ik}}(\mathbf{n_{ik}} \cdot \mathbf{v_k})\,\mathbf{n_{ik}},$$

and the vector:

$$\mathbf{R} = \sum_i \mu_i \mathbf{r}_i \left(1 + \frac{1}{2c^2}v_i^2 - \frac{1}{2c^2}\sum_{k\neq i}\frac{\mu_k}{r_{ik}}\right)$$

is such that:

$$\frac{d}{dt}\mathbf{R} = \mathbf{P},$$

to the $O(c^{-2})$ level of accuracy. Thus $\mathbf{R}$ (rescaled by the total mass) plays the role of the barycenter of the Solar System and can be used to eliminate the Sun from the equations of motion:

$$\mathbf{r}_0 = -\frac{\sum_{i\neq 0}\mu_i \mathbf{r}_i\left(1 + \frac{v_i^2}{2c^2} - \frac{U_i}{2c^2}\right)}{\mu_\odot\left(1 + \frac{v_0^2}{2c^2} - \frac{U_0}{2c^2}\right)}, \qquad U_i = \sum_{k\neq i}\frac{\mu_k}{r_{ik}}.$$

If we now take into account the PN parameters effects, we can write the linear momentum as:

$$\mathbf{P} = \mathbf{P}_0 + \mathbf{P}_\alpha, \qquad \mathbf{P}_0 = \sum_j \frac{\partial(L_{New} + L_{GR} + \eta L_\eta)}{\partial \mathbf{v}_j}, \qquad \mathbf{P}_\alpha = \sum_j \frac{\partial L_\alpha}{\partial \mathbf{v}_j}.$$

In this way, defining the center of mass of the system (rescaled by the total mass) as:

$$\mathbf{R} = \sum_i \mu_i(1 - \eta\Omega_i)\mathbf{r}_i\left(1 + \frac{1}{2c^2}v_i^2 - \frac{1}{2c^2}\sum_{k\neq i}\frac{\mu_k}{r_{ik}}\right),$$

we have

$$\frac{d}{dt}\mathbf{R} = \mathbf{P}_0,$$

and the position of the Sun in this barycentric system is now:

$$\mathbf{r}_0 = -\frac{\sum_{i \neq 0} \mu_i (1 - \eta\Omega_i)\mathbf{r}_i \left(1 + \frac{v_i^2}{2c^2} - \frac{U_i}{2c^2}\right)}{\mu_\odot (1 - \eta\Omega_0)\left(1 + \frac{v_0^2}{2c^2} - \frac{U_0}{2c^2}\right)}.$$

Finally, since we have:

$$\dot{\mathbf{P}} = 0 \Longrightarrow \dot{\mathbf{P}}_0 = -\dot{\mathbf{P}}_\alpha \Longrightarrow \ddot{\mathbf{R}} = -\dot{\mathbf{P}}_\alpha,$$

it means that the barycentric reference frame is accelerated. Thus, the equations of motion for the i-th body in this reference frame need to be corrected by the acceleration of the barycenter $\mathbf{B}$, keeping the $O(c^{-2})$ level of accuracy:

$$\mathbf{a}_i = \mathbf{a}_i^{New} + \frac{1}{\mu_i}\frac{\partial(L - L_{New})}{\partial \mathbf{r}_i} - \left[\frac{d}{dt}\frac{\partial(L - L_{New})}{\partial \mathbf{v}_i}\right]\Big|_{\mathbf{a}_i = \mathbf{a}_i^{New}} - \ddot{\mathbf{B}},$$

where:

$$\mathbf{B} = \frac{\mathbf{R}}{\sum_i \mu_i (1 - \eta\,\Omega_i)\left(1 + \frac{v_i^2}{2c^2} - \frac{U_i}{2c^2}\right)}.$$

3.3. Mercurycentric Dynamical Model

In this Section we briefly describe the models adopted to compute the Mercurycentric position of the spacecraft $\mathbf{x}_{sat}$, introduced in Section 3.1 in the expression for the range distance r.

3.3.1. Mercury Gravity Field (Static Part)

The motion of the satellite around Mercury is dominated by the gravity field of the planet. In a Mercurycentric reference frame and using spherical coordinates (r, θ, λ), the gravitational potential of the planet, intended as a static rigid mass, can be expanded in a spherical harmonics series as (see, e.g., [25]—Chapter 13):

$$V(r, \theta, \lambda) = \frac{GM_M}{r} + \sum_{\ell=2}^{+\infty} \frac{GM_M R_M^\ell}{r^{\ell+1}} \sum_{m=0}^{\ell} P_{\ell m}(\sin\theta)[C_{\ell m} \cos m\lambda + S_{\ell m} \sin m\lambda], \qquad (5)$$

where $r > 0$ is the distance from the center of the planet, $-\pi/2 < \theta < \pi/2$ the latitude and $0 \leq \lambda < 2\pi$ the longitude, M_M and R_M are Mercury's mass and mean radius, respectively, $P_{\ell m}$ the Legendre associated functions, $C_{\ell m}, S_{\ell m}$ the spherical harmonics coefficients and the summation starts from $\ell = 2$ because the potential is referred to the center of Mercury.

3.3.2. Tidal Perturbations

Mercury cannot be exhaustively described as a rigid body. The gravitational field of the Sun exerts solid tides on Mercury with the tidal bulge oriented in the direction of the Sun. This deformation can be described by adding to the Newtonian potential of Equation (5) a quantity V_L called *Love potential* [18,40,41]:

$$V_L = \frac{GM_\odot k_2 R_M^5}{r_S^3 r^3}\left(\frac{3}{2}\cos^2\psi - \frac{1}{2}\right),$$

where r_S is the Mercury-Sun distance and ψ is the angle between the Mercurycentric position of the spacecraft, $\mathbf{r}$, and the Sun Mercurycentric position. The *Love number* k_2 is the elastic constant characterizing the effect.

3.3.3. Sun and Planetary Perturbations

The solar and planetary gravitational effects on the spacecraft that orbits around Mercury can be computed as a "third-body" perturbative acceleration $\mathbf{a}_{third-body}$ in a local Mercurycentric reference frame. The N bodies acting in the perturbation are: Sun, Venus, Earth-Moon, Mars, Jupiter, Saturn, Uranus and Neptune:

$$\mathbf{a}_{third-body} = \sum_{i=0}^{N-1} GM_i \left(\frac{\mathbf{d}_i}{d_i^3} - \frac{\mathbf{r}_i}{r_i^3} \right) ,$$

where $\mathbf{d}_i$ is the position of the i-th body of mass M_i with respect to the spacecraft and $\mathbf{r}_i$ is its position with respect to Mercury (see, e.g., [35,42])

3.3.4. Rotational Dynamics

The gravity field development given by Equation (5) is valid in a body-fixed reference frame, like the Mercury body-fixed frame of reference, Ψ_{BF}, defined by the principal inertia axes, with the x-axis along the minimum inertia axis, assumed as rotational reference meridian (see [10] for details). If we define the space-fixed Mercurycentric frame, Ψ_{MC}, in which writing the equation of motion of the spacecraft, then we need to compute the rotation matrix $\mathcal{R}$ to convert the probe coordinates from Ψ_{BF} to Ψ_{MC}. To this aim, we adopt the semi-empirical model defined in [9]. Referring to that paper for an exhaustive discussion, we recall that the rotation matrix can be decomposed as $\mathcal{R} = \mathcal{R}_3(\phi)\mathcal{R}_1(\delta_2)\mathcal{R}_2(\delta_1)$, where $\mathcal{R}_i(\alpha)$ is the matrix associated with the rotation by an angle α about the i-th axis ($i = 1, 2, 3$), (δ_1, δ_2) define the space-fixed direction of the rotation axis in the Ψ_{MC} frame and ϕ is the rotation angle around the rotation axis, assuming the unit vector along the longest axis of the equator of Mercury (minimum momentum of inertia) as the rotational reference meridian.

The fundamental assumptions to describe the rotational state of Mercury in the adopted semi-empirical model are the following, as defined in [43,44]: (i) the *Cassini state theory*, defining the obliquity η with respect to the orbit normal as $\cos \eta = \cos \delta_2 \cos \delta_1$, assumed to be constant over the mission time span; (ii) addition in the description of two librations in longitude terms, the amplitude ε_1 of 88 days forced librations and the amplitude ε_2 of the Jupiter forced librations, possibly near-resonant with the free libration frequency (see, e.g., [45,46]).

3.3.5. Non-Gravitational Perturbations

The spacecraft around Mercury is perturbed significantly by non-gravitational forces such as the direct radiation pressure from the Sun, the indirect emission from the planet surface, the thermal re-emission from the spacecraft itself. The non-gravitational effects on the Mercurycentric orbit of the spacecraft are so intense that, if not properly taken into account, they would lead to a significantly biased orbit determination. Due to the general difficulty of modeling these effects, an accelerometer (ISA—Italian Spring Accelerometer) will be placed on board the spacecraft [20]. This instrument is able to measure differential accelerations between a sensitive element and its rigid frame (cage) and thus to give accurate information on the non-gravitational accelerations. During the scientific phase of the mission, the accelerometer readings will be available nearly continuously at the rate of 1 Hz.

For the purpose of simulations, we introduce a simplified model of non-gravitational perturbations in order to include the accelerometer readings among the observables. We account for the effect of direct solar radiation pressure $\mathbf{a}_{rad}$ assuming a spherical satellite with coefficient 1 (i.e., we neglect diffusive terms). The shadow of the planet is computed accurately, taking into account the penumbra effects. Moreover, we include the acceleration due to the thermal radiation from the planet, $\mathbf{a}_{th}$, assuming a zero relaxation time for the thermal re-emission of Mercury (details on the model, supplied by D.Vokrouhlicky, Charles University of Prague, can be found in [10]). The whole non-gravitational perturbations experienced by the spacecraft are, then, $\mathbf{a}_{ng} = \mathbf{a}_{rad} + \mathbf{a}_{th}$. We need to stress out that this model, although simplified, is accurate enough for the purpose of simulations. As will be detailed in Section 4.1.2, the key issue is that the accelerometer readings suffer from both random and systematic errors, which are the critical terms to deal with. We can write the accelerometer contribution to the equation of motion as: $\mathbf{a}_{ISA} = -\mathbf{a}_{ng} + \varepsilon$, where ε represents the contribution of all the error sources in the ISA readings. As already highlighted, one of the main goals of the radio science experiment is to perform a very accurate orbit determination of the Mercurycentric motion of the spacecraft. To this aim, what really matters is to remove in the most suitable way any bias introduced in the accelerometer readings by instrumental errors. For this reason, in our analysis we mainly focus on the techniques to handle these error terms instead of accurately modeling the non-gravitational perturbations themselves.

4. Simulation Scenario and Assumptions

In the following section, we outline the observational and dynamical scenario of the numerical simulations of the relativity experiment. The latest mission scenario provides for a one year orbital phase starting from 28 March 2025. The initial Mercurycentric orbit is polar and near-circular (480 × 1500 km) with the pericenter located at ~15° N. The orbital period of the spacecraft is about 2.3 h. We assume that two ground stations are available for tracking, one at the Goldstone Deep Space Communications Complex in California (USA), providing observations in the *Ka* band, and one located at the Cebreros station in Spain, supplying only *X* band observations. An average of 15–16 h of tracking per day is expected, with an average of 8 h in the *Ka* band. Range and range-rate measurements are simulated every 120 and 30 s, respectively. The propagation of the Mercurycentric dynamics in simulation stage is based on the gravity field of Mercury measured by MESSENGER [23], up to degree and order 25, with the addition of the Sun tidal effects described by the Love number k_2 and on the semi-empirical model for the planet rotation outlined in Section 3.3.4. For the Love number we adopted the value $k_2 = 0.45$ measured by MESSENGER [23]. For the rotational parameters we used the following values: the orientation of the rotation axis is defined, in our semi-empirical model, by the arbitrary angles $\delta_1 = 3$ arcmin and $\delta_2 = 1$ arcmin; the amplitudes of the librations in longitude are $\varepsilon_1 = 38.9$ arcsec, as measured by MESSENGER [47], and $\varepsilon_2 = 40$ arcsec [45]. Concerning the relativity parameters, we adopted the GR values for the PN parameters: $\beta = \gamma = 1$, $\eta = 0$, $\alpha_1 = \alpha_2 = 0$. The values of $\mu_\odot$ and $J_{2\odot}$ are taken from the DE421 ephemerides and we assume $\zeta = 0$. In the case of γ, we added the apriori $\gamma = 1 \pm 5 \times 10^{-6}$ in differential correction stage. In fact, the PN parameter γ appears both in the equations of motion for Mercury and EMB and in the equations for radio waves propagation. The delay of light propagation due to the space-time curvature, called Shapiro effect [30], is enhanced during a solar superior conjunction. Thus, a Superior Conjunction Experiment (SCE) is devised during BepiColombo cruise phase [12], with the aim of updating the constraint provided by the Cassini-Huygens mission [48]. The adopted apriori value on γ has been obtained from dedicated SCE simulations [49].

4.1. Observables Error Models

The observables we are dealing with are the tracking data (range and range-rate) and the non-gravitational accelerations, measured by the on-board accelerometer. To perform simulations in a realistic scenario, we need to properly add some measurement error to each observable.

4.1.1. Range and Range-Rate

According to [3], a nominal white noise can be associated to each tracking observation. Defining the simulated one-way range and range-rate observables as two-way measurements divided by 2 and assuming top accuracy performances of the transponder, we add a Gaussian error of $\sigma_r = 23.7$ cm to the 120 s range observables and $\sigma_{\dot{r}} = 8.7 \times 10^{-4}$ cm/s to the 30 s range-rate measurements [6].

These values represent the optimal performances in the *Ka* band, while for the *X* band we assume 10 times larger errors.

From a comparison of the accuracies in the range and range-rate, it turns out that $\sigma_r/\sigma_{\dot{r}} \sim 10^5$ s (according to Gaussian statistics, the standard deviations can be rescaled in order to be compared over the same integration time). Range measurements are, hence, more accurate than range-rate when we are observing phenomena with a period longer than 10^5 s, while the opposite is true for range-rate. We can conclude that the relativity experiment, which involves long-term periodicity phenomena, is mainly performed through the range tracking data, while gravimetry and rotation experiments mainly with the range-rate (we recall that the Mercurycentric orbital period is less than 10^4 s).

At the level of accuracy provided by the MORE relativity experiment, it could be necessary to account for additional sources of uncertainty in the range measurements. Indeed, instrumental related effects, such as residual signatures from the calibrator or residual biases after ground system calibration, can affect the observations in a non-negligible way. To account for these spurious effects, we add to the range Gaussian noise a generic systematic term, described by a bias of 3 cm and a sinusoidal trend (as already done in [12]) which reaches an amplitude of 3 cm after one year of observations. The choice of this functional behavior can be replaced by other assumptions, as done in [50]; it merely accounts for a possible scenario, which is the purpose of our simulations.

4.1.2. Accelerometer Readings and Calibration Strategy

As outlined in Section 3.3.5, we write the accelerometer contribution to the Mercurycentric motion of the spacecraft as $\mathbf{a}_{ISA} = -\mathbf{a}_{ng} + \varepsilon$, where $\mathbf{a}_{ng}$ is computed according to our simplified model. Concerning the error term ε, we assume the model provided by ISA team (private communications). It consists of a random background with some periodic terms superimposed: the main ones are a thermal term, resulting in a sinusoid at Mercury sidereal period (7.6×10^6 s) and a resonant term, resulting in a sinusoid at the orbital period of the spacecraft (8.3×10^3 s). All the details on the adopted model and the effects of the main components on the Mercurycentric orbit determination are described in [10].

The key issue in dealing with the accelerometer error term is that if we simply add it to the right hand side of the equation of motion, its detrimental effect causes a downgrading of the orbit determination of the spacecraft by orders of magnitude, vanifying the radio science experiment. In fact, this spurious instrumental effect is absorbed by the solve for parameters (like the state vector of the spacecraft at each arc) just like any other physical effect, resulting in a totally biased solution. To overcome this problem, the basic idea is to add to the right hand side of the equation of motion an additional term $c(\boldsymbol{\psi};t)$, function of a further set of parameters $\boldsymbol{\psi}$, to be added in the solve for list, and of time, such that $\varepsilon(t) - c(\boldsymbol{\psi};t) \simeq 0$. In such a way, the *calibration function* $c(\boldsymbol{\psi};t)$ absorbs most of the accelerometer error and the physical parameters of interest for the radio science experiment are, in principle, not anymore biased. In the ORBIT14 software we implemented a novel calibration strategy, in which the calibration function is represented by a C^1 cubic spline. All the details can be found in [19]. As a consequence, six additional parameters per arc (two per direction) are determined. We point out, as extensively discussed in [10], that this calibration strategy is able to absorb the low frequencies (i.e., longer than one day) error terms and the random component; in fact, the coefficients of the spline polynomials are computed once per arc, hence features with a periodicity lower than one day cannot be accounted for. This means that the resonant term, which shows a periodicity significantly lower than one day (about 2.3 h), is not absorbed by calibration at all. While this term results highly critical for what concerns the gravimetry and rotation experiments, we will see that its amplitude is not significantly detrimental for the relativity experiment.

4.2. Desaturation Maneuvres

Additional sources of perturbation on the orbit of the spacecraft around Mercury are the reaction wheels desaturation manoeuvres. We will assume as a general scenario to have one dump maneuvre during tracking and one dump maneuvre in the periods without tracking, hence a maximum amount of two dump manoeuvres per arc, as specified by the mission requirements. Each desaturation maneuvre, needed to maintain the desired attitude of the spacecraft, affects the precise Mercurycentric orbit determination. The result is a significant velocity change in the radial and out-of-plane directions and a linear momentum transfer in the transversal direction. To guarantee the expected level of accuracy in the orbit determination, these effects need to be modeled and removed from the estimation of the parameters. Each maneuvre appears in the spacecraft equation of motion as an additional acceleration acting on the probe. The downgrading effect on the spacecraft orbit determination can be significant, up to tens of meters in the range observations (see, e.g., [27]). For this reason, the velocity change $\Delta \mathbf{v}$ due to each maneuvre is added to the list of solve-for parameters, removing most of the downgrading effect from the orbit determination. The values for $\Delta \mathbf{v}$ adopted in simulations, along with all the details on the modelization and implementation of the maneuvres scenario, are given in [27]. The presence of orbital maneuvres, which are in general much larger than the desaturation maneuvres, is not considered here.

4.3. Metric Theories of Gravitation

A critical issue of the MORE relativity experiment, already discussed in [12], is that the Eddington parameter β and the Sun oblateness $J_{2\odot}$ show a near 1 correlation, as it appears from the covariance matrix obtained through the LS fit. This effect can be interpreted from a geometrical point of view considering that the main orbital effect of β is a precession of the argument of perihelion, which is a displacement taking place in the plane of the orbit of Mercury, while $J_{2\odot}$ affects the precession of the longitude of the node, thus producing a displacement in the plane of the solar equator. The angle between these two planes is only $\theta = 3.3°$, hence, being $\cos \theta \simeq 1$, we can expect such a high correlation between the two parameters. The consequence is a significant deterioration of the formal accuracies of both parameters. Since, unavoidably, the geometrical configuration cannot be changed, a possible solution to determine both parameters without a significant loss in accuracy is to add a suitable constraint on one of the involved parameter. A meaningful possibility is to link the PN parameters through the Nordtvedt equation [51]:

$$\eta = 4(\beta - 1) - (\gamma - 1) - \alpha_1 - \frac{2}{3}\alpha_2 .$$

In such a way, the knowledge on β is determined from the value of η: the correlation between β and η becomes almost 1, but that between β and $J_{2\odot}$ is greatly reduced. The introduction of the Nordtvedt equation is justified if we assume that gravitation must be described by a metric theory. In the following this becomes a basic assumption of our scenario.

4.4. Rank Deficiencies in the Mercury and EMB Orbit Determination Problem

As already stated, to perform the MORE relativity experiment we need to determine the orbit of Mercury and the EMB, that is to compute their state vector (position and velocity) at a given reference epoch. In practice, we find that we cannot solve for all the 12 components of the 2 state vectors without running into a significant deterioration of the results. The issue arises from the fact that this orbit determination problem, including simultaneously Mercury and the Earth, shows an approximate rank deficiency of order 4 (see, e.g., [12] for details).

An approximate rank deficiency of order 3 results from the breaking of an exact symmetry of the problem with respect to the full rotation group $SO(3)$. If there were only the Sun, the Earth and Mercury and if the Sun was exactly spherically symmetric (i.e., $J_{2\odot} = 0$), there would be an exact symmetry for rotation in determining both the orbits of Mercury and the Earth and therefore an exact rank deficiency of order 3. Due to the coupling with the other planets and to the asphericity of the Sun, the exact symmetry is broken but only by a small parameter (of the order of the relative size of the mutual perturbations by the other planets on the orbits of Mercury and the Earth and of the size of $J_{2\odot}$), bringing a residual approximate rank deficiency of order 3 in the problem.

A further exact symmetry would be present, if there were only the Sun, the Earth and Mercury. Changing all the lengths by a factor L, all the masses by a factor M and all the time intervals by a factor T, provided that the scaling factors are related by $L^3 = T^2 M$ (Kepler's third law), the equation of motion of the gravitational 3-body problem would remain unchanged. Again, this symmetry is broken by a small parameter, and an approximate rank deficiency of order 1 remains, leading to a total rank deficiency of order 4.

The standard solution already adopted in [12] is to solve for only 8 of the 12 components of the state vectors, assuming the remaining 4 as consider parameters. In the following we adopt the same assumption and we do not solve for the three position components of the EMB orbit (x_{EM}, y_{EM}, z_{EM}) and for the EMB velocity component perpendicular to the EMB orbital plane ($\dot{z}_{EM}$). The adopted technique is called *descoping*. A different approach to a problem of rank deficiency of order d is to add d constraint equations as apriori observations, instead of removing d parameters from the solve for list (see [25]—Chapter 6 for details). This is the technique we apply, for example, assuming the validity of the Nordtvedt equation in order to remove the degeneracy between β and $J_{2\odot}$. In ORBIT14 we have implemented also the possibility of determining all the 12 state vectors components, by adding 4 apriori constraints between the state vectors components and the Sun's mass $\mu_\odot$ in order to remove the degeneracy. A detailed discussion on this topic will be presented in a future paper by our group. In the following we assume to determine only 8 out of the 12 components.

5. Results

In this Section we will present and discuss the results of our simulations. In this review, we are mainly interested in the MORE relativity experiment: the results concerning PN and related parameters will be given in Section 5.1. For completeness, we will discuss the results concerning gravimetry and rotation in Section 5.2.

At each iteration of the differential correction process we solve for the following parameters:

- Global dynamical:

    - PN parameters: $\beta, \gamma, \eta, \alpha_1, \alpha_2$;
    - other parameters of interest for the relativity experiment: $\mu_\odot, \zeta, J_{2\odot}$;
    - the state vectors of Mercury and EMB (8 components): $(x_M, y_M, z_M; \dot{x}_M, \dot{y}_M, \dot{z}_M); (\dot{x}_{EM}, \dot{y}_{EM})$;
    - normalized harmonic coefficients of the gravity field of Mercury up to degree and order 25 and the Love number k_2;
    - rotational parameters: $\delta_1, \delta_2, \varepsilon_1, \varepsilon_2$;
    - six accelerometer calibration coefficients for each arc, plus 6+6 boundary conditions;

- Local dynamical:

    - state vector of the Mercurycentric orbit of the spacecraft, in the Ecliptic J2000 inertial reference frame, at the central time of each observed arc;
    - three dump manoeuvre components, $\Delta\mathbf{v}$, taking place during tracking, for each observed arc;

- External local dynamical:

    - three dump manoeuvre components, $\Delta\mathbf{v}$, taking place in the period without tracking between each pair of consecutive observed arcs.

5.1. The Relativity Experiment Results

The results for the PN and related parameters of interest for the relativity experiment are shown in Table 1. For each parameter we report the following quantities: (i) the formal uncertainty; (ii) the true error; (iii) the true-to-formal (T/F) error ratio; (iv) the current accuracy with which the parameter is presently known.

Table 1. Simulation results for the parameters of interest in the MORE relativity experiment (errors on $\mu_\odot$ are in cm^3/s^2, on ζ in y^{-1}).

Parameter	Formal Error	True Error	T/F Error Ratio	Current Accuracy
β	7.3×10^{-7}	2.6×10^{-6}	3.6	7×10^{-5} [52]
γ	9.3×10^{-7}	1.1×10^{-6}	1.2	2.3×10^{-5} [48]
η	2.2×10^{-6}	1.1×10^{-5}	4.9	4.5×10^{-4} [53]
α_1	4.9×10^{-7}	4.9×10^{-7}	1.0	6.0×10^{-6} [54]
α_2	8.3×10^{-8}	1.0×10^{-7}	1.2	3.5×10^{-5} [54]
$\mu_\odot$	4.2×10^{13}	4.2×10^{13}	1.0	$10^{16}, 8 \times 10^{15}$ [55,56]
ζ	2.3×10^{-14}	3.6×10^{-14}	1.5	4.3×10^{-14} [57]
$J_{2\odot}$	4.1×10^{-10}	4.1×10^{-10}	1.0	1.2×10^{-8} [52]

The formal error is obtained from the diagonal terms of the covariance matrix. The main limitation of formal analysis is that it does not account at all for any error that is non-Gaussian, like systematic errors, or time-correlated, unless they are in some way calibrated introducing further parameters in the solve-for list. Besides the formal analysis, we introduce a second quantity, which we call "true" error, to assess the expected accuracies in a more realistic way. This quantity is defined for each parameter as the difference between the nominal value of the parameter (used in simulations) and the value determined at convergence of the differential correction process. In such a way, the systematic effects are included in the computation of the accuracies. The true errors shown in Table 1 have been obtained as rms values over a number of runs carried out by changing the seed of the random numbers generator. We found that $\sim$10 runs are adequately representative to quantify systematic errors. The ideal situation would occur when T/F error ratios follow Gaussian statistics, which means either that no systematic effects are present at all or that they are accounted for, through calibration parameters, in the formal analysis. In practice, this ratio is almost always greater than 1, but what does matter is that it is limited within a maximum of T/F $\sim$ 3. Any higher value would be representative of a wrong or lacking modelization of some effects.

Analyzing the two sources of systematic effects included in simulation, i.e., the error model for accelerometer readings and the spurious effects from the ranging system, we found that T/F values higher than 1 for the relativity parameters can be only partially ascribed to non-perfectly calibrated long term components in the accelerometer error model, which are not fully absorbed by the C^1 spline calibration. The main downgrading effect turns out to be the presence of systematic terms in the range error model, which are not calibrated at all. The effect is particularly detrimental for the determination of β and η. We remind that the range error model includes a bias term of 3 cm and a sinusoidal trend up to 3 cm after one year. Analyzing individually the two error terms, we found that the bias term is responsible for most of the deterioration in estimating β and η. Adding the linear term does not further deteriorate true errors in a significant way. Moreover, as extensively discussed in [50], increasing the adopted value for the bias term results in a corresponding increase of the true errors of all the PN parameters.

A possible approach to the problem would be to introduce in the solve-for list an additional global parameter, that is a bias in modeling the range observables. Estimating the bias, it could be possible in principle to absorb most of the spurious effect in range, leading to a better estimate of the PN parameters in terms of T/F error ratios. This approach has a disadvantage that immediately appears when we take correlations into account. The correlations between PN and related parameters are shown in Table 2. As expected, the correlation between β and η is almost one. In fact we adopted the Nordvedt relation as an apriori constraint between PN parameters and we included an apriori constraint on γ from the SCE simulations during cruise phase. As a consequence, η is deduced from β

and their correlation is very high. If we add a bias in range to the solve-for list, the correlation between the bias term and, especially, β and η turns out to be almost one. This would lead to a worsening in the formal error of both β and η by more than one order of magnitude. This result would not be compliant with the scientific goals expected from MORE in terms of accuracies. In fact, the goal is to determine η at a level of, at least, 10^{-5} and β at a level of some parts in 10^{-6} [3,6,12]. In conclusion, if the systematic effects due to the ranging system remain at the level of few cm, the downgrading effect on the accuracies is still acceptable, as can be envisaged comparing the present results with current accuracies. Conversely, if the systematic terms, especially a spurious bias in the range measurements, become more significant, some suitable calibration strategy would be mandatory. As sketched in Section 4.4, the different approach of estimating all the 12, instead of only 8, components of Mercury and the EMB state vectors by adding some apriori constraints is under analysis. Preliminary attempts have shown that in such a case the correlation between the bias term in range and η and β would significantly decrease. We will report our conclusions in a future work.

Table 2. Correlations between PN and related parameters (values higher than 0.7 are highlighted in bold).

	β	γ	η	α_1	α_2	$\mu_\odot$	ζ	$J_{2\odot}$
$J_{2\odot}$	0.15	0.21	0.11	**0.90**	0.29	**0.89**	0.10	–
ζ	<0.1	0.28	<0.1	<0.1	0.17	<0.1	–	
$\mu_\odot$	0.20	0.14	0.14	**0.84**	<0.1	–		
α_2	0.35	0.28	0.36	0.27	–			
α_1	0.35	0.12	0.22	–				
η	**0.96**	0.60	–					
γ	**0.77**	–						
β	–							

A critical issue in the MORE relativity experiment, already remarked in [12] (p. 17), concerns the effects of a lacking knowledge of the Solar System model. In our simulations, we assumed that all the parameters of the SS model not included in the solve-for list (for example, the masses of the planets) are known from the ephemerides well enough that no spurious effects are introduced in the parameters estimation. An extensive discussion on this approximation has been carried out in [31] and the issue is still controversial.

Finally, we point out that the Lense-Thirring effect on the Mercury's orbit due to the Sun's angular momentum has been neglected. This choice has presently been made in order to simplify the development and implementation of the dynamical models. However, the effect is expected to be relevant [58], hence in future work we will investigate on its possible impact on the relativity parameters determination.

5.2. Results for Gravimetry and Rotation

In Section 4.1.1, we introduced one of the basic issues of the BepiColombo radio science experiment. Comparing the expected accuracies of range and range-rate, it turns out that range-rate measurements are more accurate than range data when observing phenomena with periodicity shorter than $\sim 10^5$ s, while the opposite is true for long-term periodicity phenomena. As a consequence, gravimetry and rotation experiments are mainly performed by means of range-rate data, while the relativity experiment by means of range. MORE is a comprehensive experiment in which all the parameters are solved simultaneously in the non-linear LS fit, but the expected independence between gravimetry/rotation on one side and relativity on the other suggests that, for the purpose of simulations, we can perform the experiments individually or all together and achieve the same results. We checked the validity of this statement by performing additional simulations. Referring to the solve-for list in Section 4, we ran the following simulations:

- *relativity simulations:* we removed from the solve-for list the gravimetry and rotational parameters, i.e. the gravity field spherical harmonic coefficients, Love number k_2, the angles (δ_1, δ_2), the libration amplitudes $\varepsilon_1, \varepsilon_2$;
- *gravimetry and rotation simulations:* we removed from the solve-for list the PN and related parameters.

These are mandatory tests since the chance that any further unforeseen rank deficiency between relativity and gravimetry/rotation parameters appears in the global fit needs to be verified. The results confirmed our expectations. The accuracies of PN and related parameters achieved in the global simulation, discussed in Section 5.1, and in the relativity simulations are equivalent, and the same is true for gravimetry and rotation. We have already extensively reported in [10] on the results for the MORE gravimetry and rotation experiments, together with a discussion on the achievable accuracy in the orbit determination. Therefor, we do not duplicate here the same results and we refer to that paper for a discussion on these topics. We point out that in the simulations described here we did not include among the observables the optical data from the high resolution camera HRIC, part of the SIMBIO-SYS payload [59]. In fact, camera observations significantly support range-rate measurements in the determination of the rotational parameters, while they do not contribute at all to the relativity experiment.

6. Discussion and Conclusions

In this review, we summarized all the issues concerning the BepiColombo relativity experiment. After recalling the global structure of the ORBIT14 software and the techniques to determine the parameters of interest, we detailed the essential mathematical models on which the experiment is based and the fundamental assumptions adopted. We finally presented the results of a full cycle of simulations carried out in the latest mission scenario.

At the beginning of 2000's our group performed a similar set of simulations, whose results are reported in [12], with the specific aim of dictating the mission and instrumentation requirements in order to make the BepiColombo relativity experiment feasible. Several underlying issues concerning the experiment have since been reconsidered and updated and the software has undergone significant revision. The formal results of the present paper are compared with the formal errors obtained in 2002 in [12] in Table 3, where the results reported in [6], representing the goal accuracies required for the MORE relativity experiment, have also been included. We refer to Experiment D in [12], where Nordtvedt equation has been assumed to link PN parameters. In the comparison we did not consider the ζ parameter because in [12] the quantity $\dot{G}/G$ was included instead of ζ.

Table 3. Comparison between the results in Schettino & Tommei (2016) (this paper) and previous results of the relativity experiment ($\mu_\odot$ in cm^3/s^2).

Parameter	Schettino & Tommei (2016)	Milani et al. (2002) [12]	Iess et al. (2009) [6]
β	7.3×10^{-7}	9.2×10^{-7}	2×10^{-6}
γ	9.3×10^{-7}	2×10^{-6} (SCE)	2×10^{-6}
η	2.2×10^{-6}	3.3×10^{-6}	8×10^{-6}
α_1	4.9×10^{-7}	7.1×10^{-7}	–
α_2	8.3×10^{-8}	1.9×10^{-7}	–
$\mu_\odot$	4.2×10^{13}	4.1×10^{13}	–
$J_{2\odot}$	4.1×10^{-10}	6.2×10^{-10}	2×10^{-9}

It can be seen that a slight improvement of the 2002 expectations [12] has been achieved. It can be remarked that in both cases formal errors turn out to be significantly lower than the goal accuracies of MORE. A quantitative comparison with the 2015 results in [15] is difficult because the mission time span scenario is different and, furthermore, the assumption of a metric theory of gravitation was not included in [15].

At this stage, the spacecraft is almost ready for launch and no significant modification of the mission can be addressed anymore. The aim of the set of simulations described in this review is, hence, clear: assuming the performances expected and tested for each instrument and the revised launch scenario, we want to establish the feasibility of the relativity experiment and provide the results that can be achieved in terms of accuracies. Two key issues were pointed out already during the past years: the impact on the solution from the errors in the accelerometer readings and from the aging of the transponder. Concerning the first issue, the main downgrading source was found in the thermal effects which produce periodic spurious signatures, with the periodicity of both the orbital period of the spacecraft around Mercury and the sidereal period of Mercury around the Sun. These signatures mainly affect the Mercurycentric orbit determination. An extensive discussion on the potentially downgrading effects for the gravimetry and rotation experiments have been recently discussed by our group in [10]. The results shown in Table 1 and the discussions presented in [14,50] lead us to the conclusion that, if the accelerometer error model is compliant to the one adopted, the effects on the relativity experiment are not detrimental. More critical is the question on how the ranging system affects the results. In [12] it was shown that, describing the transponder aging with a sinusoidal trend up to some tens of cm after one year, the effect was highly detrimental for the relativity parameters estimation. This issue has been tackled by introducing an on-board calibrator to account for the aging of the transponder. Nevertheless, residual spurious effects due, e.g., to the calibrator itself or to the on-ground instrumentation can still lead to a systematic error of a few cm. In our simulations, we assumed a bias of 3 cm and a sinusoidal trend up to 3 cm after one year. In such a case, the detrimental effects on the parameters are restrained, but in an unfavorable scenario in which they exceed the value of 5 cm on the one-way range, the solution would be significantly downgraded, as shown in [50]. In such a case, we envisage the need of a calibration strategy within the LS fit. In any case, in the realistic scenario presented here, we can conclude that the accuracies achievable by the BepiColombo relativity experiment for each of the PN parameter, compared with current accuracies, would represent a significant improvement of our knowledge of gravitational theories.

Acknowledgments: The results of the research presented in this paper have been performed within the scope of the Addendum n. I/080/09/1 of the contract n. I/080/09/1 with the Italian Space Agency. The authors would like to thank the three anonymous reviewers and the editors for the valuable comments and the significant improvements to the earlier version of the manuscript.

Author Contributions: Giulia Schettino and Giacomo Tommei conceived and designed the simulations; Giulia Schettino run the simulations; Giulia Schettino and Giacomo Tommei analyzed the output of the simulations; Giulia Schettino and Giacomo Tommei wrote the paper.

References

1.   Benkhoff, J.; van Casteren, J.; Hayakawa, H.; Fujimoto, M.; Laakso, H.; Novara, M.; Ferri, P.; Middleton, H.R.; Ziethe, R. BepiColombo-Comprehensive exploration of Mercury: Mission overview and science goals. *Planet. Space Sic.* **2010**, *58*, 2–20.

2.   Mukai, T.; Yamakawa, H.; Hayakawa, H.; Kasaba, Y.; Ogawa, H. Present status of the BepiColombo/Mercury magnetospheric orbiter. *Adv. Space Res.* **2006**, *38*, 578–582.

3.   Iess, L.; Boscagli, G. Advanced radio science instrumentation for the mission BepiColombo to Mercury. *Planet. Space Sci.* **2001**, *49*, 1597–1608.

4.   Milani, A.; Rossi, A.; Vokrouhlický, D.; Villani, D.; Bonanno, C. Gravity field and rotation state of Mercury from the BepiColombo Radio Science Experiments. *Planet. Space Sci.* **2001**, *49*, 1579–1596.

5.   Sanchez Ortiz, N.; Belló Mora, M.; Jehn, R. BepiColombo mission: Estimation of Mercury gravity field and rotation parameters. *Acta Astronaut.* **2006**, *58*, 236–242.

6.   Iess, L.; Asmar, S.; Tortora, P. MORE: An advanced tracking experiment for the exploration of Mercury with the mission BepiColombo. *Acta Astronaut.* **2009**, *65*, 666–675.

7. Genova, A.; Marabucci, M.; Iess, L. Mercury radio science experiment of the mission BepiColombo. *Mem. Soc. Astron. Ital. Suppl.* **2012**, *20*, 127–132.

8. Schettino, G.; Di Ruzza, S.; De Marchi, F.; Cicalò, S.; Tommei, G.; Milani, A. The radio science experiment with BepiColombo mission to Mercury. *Mem. Soc. Astron. Ital.* **2016**, *87*, 24–29.

9. Cicalò, S.; Milani, A. Determination of the rotation of Mercury from satellite gravimetry. *Mon. Not. R. Astron. Soc.* **2012**, *427*, 468–482.

10. Cicalò, S.; Schettino, G.; Di Ruzza, S.; Alessi, E.M.; Tommei, G.; Milani, A. The BepiColombo MORE gravimetry and rotation experiments with the ORBIT14 software. *Mon. Not. R. Astron. Soc.* **2016**, *457*, 1507–1521.

11. Palli, A.; Bevilacqua, A.; Genova, A.; Gherardi, A.; Iess, L.; Meriggiola, R.; Tortora, P. Implementation of an End to End Simulator for the BepiColombo Rotation Experiment. In Proceedings of the European Planetary Space Congress 2012, Madrid, Spain, 23–28 September 2012.

12. Milani, A.; Vokrouhlicky, D.; Villani, D.; Bonanno, C.; Rossi, A. Testing general relativity with the BepiColombo Radio Science Experiment. *Phys. Rev. D* **2002**, *66*, doi:10.1103/PhysRevD.66.082001.

13. Milani, A.; Tommei, G.; Vokrouhlicky, D.; Latorre, E.; Cicalò, S. Relativistic models for the BepiColombo radioscience experiment. In *Relativity in Fundamental Astronomy: Dynamics, Reference Frames, and Data Analysis*; Cambridge University Press: Cambridge, UK, 2010.

14. Schettino, G.; Cicalò, S.; Di Ruzza, S.; Tommei, G. The relativity experiment of MORE: Global full-cycle simulation and results. In Proceedings of the IEEE Metrology for Aerospace (MetroAeroSpace), Benevento, Italy, 4–5 June 2015; pp. 141–145.

15. Schuster, A.K.; Jehn, R.; Montagnon, E. Spacecraft design impacts on the post-Newtonian parameter estimation. In Proceedings of the IEEE Metrology for Aerospace (MetroAeroSpace), Benevento, Italy, 4–5 June 2015; pp. 82–87.

16. Tommei, G.; De Marchi, F.; Serra, D.; Schettino, G. On the Bepicolombo and Juno Radio Science Experiments: Precise models and critical estimates. In Proceedings of the IEEE Metrology for Aerospace (MetroAeroSpace), Benevento, Italy, 4–5 June 2015; pp. 323–328.

17. Kaula, W.M. *Theory of Satellite Geodesy: Applications of Satellites to Geodesy*; Blaisdell: Waltham, MA, USA, 1996.

18. Kozai, Y. Effects of the tidal deformation of the Earth on the motion of close Earth satellites. *Publ. Astron. Soc. Jpn.* **1965**, *17*, 395–402.

19. Alessi, E.M.; Cicalò, S.; Milani, A. Accelerometer data handling for the BepiColombo orbit determination. In *Advances in the Astronautical Science*, Proccedings of the 1st IAA Conference on Dynamics and Control of Space Systems, Porto, Portugal, 19–21 March 2012.

20. Iafolla, V.; Nozzoli S. Italian Spring Accelerometer (ISA): A high sensitive accelerometer for BepiColombo ESA Cornerstone. *Planet. Space Sci.* **2001**, *49*, 1609–1617.

21. Will, C.M. *Theory and Experiment in Gravitational Physics*; Cambridge University Press: Cambridge, UK, 1993.

22. Tommei, G.; Milani, A.; Vokrouhlicky, D. Light-time computations for the BepiColombo Radio Science Experiment. *Celest. Mech. Dyn. Astron.* **2010**, *107*, 285–298.

23. Mazarico, E.; Genova, A.; Goossens, S.; Lemoine, F.G.; Neumann, G.A.; Zuber, M.T.; Smith, D.E.; Solomon, S.C. The gravity field, orientation, and ephemeris of Mercury from MESSENGER observations after three years in orbit. *J. Geophys. Res.* **2014**, *119*, 2417–2436.

24. Jehn, R.; Rocchi, A. *BepiColombo Mercury Cornerstone Mission Analysis: The April 2018 Launch Option*; MAS Working Paper No. 608; BC-ESC-RP-50013; 24 March 2016; Issue 2.1. Available online: http://issfd.org/ISSFD_2014/ISSFD24_Paper_S6-5_jehn.pdf (accessed on 5 August 2016).

25. Milani, A.; Gronchi, G.F. *Theory of Orbit Determination*; Cambridge University Press: Cambridge, UK, 2010.

26. Folkner, W.M.; Williams, J.G.; Boggs, D.H. *The Planetary and Lunar Ephemeris DE 421*; JPL Publication: Pasadena, CA, USA, 2008.

27. Alessi, E.M.; Cicalò, S.; Milani, A.; Tommei, G. Desaturation Manoeuvres and Precise Orbit Determination for the BepiColombo Mission. *Mon. Not. R. Astron. Soc.* **2012**, *423*, 2270–2278.

28. Bonanno, C.; Milani, A. Symmetries and rank deficiency in the orbit determination around another planet. *Celest. Mech. Dyn. Astron.* **2002**, *83*, 17–33.

29. Will, C.M. The Confrontation between General Relativity and Experiment. *Living Rev. Relativ.* **2014**, *17*, 1–117.

30. Shapiro, I.I. Fourth Test of General Relativity. *Phys. Rev. Lett.* **1964**, *13*, 789–791.

31. De Marchi, F.; Tommei, G.; Milani, A.; Schettino, G. Constraining the Nordtvedt parameter with the BepiColombo Radioscience experiment. *Phys. Rev. D* **2016**, *93*, 123014.

32. Peebles P.J.E. *Principles of Physical Cosmology*; Princeton University Press: Princeton, NJ, USA, 1993.

33. Klioner, S.A. Relativistic scaling of astronomical quantities and the system of astronomical units. *Astron. Astrophys.* **2008**, *478*, 951–958.

34. Klioner, S.A.; Capitaine, N.; Folkner, W.; Guinot, B.; Huang, T.Y.; Kopeikin, S.; Petit, G.; Pitjeva, E.; Seidelmann, P.K.; Soffel, M. Units of Relativistic Time Scales and Associated Quantities. In *Relativity in Fundamental Astronomy: Dynamics, Reference Frames, and Data Analysis*; Cambridge University Press: Cambridge, UK, 2010.

35. Moyer, T.D. *Formulation for Observed and Computed Values of Deep Space Network Data Types for Navigation, NASA-JPL, Deep Space Communication and Navigation Series, Monograph 2*; John Wiley & Sons: Hoboken, NJ, USA, 2000.

36. Soffel, M.; Klioner, S.A.; Petit, G.; Wolf, P.; Kopeikin, S.M.; Bretagnon, P.; Brumberg, V.A.; Capitaine, N.; Damour, T.; Fukushima, T. The IAU Resolutions for Astrometry, Celestial Mechanics, and Metrology in the Relativistic Framework: Explanatory Supplement. *Astron. J.* **2003**, *126*, 2687–2706.

37. Irwin, A.W.; Fukushima, T. A numerical time ephemeris of the Earth. *Astron. Astrophys.* **1999**, *348*, 642–652.

38. Damour, T.; Soffel, M.; Hu, C. General-relativistic celestial mechanics. IV. Theory of satellite motion. *Phys. Rev. D* **1994**, *49*, 618–635.

39. Archinal, B.A.; A'Hearn, M.F.; Bowell, E.; Conrad, A.; Consolmagno, G.J.; Courtin, R.; Fukushima, T.; Hestroffer, D.; Hilton, J.L.; Krasinskyet, G.A.; et al. Report of the IAU Working Group on Cartographic Coordinates and Rotational Elements: 2009. *Celest. Mech. Dyn. Astron.* **2011**, *2*, 101–135.

40. Montenbruck, O.; Gill, E. *Satellite Orbits: Models, Methods and Applications*; Springer: Berlin, Germany, 2005.

41. Padovan, S.; Margot, J.-L.; Hauck, S.A., II; Moore, W.B.; Solomon, S.C. The tides of Mercury and possible implications for its interior structure. *J. Geophys. Res.* **2014**, *119*, 850–866.

42. Roy, A.E. *Orbital Motion*, 4th ed.; CRC Press: Boca Raton, FL, USA, 2005.

43. Peale, S.J. *The Rotational Dynamics of Mercury and the State of Its Core*; University Arizona press: Tucson, AZ, USA, 1988; pp.461–493.

44. Peale, S.J. The proximity of Mercury's spin to Cassini state 1 from adiabatic invariance. *Icarus* **2006**, *181*, 338–347.

45. Yseboodt, M.; Margot, J.L.; Peale, S.J. Analytical model of the long-period forced longitude librations of Mercury. *Icarus* **2010**, *207*, 536–544.

46. Yseboodt, M.; Rivoldini, A.; Van Hoolst, T.; Dumberry, M. Influence of an inner core on the long-period forced librations of Mercury. *Icarus* **2013**, *226*, 41–51.

47. Stark, A.; Oberst, J.; Preusker, F.; Peale, S.J.; Margot, J.-L.; Phillips, R.J.; Neumann, G.A.; Smith, D.E.; Zuber, M.T.; Solomon, S.C. First MESSENGER orbital observations of Mercury's librations. *Geophys. Res. Lett.* **2015**, *42*, 7881–7889.

48. Bertotti, B.; Iess, L.; Tortora, P. A test of general relativity using radio links with the Cassini spacecraft. *Nature* **2003**, *425*, 374–376.

49. Imperi, L.; Iess, L. Testing general relativity during the cruise phase of the BepiColombo mission to Mercury. In Proceedings of the IEEE Metrology for Aerospace (MetroAeroSpace), Benevento, Italy, 4–5 June 2015.

50. Schettino, G.; Imperi, L.; Iess, L.; Tommei, G. Sensitivity study of systematic errors in the BepiColombo relativity experiment. In Proceedings of the Metrology for Aerospace (MetroAeroSpace), Florence, Italy, 22–23 June 2016, in press.

51. Nordtvedt, K.J. Post-Newtonian Metric for a General Class of Scalar-Tensor Gravitational Theories and Observational Consequences. *Astrophys. J.* **1970**, *161*, 1059–1067.

52. Fienga, A.; Laskar, J.; Exertier, P.; Manche, H.; Gastineau, M. Numerical estimation of the sensitivity of INPOP planetary ephemerides to general relativity parameters. *Celest. Mech. Dyn. Astron.* **2015**, *123*, 325–349.

53. Williams, J.G.; Turyshev, S.G.; Boggs, D.H. Lunar Laser Ranging Tests of the Equivalence Principle with the Earth and Moon. *Int. J. Mod. Phys. D* **2009**, *18*, 1129–1175.

54. Iorio, L. Constraining the Preferred-Frame α_1, α_2 Parameters from Solar System Planetary Precessions. *Int. J. Mod. Phys. D* **2014**, *23*, 1450006.

55. Pitjeva, E.V. Determination of the Value of the Heliocentric Gravitational Constant ($GM_\odot$) from Modern Observations of Planets and Spacecraft. *J. Phys. Chem. Ref. Data* **2015**, *44*, 031210.

56. The Value has Been Obtained from Latest JPL Ephemerides and Has Been Published on the NASA HORIZONS Web-Interface. Available online: http://ssd.jpl.nasa.gov/?constants (accessed on 5 August 2016).

57. Pitjeva, E.V.; Pitjev, N.P. Relativistic effects and dark matter in the Solar system from observations of planets and spacecraft. *Mon. Not. R. Astron. Soc.* **2013**, *432*, 3431–3437.

58. Iorio, L.; Lichtenegger, H.I.M.; Ruggiero, M.L.; Corda, C. Phenomenology of the Lense-Thirring effect in the solar system. *Astrophys. Space Sci.* **2011**, *331*, 351–395.

59. Flamini, E.; Capaccioni, F.; Colangeli, L.; Cremonese, G.; Doressoundiram, A.; Josset, J.L.; Langevin, L.; Debei, S.; Capria, M.T.; De Sanctis, M.C.; et al. SIMBIO-SYS: The spectrometer and imagers integrated observatory system for the BepiColombo planetary orbiter. *Planet. Space Sci.* **2010**, *58*, 125–143.

6

Autoparallel *vs.* Geodesic Trajectories in a Model of Torsion Gravity

Luis Acedo

Instituto Universitario de Matemática Multidisciplinar, Universitat Politècnica de València, Building 8G, 2°
Floor, Camino de Vera 46022, Valencia, Spain; luiacrod@imm.upv.es

Academic Editor: Lorenzo Iorio

Abstract: We consider a parametrized torsion gravity model for Riemann–Cartan geometry around a
rotating axisymmetric massive body. In this model, the source of torsion is given by a circulating
vector potential following the celestial parallels around the rotating object. Ours is a variant of the
Mao, Tegmark, Guth and Cabi (MTGC model) in which the total angular momentum is proposed as
a source of torsion. We study the motion of bodies around the rotating object in terms of autoparallel
trajectories and determine the leading perturbations of the orbital elements by using standard
celestial mechanics techniques. We find that this torsion model implies new gravitational physical
consequences in the Solar system and, in particular, secular variations of the semi-major axis of
the planetary orbits. Perturbations on the longitude of the ascending node and the perihelion of
the planets are already under discussion in the astronomical community, and if confirmed as truly
non-zero effects at a statistically significant level, we might be at the dawn of an era of torsion
phenomenology in the Solar system.

Keywords: Solar system anomalies; Riemann–Cartan spacetime; gravitation models; autoparallel
curves; geodesic curves

1. Introduction

After one hundred years since its proposal [1], gravitation is still understood in terms of the
theory of general relativity (GR). This theory is considered as the pinnacle of classical physics, and the
status of its agreement with experiments is very good, although the progress in its verification has been
painfully slow [2] due to the weakness of the gravitational interaction and the technical difficulties
in measuring/observing its predicted effects due to their smallness [3–6]. In the last few years, an
important advance has been achieved with the confirmation of the geodetic and the frame-dragging
effects upon a gyroscope mounted on an artificial satellite orbiting the Earth. This is known as the
Gravity Probe B experiment [7]. With this outstanding result, most of the major deviations from
Newtonian gravity, as predicted by GR in the Solar system, are already experimentally checked. Efforts
to test a few other ones, such as the Lense–Thirring [8–12] and the post-Newtonian quadrupolar orbital
precessions [13,14], are ongoing. Moreover, the discovery of exoplanets orbiting other stars provides an
opportunity to obtain additional substantiation of GR [15–17]. In particular, some authors have claimed
that the relativistic precession of periastra in exoplanets could be detectable in the near future [18,19].
Zhao and Xie have also studied the influence of parametrized post-Newtonian dynamics in their
transit times and the possibility of testing GR to a 6% level [20]. Even testing a putative fifth-force
have also been considered [21]. The hot exoplanet WASP-33b also constitutes an excellent natural
laboratory for GR [22], because the predicted Lense–Thirring node precession is 3.25×10^5 larger than
that of Mercury [23], and this is only one order of magnitude below the measurability threshold for
these systems.

On the other hand, since the development of GR in its standard form, there have been many attempts to propose modified theories of gravity capable of predicting new testable phenomena. From the 1920s to the 1950s, the main drive of this research was to find a way of unifying gravitation and electromagnetism, although this objective was slowly being abandoned, except for Einstein himself and his collaborators [24]. Many of these theories were characterized by including a nonzero torsion tensor field, as well as the curvature tensor of standard Riemannian geometry. Extended geometries including both curvature and torsion have been used in physics since the early work of Einstein and Cartan [24]. The resulting Einstein–Cartan theory, much later improved by Sciama [25] and Kibble [26], is still considered a viable alternative to standard GR. In fact, it is still actively investigated as attested from the papers, conferences and even books published on this topic [27–29]. The Einstein–Cartan–Sciama–Kibble theory (ECSK) has also an interesting structure, as it can be consistently described as a gauge theory of the Poincaré group [30]. This way, it was put in correspondence with the very successful gauge theory approach to other interactions.

As beautiful as it could be, ECSK theory has not received any experimental support yet. This is not sufficient to dismiss any gravitation theory beforehand, because, as has happened with GR, the experiments are very difficult to design and carry out. Net spin densities are very small in most substances, as alignment of individual atomic spins is random, but it can be large in some elements, such as helium three (^{3}He) or dysprosium-iron compounds (Dy_6Fe_{23}). Ni has suggested to use these elements to build gyroscopes capable of testing PGtheory [31]. Apart from small spin densities, there is also the peculiarity that ECSK predicts null torsion outside macroscopic bodies. The reason for that behavior is that the field equation for torsion relates it linearly with spin density. Consequently, in a vacuum, where spin densities are null, torsion is also zero. Another important application of ECSK theory has been recently found by Popławski, who showed that torsion in the early Universe generates repulsion, and this could solve the flatness and horizon problems without resorting to an *ad hoc* inflation scenario [32,33]. A class of Poincaré gauge theories with extended Lagrangians quadratic in curvature and torsion have also been studied in the last few decades. These theories follow closely the analogy of the gauge paradigms of Weyl and Yang–Mills and, in some cases, predict a propagating torsion [34].

However, exploring alternative theories and models to standard PG is still both viable and timely. Theories in which torsion propagates outside macroscopic bodies can also be developed consistently [35]. As early as 1979, Hojman *et al.* proposed a model in which torsion is connected with a massless scalar field [36,37]. In this theory, torsion propagates in a vacuum, and torsion waves can be generated by sources with variable spin. At the same time, Hayashi and Shirafuji discussed an alternative to GR in which they revived the notion of Einstein's teleparallelism [38]. In the Hayashi–Shirafuji theory the fundamental entities are the tetrads instead of the metric, and the action is varied with respect to them to obtain the field equations. Interestingly, Hayashi and Shirafuji found a static spherically-symmetric vacuum solution in Weitzenböck spacetime (characterized by a null curvature tensor and a nonzero torsion [39,40]), which replaces the Schwarzschild solution in standard GR. The so-called new general relativity agrees with the classical tests for light bending, the anomalous advance of the perihelion of Mercury and Shapiro's delay of radar signals [38]. However, this theory fails to predict the geodetic and frame-dragging effects already checked in the Gravity Probe B experiment [7,41].

In 2007, Mao, Tegmark, Guth and Cabi proposed a phenomenological parametrized model for torsion in the Solar system (MTGC model). In this model, the source of torsion is assumed to be the rotational angular momentum of the planets and the Sun [41]. This is not the case in standard PG theory in which only the microscopic spin of elementary particles can generate torsion. The MTGC model does not depend on any specific theoretical framework, as the authors deduce the form of the torsion tensor from symmetry principles as the invariance under rotation, the antisymmetry of the torsion tensor in its covariant indices and the behavior of the angular momentum vector under parity transformations [41]. By using autoparallel or extreme schemes for the spin four vector S^μ or spin fourth tensor $S^{\mu\nu}$, these authors calculate extra contributions to the geodetic and frame-dragging precessions of a gyroscope's spin orbiting around the Earth. They claim that a refined version of Gravity Probe B experiment could be used to determine the values of some combinations of the seven constant parameters used to parametrize torsion. For the time being, error bars in the GPBexperiment are so large that we can only give some estimates on the bounds of the torsion parameters, but compatibility with standard GR is still not excluded by observations. Applying the planetary equations of Lagrange in the Gauss form, March *et al.* calculated the secular variations of the orbital elements for the planets and the Earth's geodynamics satellites [42,43]. In particular, they have found extra precessions rates for the longitude of the ascending node, *i.e.*, an anomalous Lense–Thirring effect, and also a contribution to the precession of the perihelion. Unfortunately, the precision in the determination of the Lense–Thirring effect or the perihelion precession, even for the highly-accurate measurements of the geodynamics satellites, such as LAGEOS, is still not sufficient to evince an irrefutable discrepancy with the predictions of standard GR. It is hoped that the newly-launched LARES [44] satellite may yield an improvement in the accuracy of the ongoing and forthcoming tests of fundamental physics, although also, such a possibility is currently debated [45–48]. Although the authors of these works have given some bounds on the values of the torsion parameters, based on the known error bars from the most recent ephemerides, it is still premature to draw any firm conclusion on the need of modified theories of gravity to explain the data.

The approaches of Mao *et al.* [41] and the subsequent spin-off applications by March *et al.* [42,43] have been heavily criticized by advocates of the PG theory. Hehl *et al.* have argued the following [49,50]: (i) Postulating that structureless test bodies follow autoparallel trajectories is incorrect in a general relativistic setup. In standard torsion theories test, bodies follow extremal trajectories, as is also the case in GR. The extremal trajectory is derived from the field equations themselves, and this is a theoretical feature of general relativity that should be preserved in any future theory. (ii) The net orbital angular momentum is not an integral over a local density, and consequently, it cannot be the source of torsion in a local field theory of gravity. Concerning the first objection, Kleinert and Pelster argued that in a spacetime with torsion, we must notice that parallelograms are, in general, not closed [51], the closure failure being proportional to the torsion tensor. This implies that the variational principle for finding the extrema of the action must take into account that the variation at the final point is nonzero as a consequence of the closure failure [52]. Using this modified variational principle, Kleinert and Pelster found that the equation of motion of structureless test bodies is given by autoparallel trajectories instead of extremal trajectories [51]. Bel has also shown that an analogy can be established among geodesics in Riemannian spacetime and autoparallels of a Weitzenböck connection [40]. In the absence of torsion extremal and autoparallel trajectories coincides as happens in standard GR. Hehl and Obukhov criticized this approach, because the autoparallel trajectories were not derived from the energy-momentum conservation laws, as is done in the GR and PG theories [50]. However, the closure failure also implies that the energy-momentum tensor of spinless point particles satisfies a different conservation law, as shown by Kleinert [53]. The second objection is, however, lethal to the MTGC model and its consequences. Any consistent theory of gravity must admit only local quantities or quantities obtained as the integration of local densities as sources of the tensor fields.

For these reasons, we investigate in this paper an alternative source for torsion around a rotating sphere. In our model, torsion is related to an axial vector field following the celestial parallels, $\mathbf{A}(r,\theta,\phi) = A(r,\theta,\phi)\hat{\boldsymbol{\phi}}$. This field structure could be obtained from the solution of a local Laplacian equation relating the vector potential with the energy-momentum flux of the rotating body in analogy to the corresponding equation for the magnetic field around a charged rotating sphere. All quantities in this model are local, and the non-locality induced by considering the total angular momentum as the source of torsion is removed. We study the secular evolution of the orbital elements for a test particle orbiting around a rotating central body. Some new effects unknown in GR are found: (i) a secular variation of the semi-major axis of the orbit; and (ii) a secular variation of the orbital eccentricity. As the increase of the astronomical unit is currently being discussed and no conventional explanation has still been found, our model could provide such an explanation, and moreover, we can give estimations on the torsion parameters from the preliminary data on these anomalies [54–56]. If these anomalies are confirmed, torsion fields generated by a circulating potential vector around rotating bodies could provide a parsimonious explanation of these phenomena, and they would stimulate further research in torsion gravity.

The structure of the paper is as follows: In Section 2, we provide a brief review on Riemann–Cartan spacetime as a quick reference for the rest of the paper. Our proposal for the torsion around spherical rotating bodies is discussed in Section 3 by following the symmetry arguments of Mao *et al.* [41]; autoparallel trajectories and orbital equations for perturbation theory are derived in Section 4. Results for the secular variation of the elements and comparison with Solar system anomalies are used in Section 5 to estimate the torsion parameters of our model. The discussion and conclusions are given in Section 6. Appendix A A is also included, in which the relation among the perturbing forces in the Sun's and the orbital system of reference is derived.

2. The Torsion and Contortion Tensors in Riemann–Cartan Spacetime

In this section, we remind about the main definitions and relations among tensors and the affine connection in Riemann–Cartan spacetime [57]. This spacetime is characterized by a non-zero curvature tensor and a torsion tensor defined as follows:

$$S_{jk}{}^{i} = \frac{1}{2}\left(\Gamma_{jk}^{i} - \Gamma_{kj}^{i}\right). \tag{1}$$

Therefore, a nonvanishing torsion implies that the affine connection is not symmetrical in the two lower indices in contrast with the postulates of ordinary Riemannian geometry. Christoffel's symbols are defined in terms of the metric tensor by the same expression found in Riemannian geometry. However, as we will find below, they do not coincide with the affine connection. Therefore, we have for Christoffel's symbols:

$$\left\{ \begin{array}{c} i \\ jk \end{array} \right\} = \frac{1}{2}g^{il}\left(g_{lk,j} + g_{jl,k} - g_{jk,l}\right), \tag{2}$$

where the commas denote, as usual, ordinary derivatives with respect to the coordinates. However, covariant and contravariant derivatives of a vector field must be defined in terms of the affine connection by the following relations:

$$A_{i|j} = A_{i,j} - A_k\Gamma_{ji}^{k}, \tag{3}$$

$$A^{i}{}_{|j} = A^{i}{}_{,j} + A^{k}\Gamma_{jk}^{i}. \tag{4}$$

The metric condition is given as usual:

$$g_{ij|k} = g_{ij,k} - g_{hj}\Gamma_{ki}^{h} - g_{ih}\Gamma_{kj}^{h} = 0, \tag{5}$$

so the non-metricity is null [41]. By adding up the equivalent equations resulting from Equation (5) by the cyclic permutation of the the three indices, we have:

$$\left\{ \begin{array}{c} i \\ jk \end{array} \right\} = \Gamma_{jk}^{i} + K_{jk}{}^{i}, \tag{6}$$

where $K_{jk}{}^{i}$ is the contortion tensor defined in terms of the torsion as follows:

$$\begin{aligned} K_{jk}{}^{i} &= -S_{jk}{}^{i} + g^{il}g_{hk}S_{jl}{}^{h} + g^{il}g_{jh}S_{kl}{}^{h} \\ K_{jk}{}^{i} &= -S_{jk}{}^{i} - S^{i}{}_{jk} + S_{k}{}^{i}{}_{j}, \end{aligned} \tag{7}$$

where we have used the metric tensor to raise and lower the indices. Similarly, in terms only of covariant indices, we have:

$$K_{ijk} = S_{jik} - S_{kij} + S_{jki} ,\tag{8}$$

from which we deduce the following antisymmetry property:

$$K_{ijk} = -K_{ikj} ,\tag{9}$$

which also implies $K_{ij}{}^j = 0$. In the case of the Riemann–Cartan spacetime, we have a generalization of a Ricci identity involving the torsion tensor as follows:

$$A^i{}_{|jk} - A^i{}_{|kj} = -A^h R_{kjh}{}^i - 2Q_{kj}{}^h A^i{}_{|h} ,\tag{10}$$

where both the curvature tensor and the torsion tensor appear.

3. Parametrization of Torsion in Spherically-Symmetric and Axisymmetric Spacetimes

The main idea of the MTGC model consist of a parametrization of torsion for both the static, spherical and parity symmetric case and the stationary, spherically-axisymmetric spacetime by using dimensional and symmetry arguments [41].

In the first case, we expect torsion to be invariant under the group of spatial rotations, $O(3)$, and, consequently, to involve only invariant quantities, such as the radio vector, x^i, $i = 1, 2, 3$, the Kronecker δ-function and the mass of the spherical object generating the field. The most general torsion tensor with these conditions becomes:

$$S_{0i}{}^0 = t_1 \frac{m}{2r^3} x^i ,\tag{11}$$

$$S_{jk}{}^i = t_2 \frac{m}{2r^3} \left(x^j \delta_{ki} - x^k \delta_{ji} \right) ,\tag{12}$$

where i, j and k are spatial indices and t_1, t_2 are functions of r alone to be treated as constants for an orbit of fixed radius. For perturbation calculations, it is highly convenient to transform this result to spherical coordinates by using the identities: $\partial x^i / \partial r = \hat{e}^i_r$, $\partial x^i / \partial \theta = r\hat{e}^i_\theta$ and $\partial x^i / \partial \phi = r \sin\theta \hat{e}^i_\phi$, which yields for the nonvanishing components:

$$S_{tr}{}^t = S_{tr}{}^t \frac{\partial x^i}{\partial r} = t_1 \frac{m}{2r^2} \tag{13}$$

$$S_{r\theta}{}^\theta = S_{jk}{}^i \frac{\partial x^j}{\partial r} \frac{\partial x^k}{\partial \theta} \frac{\partial \theta}{\partial x^i} = t_2 \frac{m}{2r^2} ,\tag{14}$$

$$S_{r\phi}{}^\phi = S_{r\theta}{}^\theta .\tag{15}$$

Now, we consider the stationary spherically-axisymmetric spacetime whose metric is given, to first order, as:

$$\begin{aligned} ds^2 =& -\left[1 - \frac{\rho_S}{r}\right] c^2 dt^2 + \left[1 + \gamma \frac{\rho_S}{r}\right] \\ &+ r^2\left(d\theta^2 + \sin^2\theta d\phi^2\right) \\ &- (1 + \gamma + \alpha_1/4)\frac{\rho_S \rho_J}{r} c dt d\phi , \end{aligned}\tag{16}$$

where $\rho_S = 2GM/c^2$ is the Schwarzschild radius and $\rho_J = J/(Mc)$ is a distance given in terms of the total angular momentum of the rotating object, its mass, M, and the speed of light, c. The constant parameters γ and α_1 are the parametrized post-Newtonian mechanics (PPN) parameters whose values in the case of standard general relativity are $\gamma = 1$, $\alpha_1 = 0$ as we will assume in this paper. The non-zero first-order contributions to Christoffel's symbols in Equation (2) are listed below:

$$\left\{ \begin{matrix} t \\ rt \end{matrix} \right\} = \left\{ \begin{matrix} r \\ tt \end{matrix} \right\} = \frac{\rho_S}{2r} ,\tag{17}$$

$$\begin{Bmatrix} t \\ r\phi \end{Bmatrix} = -\frac{3}{2}\sin^2\theta\,\frac{\rho_S\rho_J}{r^2}\,, \tag{18}$$

$$\begin{Bmatrix} r \\ \phi t \end{Bmatrix} = -\sin^2\theta\,\frac{\rho_S\rho_J}{2r^2}\,, \tag{19}$$

$$\begin{Bmatrix} r \\ rr \end{Bmatrix} = -\frac{\rho_S}{2r^2}\,, \tag{20}$$

$$\begin{Bmatrix} r \\ \theta\theta \end{Bmatrix} = \rho_S - r = \begin{Bmatrix} r \\ \phi\phi \end{Bmatrix}/\sin^2\theta\,, \tag{21}$$

$$\begin{Bmatrix} \theta \\ t\phi \end{Bmatrix} = \frac{\rho_S\rho_J}{r^3}\sin\theta\cos\theta\,, \tag{22}$$

$$\begin{Bmatrix} \theta \\ r\theta \end{Bmatrix} = \begin{Bmatrix} \phi \\ r\phi \end{Bmatrix} = \frac{1}{r}\,, \tag{23}$$

$$\begin{Bmatrix} \theta \\ \phi\phi \end{Bmatrix} = -\sin\theta\cos\theta\,, \tag{24}$$

$$\begin{Bmatrix} \phi \\ tr \end{Bmatrix} = \frac{\rho_S\rho_J}{2r^4}\,, \tag{25}$$

$$\begin{Bmatrix} \phi \\ t\theta \end{Bmatrix} = -\frac{\cos\theta}{\sin\theta}\frac{\rho_S\rho_J}{r^3}\,, \tag{26}$$

$$\begin{Bmatrix} \phi \\ \theta\phi \end{Bmatrix} = \frac{\cos\theta}{\sin\theta}\,, \tag{27}$$

and the corresponding symbols with permutated lower indices, which coincide with the listed ones by symmetry. First-order refers to the fact that we only consider terms proportional to ρ_S, ρ_J and $\rho_S\rho_J$ and ignore any higher-order power.

Finally, we will consider the torsion tensor for the stationary spherically-axisymmetric case. Torsion will be associated with an axial vector field A^k instead of the angular momentum, as proposed in the MTGC model [41]. This vector field reverses under time reversal and improper rotations. This requires that only those components with a single temporal index are nonvanishing. Moreover, to cancel the minus sign arising in improper rotations, we must include the Levi–Cività symbol, ϵ_{ijk}, because, being a pseudotensor, it also changes sign in improper rotations. With these conditions, it was found that:

$$\begin{aligned} S_{ij}{}^t &= \frac{f_1}{2r^3}\epsilon_{ijk}A^k + \frac{f_2}{2r^5}A^k x^l\left(\epsilon_{ikl}x^j - \epsilon_{jkl}x^i\right) \\ &+ \frac{f_6}{2r^5}A^k x_k \epsilon_{ijl}x^l\,, \\ S_{tij} &= \frac{f_3}{2r^3}\epsilon_{ijk}A^k + \frac{f_4}{2r^5}A^k x^l \epsilon_{ikl}x^j \\ &+ \frac{f_5}{2r^5}A^k x^l \epsilon_{jkl}x^i + \frac{f_7}{2r^5}A^k x_k \epsilon_{ijl}x^l\,, \end{aligned} \tag{28}$$

where $f_1, \dots, f_7$ are constants by dimensional analysis. Notice that Mao *et al.* [41] identified the vector A^k with the angular momentum J^k, but in the model presented in this paper, it could be a more general vector field. Save for constant prefactors, we chose a vector field with the same structure that the vector potential of a rotating charged sphere is as follows:

$$\mathbf{A} = \frac{4GM\Omega R^2}{5c^3}\sin\theta(-\sin\phi\,\hat{\mathbf{m}}_1 + \cos\phi\,\hat{\mathbf{m}}_2)\,, \tag{29}$$

where Ω is the angular velocity of the rotating central body, m is the mass and R is its radius. The prefactor is, then, proportional to the modulus of the total angular momentum. Notice that we can also write it as $4GM\Omega R^2/(5c^3) = \rho_S\rho_J$, where $\rho_S = 2GM/c^2$ is the Schwarzschild radius of the central body and $\rho_J = J/(Mc) = 2/5(\Omega R/c)R$. The unit vector $\hat{\mathbf{m}}_1$ points towards the ascending node of the Sun's axial rotation, and $\hat{\mathbf{m}}_2$ is perpendicular to it in the equatorial plane of the Sun or the rotating body that we are considering. The nonvanishing components of the torsion tensor corresponding to the static spherically-symmetric case are derived from Equation (13) as given in [41] as follows:

$$S_{tr}{}^t = t_1\frac{m}{2r^2} \; , \; S_{r\theta}{}^\theta = S_{r\phi}{}^\phi = t_2\frac{m}{2r^2} \; . \tag{30}$$

Additionally, those components, opposite in sign, correspond to the permutation of the covariant indices. Similarly, we find from Equations (28) and (29) that additional nonzero components of the torsion tensor for the stationary spherically-axisymmetric case are given by:

$$\begin{aligned}
S_{r\theta}{}^t = -S_{\theta r}{}^t &= \chi_1\frac{\rho_S\rho_J}{2r^2}\sin\theta \; , \\
S_{t\theta}{}^r = -S_{\theta t}{}^r &= \chi_2\frac{\rho_S\rho_J}{2r^2}\sin\theta \; , \\
S_{tr}{}^\theta = -S_{rt}{}^\theta &= \chi_3\frac{\rho_S\rho_J}{2r^4}\sin\theta \; ,
\end{aligned} \tag{31}$$

where χ_1, χ_2 and χ_3 are constant parameters. Finally, we must tabulate the values of the nonvanishing components of the contortion tensor (up to first-order in ρ_S, ρ_J and $\rho_S\rho_J$) by using the relation with the torsion in Equation (7) and Equations (30) and (31). The results are listed below:

$$K_{01}{}^0 = K_{00}{}^1 = -\frac{\rho_S t_1}{r^2} \; , \tag{32}$$

$$K_{22}{}^1 = K_{33}{}^1/\sin^2\theta = -\rho_S t_2 \; , \tag{33}$$

$$K_{21}{}^2 = K_{31}{}^3 = \frac{\rho_S t_2}{r^2} \; , \tag{34}$$

$$K_{21}{}^0 = (\chi_1 + \chi_2 + \chi_3)\frac{\rho_S\rho_J}{2r^2}\sin\theta \; , \tag{35}$$

$$K_{12}{}^0 = (\chi_1 - \chi_2 - \chi_3)\frac{\rho_S\rho_J}{2r^2}\sin\theta \; , \tag{36}$$

$$K_{20}{}^1 = (\chi_1 + \chi_2 + \chi_3)\frac{\rho_S\rho_J}{2r^2}\sin\theta \; , \tag{37}$$

$$K_{02}{}^1 = (\chi_1 - \chi_2 + \chi_3)\frac{\rho_S\rho_J}{2r^2}\sin\theta \; , \tag{38}$$

$$K_{10}{}^2 = (\chi_2 + \chi_3 - \chi_1)\frac{\rho_S\rho_J}{2r^4}\sin\theta \; , \tag{39}$$

$$K_{01}{}^2 = (\chi_2 - \chi_1 - \chi_3)\frac{\rho_S\rho_J}{2r^4}\sin\theta \; . \tag{40}$$

In the next section, we will study the autoparallel trajectories in the spacetime with this contortion tensor. Notice that the torsion and contortion tensor fields are determined by five constant parameters: t_1, t_2, χ_1, χ_2 and χ_3. The effect of the first two, t_1 and t_2, has already been analyzed by March *et al.* [42,43], and we will be concerned in this paper with the bounds or estimated values of χ_1, χ_2 and χ_3.

4. Autoparallel Trajectories and Perturbation Theory

As in previous models [41–43], we will assume that structureless point particles move along autoparallel trajectories of the Riemann–Cartan spacetime. Therefore, we have that:

$$\frac{d^2x^\alpha}{d\tau^2} + \left\{\begin{matrix}\alpha \\ \mu\nu\end{matrix}\right\}\frac{dx^\mu}{d\tau}\frac{dx^\nu}{d\tau} = K_{\mu\nu}{}^\alpha\frac{dx^\mu}{d\tau}\frac{dx^\nu}{d\tau} \; , \tag{41}$$

where τ is proper time measured along the trajectory. Notice that in a purely Riemannian spacetime, the contortion tensor is null, and Equation (41) is also found for geodesic trajectories. It is usually claimed that only test bodies with a microstructure can couple to torsion [58] and that spinless particles should follow the geodesic trajectories defined by:

$$\frac{d^2x^\alpha}{d\tau^2} + \left\{ \begin{matrix} \alpha \\ \mu\nu \end{matrix} \right\} \frac{dx^\mu}{d\tau}\frac{dx^\nu}{d\tau} = 0 \,. \tag{42}$$

In standard general relativity, one finds this equation of motion in two ways: as the extremal of the integral of the spacetime element, ds, or as a consequence of the field equations, *i.e.*, as the condition that the covariant divergence of the stress-energy tensor has zero value [59]. Both ways are equivalent and lead to the geodesic equation of motion in Equation (42). Mathematically, this coincides with the autoparallels in Equation (41) for zero torsion. If the calculations are translated to a Riemann-Cartan spacetime using holonomic constraints at the initial and final points (with zero variation in these points) of the trajectory, we find again the geodesic trajectories, because only the Christoffel symbols enter into the analysis [53].

However, geodesics are a global concept defined, as they are, as the shortest paths between two points. On the contrary, autoparallels can be defined locally as the straightest paths in spacetime. Therefore, from the fundamental principle of locality in classical field theory, it seems more natural to ascribe physical significance to autoparallels instead of geodesics.

Kleinert and collaborators [51–53] have studied a new nonholonomic mapping principle from flat spacetime to curved spacetime with torsion in which curvature is described as a disclination and torsion as a dislocation of the spacetime fabric. This implies that a closure failure appears in parallelograms, and the endpoints of a variational trajectory are displaced by:

$$\delta^S b^\mu = \delta^S q^\mu - \delta q^\mu \,, \tag{43}$$

where δ^S denotes the nonholonomic variations and δ are the auxiliary variations vanishing at the endpoints [53]. Starting from the action:

$$\mathcal{A} = -\frac{1}{2} \int_{\sigma_1}^{\sigma_2} d\sigma\, g_{\mu\nu}(q(\sigma))\dot{q}^\mu(\sigma)\dot{q}^\nu(\sigma) \,, \tag{44}$$

and applying the nonholonomic variations, we get:

$$\delta^S \mathcal{A} = -\int_{\sigma_1}^{\sigma_2} d\sigma \left(g_{\mu\nu}\dot{q}^\nu \delta^S \dot{q}^\mu + \frac{1}{2}\partial_\mu g_{\lambda\kappa}\delta^S q^\mu \dot{q}^\lambda \dot{q}^\kappa \right) \,, \tag{45}$$

where the dot denotes a derivative with respect the proper time, σ. In terms of the multivalued tetrads $e^i{}_\nu$, $i = 0,\dots,3$, $\mu = 0,\dots,3$, the metric tensor is defined as follows $g_{\mu\nu} = e^i{}_\mu e^i{}_\nu$, and the affine connection is given by $\Gamma_{\mu\nu}{}^\lambda = e_i{}^\lambda \partial_\mu e^i{}_\nu$. Taking also into account that $\partial_\mu g_{\nu\lambda} = \Gamma_{\mu\nu\lambda} + \Gamma_{\mu\lambda\nu}$ and integrating by parts, we obtain:

$$\begin{aligned} \delta^S \mathcal{A} &= -\int_{\sigma_1}^{\sigma_2} d\sigma \left[-g_{\mu\nu}\left(\ddot{q}^\nu + \left\{ \begin{matrix} \nu \\ \lambda\kappa \end{matrix} \right\}\dot{q}^\lambda \dot{q}^\kappa \right)\delta q^\mu \right. \\ &\quad + \left. \left(g_{\mu\nu}\dot{q}^\nu \frac{d}{d\sigma}\delta^S b^\mu + \Gamma_{\mu\lambda\kappa}\delta^S b^\mu \dot{q}^\lambda \dot{q}^\kappa \right) \right] \,. \end{aligned} \tag{46}$$

By using the differential equation for the nonholonomic variation $\delta^S b^\mu$:

$$\frac{d}{d\sigma}\delta^S b^\mu = -\Gamma_{\lambda\nu}{}^\mu \delta^S b^\lambda \dot{q}^\nu + 2S_{\lambda\nu}{}^\mu \dot{q}^\lambda \delta q^\nu \,, \tag{47}$$

Kleinert finds:

$$\delta^S \mathcal{A} = \int_{\sigma_1}^{\sigma_2} d\sigma g_{\mu\nu} \left(\ddot{q}^\nu + \Gamma_{\lambda\kappa}^\nu \dot{q}^\lambda \dot{q}^\kappa \right) \delta q^\mu = 0 \,. \tag{48}$$

As the auxiliary variations are arbitrary, but null at the endpoints, σ_1 and σ_2, we get from this the autoparallel equation instead of the geodesic. A similar derivation was discussed also by Kleinert and Pelster [51]. Further details are also given in several textbooks [60,61]. Autoparallels also satisfy a gauge invariance relating the autoparallel equations of motion in different Riemann–Cartan spacetimes [62]. This property could find a deeper physical meaning, apart from its mathematical interest, in future theories of gravity involving curvature and torsion.

The question of the relevance for the physics of the standard method or the method based on nonholonomic variations for the derivation of the equations of motion in Riemann–Cartan spacetime should be decided experimentally if, as we suggest, there is an additional torsion structure in spacetime. This question cannot be answered in the context of standard general relativity, because, in this case, autoparallel and geodesic equations coincide.

In Riemann–Cartan spacetime, the right-hand side of Equation (41) represents a perturbing force whose effects can be calculated by standard perturbation theory in celestial mechanics. We should calculate the first-order perturbation terms arising from the contortion tensor field in Equations (32)–(40). As the planets in the Solar system move with velocities much smaller than the speed of light, we can identify the proper time with the ephemeris time or the atomic time used by astronomers, *i.e.*, $\tau = t$. We will also consider that the torsion parameters $t_1 = t_2 = 0$, because it has been shown that t_1 can be absorbed in a redefinition of the mass of the source, and t_2 does not appear in the equations of the trajectories [42,43].

The components of the perturbing torsion terms are then given by:

$$\delta F^i = K_{\mu\nu}{}^i \frac{dx^\mu}{d\tau} \frac{dx^\nu}{d\tau} \,, \tag{49}$$

with $i = 1, 2, 3$ corresponding to the radial, polar and azimuthal coordinates, respectively. The leading term of the radial perturbing acceleration is found by direct substitution of Equations (32)–(40) into Equation (49):

$$\begin{aligned} \delta a_r &= \delta F^r = c \left(K_{02}{}^1 + K_{20}{}^1 \right) \dot{\theta} \\ &= c(\chi_1 + \chi_3) \frac{\rho_S \rho_J}{r^2} \sin\theta \, \dot{\theta} \,, \end{aligned} \tag{50}$$

ignoring corrections $\mathcal{O}\left(\dot{\theta}^2\right)$ and $\mathcal{O}\left(\dot{\phi}^2\right)$, c being the speed of light in a vacuum, $dx^0/dt = c$ and $\dot{\theta} = d\theta/dt$. Similarly, for the polar component of the perturbing force, we find:

$$\begin{aligned} \delta a_\theta &= r\delta F^\theta = r \left(K_{01}{}^2 + K_{10}{}^2 \right) \dot{r} \\ &= c(\chi_2 - \chi_1) \frac{\rho_S \rho_J}{r^3} \sin\theta \, \dot{r} \,. \end{aligned} \tag{51}$$

On the other hand, the azimuthal component of the perturbing torsion force is zero, as expected for an axisymmetric source, $\delta a_\phi = 0$. The perturbing accelerations in Equations (50) and (51) are corrections to the accelerations in spherical coordinates in the system of reference of the Sun (see Appendix A 6), which are obtained from the left-hand side of the equations for the autoparallel trajectories and the first-order Christoffel symbols in Equations (17)–(27) as follows:

$$a_r = \ddot{r} - r\dot{\theta}^2 - r\dot{\phi}^2 \sin^2\theta \,, \tag{52}$$

$$a_\theta = r\ddot{\theta} + 2\dot{r}\dot{\theta} - r\dot{\phi}^2 \sin\theta\cos\theta \,, \tag{53}$$

$$a_\phi = r\ddot{\phi}\sin\theta + 2\dot{r}\dot{\phi}\sin\theta + 2r\dot{\theta}\dot{\phi}\cos\theta \,. \tag{54}$$

As our objective is to apply celestial mechanics perturbation techniques, we will find it convenient to calculate the radial, $\mathcal{R}$, tangential to the orbit, $\mathcal{T}$, and normal to the orbital plane, $\mathcal{N}$, components of the perturbing torsion force per unit mass. In Equations (88)–(90) in the Appendix A 6, we have found those components in terms of the accelerations in spherical coordinates in a system of reference whose z axis is the rotation axis of the Sun. The relation is expressed in terms of a transformation matrix α_{ij}, $i, j = 1, 2, 3$ obtained as the set of scalar products among the vectors in the orbital system of reference and the Sun's system of reference, as given in Equation (79).

Direct substitution of Equations (50) and (51) and $\delta a_\phi = 0$ into Equations (88)–(90) yields:

$$\mathcal{R} = c(\chi_1 + \chi_3)\frac{\rho_s \rho_J}{r^2}(\alpha_{13} \sin v \tag{55}$$

$$- \alpha_{23} \cos v)\dot{v} ,$$

$$\mathcal{T} = c(\chi_2 - \chi_1)\frac{\rho_s \rho_J}{r^3}(\alpha_{13} \sin v \tag{56}$$

$$- \alpha_{23} \cos v)\dot{r} ,$$

$$\mathcal{N} = c(\chi_2 - \chi_1)\frac{\rho_s \rho_J}{r^3} \sin\theta$$

$$(\alpha_{31} \cos\theta \cos\phi + \alpha_{32} \cos\theta \sin\phi - \alpha_{33} \sin\theta)\dot{r} ,$$

$$= c\,\alpha_{33}(\chi_1 - \chi_2)\frac{\rho_s \rho_J}{r^3}\dot{r} . \tag{57}$$

Notice that all terms in Equations (55)–(57) are constants or can be written in terms of the true anomaly. The relations among the polar, θ, and azimuthal, ϕ, angles and the true anomaly, v, are given in Equations (80)–(82). Using these relations and the identity in Equation (83), we have found the simplifications for $\mathcal{T}$ and $\mathcal{N}$.

For the radio vector and radial velocity, we have [63–65]:

$$r = \frac{p}{1 + \epsilon \cos(v - \omega)} , \tag{58}$$
$$\dot{r} = \sqrt{\frac{\mu}{p}}\epsilon \sin(v - \omega) ,$$

where $p = a(1 - \epsilon^2)$ is the semilatus rectum, ϵ is the eccentricity, a is the semimajor axis, ω is the argument of the perihelion and $\mu = GM$ is the product of the gravitational constant and the mass of the Sun. The relation among time, t, and the true anomaly, v, will also be useful in the following perturbation calculations:

$$dt = \frac{T}{2\pi}\frac{(1 - \epsilon^2)^{3/2}}{(1 + \epsilon \cos(v - \omega))^2} dv , \tag{59}$$

where the orbital period, T, is given by Kepler's third law: $T = 2\pi a^{3/2}/\mu^{1/2}$. We are now ready for calculating the perturbations in the orbital elements as a consequence of the torsion force in Equations (55)–(57). Following the classical treatment of Burns [65], we can write for the semimajor axis:

$$\frac{\dot{a}}{a} = \frac{2a}{\mu}\dot{E} = \frac{2a}{\mu}(\dot{r}\mathcal{R} + r\dot{v}\mathcal{T}) , \tag{60}$$

where E is the total energy. After some simplifications using Equations (55), (56), (58) and (59), we have:

$$da = \frac{cT}{\pi}\frac{\rho_s \rho_J}{a^2}\frac{\epsilon}{(1 - \epsilon^2)^{5/2}}(\chi_2 + \chi_3)\sin(v - \omega)$$
$$(1 + \epsilon \cos(v - \omega))^2(\alpha_{13} \sin v - \alpha_{23} \cos v)dv . \tag{61}$$

For the instantaneous variation of the eccentricity, we start from:

$$\dot{\epsilon} = \frac{\epsilon^2 - 1}{2\epsilon}\left[-\frac{\dot{a}}{a} + \frac{2rT}{H}\right] , \tag{62}$$

where $H = \sqrt{\mu p} = \sqrt{GMa(1-\epsilon^2)}$ is the angular momentum per unit mass. By using Equations (61), (56) and (59) conjointly with Equation (62), we arrive at:

$$\begin{aligned} d\epsilon &= \frac{cT}{2\pi a}\frac{\rho_S\rho_J}{a^2}\frac{\sin(\nu-\omega)}{(1-\epsilon^2)^{1/2}}(\alpha_{13}\sin\nu - \alpha_{23}\cos\nu) \\ &\left[\frac{(\chi_2+\chi_3)}{1-\epsilon^2}(1+\epsilon\cos(\nu-\omega))^2 + \chi_1 - \chi_2\right]d\nu . \end{aligned} \tag{63}$$

Torsion also induces an extra precession of the longitude of the ascending node of the planetary orbits in addition to the Lense–Thirring effect arising in standard general relativity. This precession rate is proportional to the normal component to the planetary orbits of the perturbing force, as found in perturbation theory [63]:

$$\frac{d\Omega}{dt} = \frac{r\mathcal{N}\sin\nu}{H\sin I} = \sqrt{\frac{p}{GM}}\frac{\mathcal{N}}{\sin I}\frac{\sin\nu}{1+\epsilon\cos(\nu-\omega)} , \tag{64}$$

where I is the orbital inclination and $p = a(1-\epsilon^2)$ is the semi-latus rectum. From the expression of the normal component of the perturbing force in Equation (57) and the contribution to the precession of the longitude of the ascending node in Equation (64), we obtain:

$$d\Omega = \alpha_{33}(\chi_1 - \chi_2)\frac{cT}{2\pi a}\frac{\rho_S\rho_J}{a^2}\frac{\epsilon}{(1-\epsilon^2)^{3/2}}\frac{\sin\nu}{\sin I}\sin(\nu-\omega)d\nu . \tag{65}$$

Finally, for the extra precession of the perihelion contributed by spacetime torsion, we find:

$$\begin{aligned} d\omega &= (\chi_2 - \chi_1)\frac{cT}{2\pi a}\frac{\rho_S\rho_J}{a^2}\frac{1}{(1-\epsilon^2)^{3/2}} \\ &\left(\sin^2(\nu-\omega)\frac{2+\epsilon\cos(\nu-\omega)}{(1+\epsilon\cos(\nu-\omega))^3}(\alpha_{13}\sin\nu - \alpha_{23}\cos\nu)\right. \\ &\left. +\epsilon\alpha_{33}\cot I\sin\nu\sin(\nu-\omega)\right)d\nu \\ &-c(\chi_1+\chi_3)\sqrt{\frac{a}{GM}}\frac{\rho_S\rho_J}{a^2}\frac{1}{\epsilon(1-\epsilon^2)^{3/2}} \\ &\cos(\nu-\omega)(\alpha_{13}\sin\nu - \alpha_{23}\cos\nu)d\nu . \end{aligned} \tag{66}$$

In the next section, we will discuss the predictions of Equations (61)–(66) for the variation of the orbital elements of the planet and its possible connection with certain anomalies recently found by astronomers.

5. Results

As we cannot give to the parameters χ_1, χ_2 and χ_3 definite values on a theoretical basis, it is not possible, in principle, to make predictions on the variation of the orbital elements as a consequence of torsion in our model. It is reasonable to assume that these parameters are of the order of unity, if there is a theory consistent with the torsion model, but this is not sufficient to suggest any reliable prediction. However, we can use an inductive approach by assuming that some anomalies recently found for the planetary orbits are the consequence of torsion gravity arising in a theory that includes the phenomenological model discussed in this paper as a particular case. Specifically, we refer to the anomalous secular increase of the astronomical unit (AU) first reported by Krasinsky and Brumberg in 2004 [66]. The analysis of databases of radar and laser ranging and spacecraft observations in the last few decades showed that the astronomical unit increases by 15 ± 4 meters per century. An independent study by Standish reduced this figure to 7 ± 2 meters per century [67]. This problem has even motivated the International Astronomical Union to redefine the AU as a constant and, as a consequence, to remove the Gaussian constant of mass from the list of astronomical constants [56]. However, these redefinitions of constants do not solve the problem pointed out by the aforementioned astronomers. However, we should notice that, according to this new definition, it is not rigorous to compare the rates of the semi-major axes of the planets with the rates of the astronomical unit, because this is fixed. On the other hand, we can use the recent reports on a secular decrease of the mass parameter $\mu = GM$ of the Sun [56]. The average of the rates determined with EPM2008, EPM2010 and EPM2011 ephemerides is $\mu = (5.73 \pm 4.27) \times 10^{-14}$ yr^{-1}. It is known from the perturbation analysis of the planetary orbits in a scenario of a diminishing gravitational constant that the semi-major axis increase as [68]:

$$\frac{\dot{a}}{a} = -\frac{\dot{\mu}}{\mu} , \tag{67}$$

which implies a rate of 0.86 ± 0.64 meters per century. It is important to point out that the recent analyses with the INPOP13cephemerides are statistically compatible with a zero variation of the Sun's mass parameter [56]. Some conventional and unconventional attempts for an explanation of the variation rates for the semi-major axes of the planets have been suggested, but there is still no convincing solution of the problem [69–73].

It is interesting to notice that Equation (61) implies a variation of the semimajor orbital axes for $\eta = \chi_2 + \chi_3 \neq 0$. Moreover, if we assume that:

$$\chi_1 + \chi_3 = 0 , \tag{68}$$

it is found from Equations (61)–(66) that the variations of a, ϵ, Ω and ω depend only on the single parameter $\eta = \chi_2 - \chi_1$. We will choose the condition in Equation (68) without losing the perspective that other possibilities are compatible with our model, even the case in which no secular change is found for the semimajor planetary axes, i.e., the case $\eta = 0$.

Using this condition and averaging Equations (61), (63), (65) and (66) over a whole orbit, we obtain the following results for the variation of the elements in one year:

$$\frac{\Delta a}{a} = \eta \frac{Ly}{a} \frac{\Omega R}{c} \left(\frac{R}{a}\right)^2 \frac{\epsilon}{(1-\epsilon^2)^{5/2}} \tag{69}$$

$$\left(1 + \frac{\epsilon^2}{4}\right)(\alpha_{13} \cos \omega + \alpha_{23} \sin \omega) ,$$

$$\Delta \epsilon = \eta \frac{5Ly}{8a} \frac{\rho_s \rho_J}{a^2} \frac{\epsilon^2}{(1-\epsilon^2)^{3/2}} \tag{70}$$

$$(\alpha_{13} \cos \omega + \alpha_{23} \sin \omega) ,$$

$$\Delta \Omega = -\eta \, \alpha_{33} \frac{Ly}{2a} \frac{\rho_s \rho_J}{a^2} \frac{\epsilon}{(1-\epsilon^2)^{3/2}} \frac{\cos \omega}{\sin I} , \tag{71}$$

$$\Delta \omega = \eta \frac{Ly}{2a} \frac{\rho_s \rho_J}{a^2} \frac{\epsilon}{(1-\epsilon^2)^{3/2}} \tag{72}$$

$$\left[\alpha_{33} \cot I \cos \omega - \frac{5}{4}(\alpha_{13} \sin \omega - \alpha_{23} \cos \omega)\right] ,$$

where Ly stands for a light year. By assuming that the increase of the astronomical unit is obtained as an average over the inner planets, Mercury, Venus, the Earth and Mars, and taking the value reported by Standish, $\Delta AU = 7 \pm 2$ meters per century, we find that $\eta = -0.154$. If we take the values deduced from the variation of the mass parameter GM of the Sun as reported in the EPM2008–2011 ephemerides, a smaller value $\eta = -0.019$ is found. In these calculation, the elements of the planets as given in [74,75] were used. We can also make some predictions for the variation of the other orbital elements and check if they are consistent with the presented observations. We should see that, at least, our model is not inconsistent in this sense.

For the secular variation of the eccentricity for this value of η, we find $\Delta\epsilon = 4.57 \times 10^{-13}$ in one year in the case of Mercury and even lower values for the other planets. This is below the precision threshold for present determinations of this magnitude [69].

Possible anomalous contributions to the secular node precessions are currently considered in recent ephemerides. For the moment, no statistically-significant results have been obtained with the attained precisions, but these values set an upper limit to any prediction by theoretical models. For INPOP10a, these corrections are listed in milliarcseconds per century and compared to our predictions from Equation (71) in Table 1. The most recent ephemerides, INPOP10a and EPM2011, in connection with possible Solar system anomalies have been discussed by Iorio [56].

Table 1. Corrections to the secular node precessions as obtained in the INPOP10a ephemeris and predictions of the torsion gravity model discussed in this paper in milliarcseconds per century.

Planet	$\Delta\dot{\Omega}$ (INPOP10a)	$\Delta\dot{\Omega}$ (Torsion Gravity)
Mercury	1.4 ± 1.8	0.30
Venus	0.2 ± 1.5	2.14×10^{-2}
Earth	0.0 ± 0.9	-1.70×10^{-4}
Mars	-0.05 ± 0.13	2.62×10^{-3}
Jupiter	-40 ± 42	1.51×10^{-5}
Saturn	-0.1 ± 0.4	1.31×10^{-5}

The corrections to the standard secular perihelion precessions are given in Table 2 and compared to the predictions of our torsion gravity model.

Inspection of Tables 1 and 2 shows that the predictions of the model are compatible with both ephemerides at the 2σ level. Therefore, in the context of the torsion gravity model, the anomalous increase of the astronomical unit is consistent with the rest of measurements on possible corrections to other orbital elements. On the other hand, we have shown that only for Mercury, it seems that an improved ephemeris could detect a statistically-significant nonzero correction for $\Delta\dot{\Omega}$ and $\Delta\dot{\omega}$ in the foreseeable future.

Table 2. Corrections to the secular perihelion precessions as obtained in the INPOP10a and EPM2011ephemerides and predictions of the torsion gravity model discussed in this paper in milliarcseconds per century.

Planet	$\Delta\dot{\omega}$ (INPOP10a)	$\Delta\dot{\omega}$ (EPM2011)	$\Delta\dot{\omega}$ (Torsion Gravity)
Mercury	0.4 ± 0.6	-2.0 ± 3.0	-0.622
Venus	0.2 ± 1.5	2.6 ± 1.6	-4.28×10^{-3}
Earth	-0.2 ± 0.9	0.19 ± 0.19	-1.03×10^{-4}
Mars	-0.04 ± 0.15	-0.02 ± 0.037	-4.9×10^{-3}
Jupiter	-41 ± 42	58.7 ± 28.3	-3.48×10^{-5}
Saturn	0.15 ± 0.65	-0.32 ± 0.47	-2.69×10^{-5}

6. Conclusions

The advancement of physics can only proceed by a continuous interplay between theory and experiment. This healthy interaction allows for a selection of the most promising hypotheses among the different proposals. Gravity theory has been an exception to this methodological rule for the most part of the 20th century, because experiments are very difficult to develop, as they usually imply very accurate devices, and these must be set into orbit to perform the measurements [7,8]. Gravity being the weakest of all interactions is also the one we know least, because accurate tests of all general relativity predictions are still lacking [2]. For example, Lense–Thirring precession of orbital nodes is only known with a wide error bar from the laser range monitoring of the geodynamic satellites [8–10,76–78].

A further, non-negligible source of difficulty in gravitational experiments resides in the extremely long times required either to collect data or to analyze them: the Gravity Probe B is a case-study example [7].

The experimental situation has been complicated in recent years because of the discovery of a set of anomalies that, apparently, cannot be explained conventionally (see, e.g., the recent review by Iorio [56]). Similarly, the most recent ephemerides has allowed the determination of upper bounds on the variation of the orbital elements of the planets beyond the predictions of classical perturbation theory and general relativity [56,75]. Although some of these anomalies may lose their statistical significance in the more or less near future in view of further observations and related analyses, it is nonetheless important to discuss the possibility that they may constrain theories and extensions beyond general relativity.

Many extensions of general relativity, some of them proposed by Einstein himself, have made use of the concept of torsion [24]. These ideas coalesced in the 1960s and 1970s in the so-called Einstein–Cartan–Sciama–Kibble (ECSK) theory in which torsion is connected with the microscopic spin density and does not propagate outside massive bodies [30]. This theory is still considered a viable alternative to general relativity, but it suffers from a total lack of experimental support, despite some claims that it could explain inflation [32,33]. This situation leaves room for the study of more alternatives without restricting to a given mathematical formalism. In such a spirit, Mao *et al.* proposed in 2007 the MTGC parametrized model in which torsion is connected to macroscopic angular momentum [41]. It was shown that this model can be constrained by perturbations in the orbital elements of the planets and geodynamic satellites [42,43].

Hehl *et al.* [49,50] have pointed out that it is inconsistent to use total angular momentum, *i.e.*, a quantity not obtained by integration over local densities, as the source of a local field quantity, such as torsion. To avoid this inconsistency, we have modified the approach of the original MTGC model by connecting torsion with a local circulating vector potential as the one obtained in classical electromagnetism for a rotating charged sphere. This way, we have shown that a new phenomenon, qualitatively distinct from those obtained in general relativity, is predicted. Namely, a secular increase of the semi-major axes of the planets [66,67]. This problem has attracted the attention of several authors in recent years, who have tried to find explanations in terms of nonstandard and conventional hypotheses [56,69,70,72]. This observation remains unexplained and, although it could be dismissed by more precise analyses in the future, it deserves further attention. We have shown that our torsion model is compatible with these observations for a value of the parameters of the order of unity.

Moreover, planetary ephemerides are becoming increasingly precise year after year. It is still premature to state that statistically-significant anomalies have been revealed in the secular precessions of the longitude of the ascending node and the argument of the perihelion, but the uncertainty intervals are promisingly small [56]. Anyway, it is now clear that any deviation from the predictions of general relativity (once we take into account standard perihelion precessions and the gravitomagnetic Lense–Thirring effect) is expected only in the range of a few milliarcseconds per century.

We have also shown that extra secular precessions of these elements are the consequence of torsion, but for any planet, they are very small in relation to the confidence intervals of the INPOP10a and EPM2011 ephemerides. However, for the case of Mercury, they could be detected in the foreseeable future, because they lie in the range of a few tenths of milliarcseconds per century. The important fact is that this agreement is achieved in consistency with an increase of the astronomical unit of a few meters per century. This means that testing a torsion gravity extension of general relativity as the one discussed in this paper is within the reach of modern observation techniques and data analyses in astronomy.

We conclude that further experiments and observations are required, achieving the maximum precision possible with present-day technology, to confirm or dismiss possible anomalies beyond general relativity in the secular evolution of the elements of the planets and spacecraft. From these future observations, the model proposed in this paper could receive further support. In such a case, it could serve as the basis for a consistent theory of torsion gravity obtained by the scientific method of induction from experience.

Acknowledgments: The author gratefully acknowledges Lluís Bel for many useful comments and discussions.

Appendix A. Relation among the Sun's System of Reference and the Orbital Coordinates

Firstly, we define the ecliptic system of reference: $\hat{k}$ is a unit vector perpendicular to the ecliptic plane; $\hat{i}$ points towards the point of Aries; and $\hat{j}$ is perpendicular to the preceding unit vectors in such a way that we have a right-handed Cartesian coordinate system [75].

The orbital plane of a given planet is then characterized by two angles: the inclination ι with respect to the ecliptic plane and the angle Ω among the line of nodes (*i.e.*, the intersection among the two planes) and the point of Aries. Consequently, we can write the unit vector of the orbital system of reference as follows:

$$\hat{n}_1 = \cos\Omega\,\hat{i} + \sin\Omega\,\hat{j}, \tag{73}$$

$$\hat{n}_2 = -\cos\iota\sin\Omega\,\hat{i} + \cos\iota\cos\Omega\,\hat{j} + \sin\iota\,\hat{k}, \tag{74}$$

$$\hat{n}_3 = \sin\iota\sin\Omega\,\hat{i} - \sin\iota\cos\Omega\,\hat{j} + \cos\iota\,\hat{k}. \tag{75}$$

Similarly, we can define the inclination of the Sun's axis and the longitude of the ascending node of its equator. These two angles determine the orientation of the Sun's rotation axis on space and were obtained in the 19th century by careful observations of Carrington. Carrington's elements, as they are called, are given by [79,80]:

$$\iota_c = 7.25°, \tag{76}$$

$$\Omega_c = 73.67° + 0.013958°\,(t - 1850), \tag{77}$$

where t is the year of observation. We define the Sun's system of reference as the Cartesian system obtained by the three unit vectors $\hat{m}_i$, $i = 1, 2, 3$ whose expression in terms of the Carrington elements is also given by Equation (73).

The planet's orbital radius vector is usually written as [63,64]:

$$\mathbf{r} = p\,\frac{\cos\nu\,\hat{n}_1 + \sin\nu\,\hat{n}_2}{1 + \epsilon\cos(\nu - \omega)}, \tag{78}$$

where $p = a(1 - \epsilon^2)$ is the semilatus rectum, a the semi-major axis, ϵ the orbital eccentricity, ν is the true anomaly and ω is the argument of the perihelion. Notice that we use an unconventional definition of the true anomaly as the angle among the radius vector and the ascending node of the planet instead of measuring it from the perihelion as usual.

We now introduce the transformation matrix α_{ij} as the scalar products of the unit vectors of the orbital and the Sun's system of reference:

$$\alpha_{ij} = \hat{n}_i\cdot\hat{m}_j\,,i\,,j = 1,2,3. \tag{79}$$

If θ, ϕ are, respectively, the polar angle and azimuthal angle in the Sun's system of reference, we find from Equation (78) and using the definition in Equation (79) the relation among them and the true anomaly, ν, in the following form:

$$\cos\theta = \tfrac{\mathbf{r}}{r}\cdot\hat{m}_3 = \alpha_{13}\cos\nu + \alpha_{23}\sin\nu, \tag{80}$$

$$\sin\theta\cos\phi = \tfrac{\mathbf{r}}{r}\cdot\hat{m}_1 = \alpha_{11}\cos\nu + \alpha_{21}\sin\nu, \tag{81}$$

$$\sin\theta\cos\phi = \tfrac{\mathbf{r}}{r}\cdot\hat{m}_2 \tag{82}$$

$$= \alpha_{12} \cos \nu + \alpha_{22} \sin \nu \ .$$

The coefficients of the transformation matrix satisfy some useful identities:

$$\sum_{k=1}^{3} \alpha_{ik} \alpha_{jk} = \delta_{ij} \ , \ i, j = 1, 2, 3 \ , \tag{83}$$

where δ_{ij} is Kronecker's delta. The spherical unit vectors in the Sun's system of reference are given as:

$$\begin{aligned} \hat{\mathbf{r}} &= \sin \theta \cos \phi \, \hat{\mathbf{m}}_1 \\ &+ \sin \theta \cos \phi \, \hat{\mathbf{m}}_2 + \cos \theta \, \hat{\mathbf{m}}_3 \ , \end{aligned} \tag{84}$$

$$\begin{aligned} \hat{\boldsymbol{\theta}} &= \cos \theta \cos \phi \, \hat{\mathbf{m}}_1 \\ &+ \sin \theta \sin \phi \, \hat{\mathbf{m}}_2 - \sin \theta \, \hat{\mathbf{m}}_3 \ , \end{aligned} \tag{85}$$

$$\hat{\boldsymbol{\phi}} = - \sin \phi \, \hat{\mathbf{m}}_1 + \cos \phi \, \hat{\mathbf{m}}_2 \ . \tag{86}$$

From Equations (80) and (84), we can also find the tangential unit vector to the planetary orbit:

$$\begin{aligned} \hat{\nu} = \tfrac{d\hat{\mathbf{r}}}{d\nu} &= (-\alpha_{11} \sin \nu + \alpha_{21} \cos \nu) \, \hat{\mathbf{m}}_1 \\ &+ (-\alpha_{12} \sin \nu + \alpha_{22} \cos \nu) \, \hat{\mathbf{m}}_2 \\ &+ (-\alpha_{13} \sin \nu + \alpha_{23} \cos \nu) \, \hat{\mathbf{m}}_3 \ . \end{aligned} \tag{87}$$

Orthogonality with $\hat{\mathbf{r}}$ and normalization to the unity modulus can be shown by applying the identities in Equation (83).

Now, we can find the radial, $\mathcal{R}$, tangential to the orbit, $\mathcal{T}$, and normal to the orbital plane, $\mathcal{N}$, components of the perturbing force per unit mass in terms of the accelerations, a_r, a_θ and a_ϕ, in the Sun's spherical system of reference. It is obvious that:

$$\mathcal{R} = a_r \ . \tag{88}$$

For the tangential component, we have:

$$\begin{aligned} \mathcal{T} &= \left(a_\theta \, \hat{\boldsymbol{\theta}} + a_\phi \, \hat{\boldsymbol{\phi}} \right) \cdot \hat{\nu} \\ &= \tfrac{a_\theta}{\sin \theta} (\alpha_{13} \sin \nu - \alpha_{23} \cos \nu) \\ &+ \tfrac{a_\phi}{\sin \theta} (\alpha_{11} \alpha_{22} - \alpha_{12} \alpha_{21}) \ , \end{aligned} \tag{89}$$

after some simplifications using Equations (80)–(87). Finally, for the normal component, we find:

$$\begin{aligned} \mathcal{N} &= \left(a_\theta \, \hat{\boldsymbol{\theta}} + a_\phi \, \hat{\boldsymbol{\phi}} \right) \cdot \hat{\mathbf{n}}_3 \\ &= a_\theta (\alpha_{31} \cos \theta \cos \phi + \alpha_{32} \cos \theta \sin \phi - \alpha_{33} \sin \theta) \\ &+ a_\phi (-\alpha_{31} \sin \phi + \alpha_{32} \cos \phi) \ . \end{aligned} \tag{90}$$

Notice that the angles θ and ϕ can be formally expressed in terms of the true anomaly by using Equation (80).

References

1. Iorio, L. Editorial for the Special Issue 100 Years of Chronogeometrodynamics: The Status of the Einstein's Theory of Gravitation in Its Centenial Year. *Universe* **2015**, *1*, 38–81. [CrossRef]
2. Will, C.M. The Confrontation between General Relativity and Experiment. *Living Rev. Relativ.* **2006**, *9*, 3. [CrossRef]
3. Lämmerzahl, C.; Ciufolini, I.; Dittus, H.; Iorio, L.; Müller, H.; Peters, A.; Samain, E.; Scheithauer, S.; Schiller, S. OPTIS–An Einstein Mission for Improved Tests of Special and General Relativity. *Gen. Relativ. Gravit.* **2004**, *36*, 2373–2416. [CrossRef]

4.   Iorio, L.; Ciufolini, I.; Pavlis, E.C.; Schiller, S.; Dittus, H.; Lämmerzahl, C. On the possibility of measuring the Lense-Thirring effect with a LAGEOS LAGEOS II OPTIS mission. *Classical. Quant. Grav.* **2004**, *21*, 2139–2151. [CrossRef]

5.   Schiller, S.; Tino, G.M.; Gill, P.; Salomon, C.; Sterr, U.; Peik, E.; Nevsky, A.; Görlitz, A.; Svehla, D.; Ferrari, G. Einstein Gravity Explorer-a medium-class fundamental physics mission. *Exp. Astron.* **2009**, *23*, 573–610. [CrossRef]

6.   Turyshev, S.G.; Turyshev, S.G.; Sazhin, M.V.; Toth, V.T. General relativistic laser interferometric observables of the GRACE-Follow-On mission. *Phys. Rev. D* **2014**, *89*, 105029. [CrossRef]

7.   Everitt, C.W.F.; de Bra, D.B.; Parkinson, B.W.; Turneaure, J.P.; Conklin, J.W.; Heifetz, M.I.; Keiser, G.M.; Silbergleit, A.S.; Holmes, T.; Kolodziejczak, J.; *et al.* Gravity Probe B: Final Results of a Space Experiment to Test General Relativity. *Phys. Rev. Lett.* **2011**, *106*, 221101. [CrossRef] [PubMed]

8.   Ciufolini, I.; Pavlis, E.C. A confirmation of the general relativistic prediction of the Lense-Thirring effect. *Nature* **2004**, *431*, 958–960. [CrossRef] [PubMed]

9.   Ciufolini, I.; Paolozzi, A.; Pavlis, E.C.; Ries, J.C.; Koenig, R.; Matzner, R.A.; Sindoni, G.; Neumayer, H. Towards a One Percent Measurement of Frame Dragging by Spin with Satellite Laser Ranging to LAGEOS, LAGEOS 2 and LARES and GRACE Gravity Models. *Space Sci. Rev.* **2009**, *148*, 71–104. [CrossRef]

10.   Iorio, L.; Lichtenegger, H.I.M.; Ruggiero, M.L.; Corda, C. Phenomenology of the Lense-Thirring effect in the solar system. *Astrophys. Space Sci.* **2011**, *331*, 351–395. [CrossRef]

11.   Ciufolini, I. Frame Dragging and Lense-Thirring Effect. *Gen. Relativ. Gravit.* **2004**, *36*, 2257–2270. [CrossRef]

12.   Renzetti, G. History of the attempts to measure orbital frame-dragging with artificial satellites. *Cent. Eur. J. Phys.* **2013**, *11*, 531–544. [CrossRef]

13.   Soffel, M.; Wirrer, R.; Schastok, J.; Ruder, H.; Schneider, M. Relativistic effects in the motion of artificial satellites. I—The oblateness of the central body. *Celest. Mech.* **1988**, *42*, 81–89. [CrossRef]

14.   Iorio, L. A possible new test of general relativity with Juno. *Classical Quant. Grav.* **2013**, *30*, 195011. [CrossRef]

15.   Iorio, L. Are we far from testing general relativity with the transitting extrasolar planet HD 209458b "Osiris"? *New Astron.* **2006**, *11*, 490–494. [CrossRef]

16.   Iorio, L. Classical and relativistic long-term time variations of some observables for transiting exoplanets. *Mon. Not. R. Astron. Soc.* **2011**, *411*, 167–183. [CrossRef]

17.   Adams, F.C.; Laughlin, G. Relativistic Effects in Extrasolar Planetary Systems. *Int. J. Mod. Phys. D* **2006**, *15*, 2133–2140. [CrossRef]

18.   Pal, A.; Kocsis, B. Periastron precession measurements in transiting extrasolar planetary systems at the level of general relativity. *Mon. Not. R. Astron. Soc.* **2008**, *389*, 191–198. [CrossRef]

19.   Jordan, A.; Bakos, G. Observability of the General Relativistic Precession of Periastra in Exoplanets. *Astrophys. J.* **2008**, *685*, 543–552. [CrossRef]

20.   Zhao, S.; Xie, Y. Parametrized post-Newtonian secular transit timing variations for exoplanets. *Res. Astron. Astrophys.* **2013**, *13*, 1231–1239. [CrossRef]

21.   Xie, Y.; Deng, X.-M. On the (im)possibility of testing new physics in exoplanets using transit timing variations: deviation from inverse-square law of gravity. *Mon. Not. R. Astron. Soc.* **2014**, *438*, 1832–1838. [CrossRef]

22.   Iorio, L. Accurate characterization of the stellar and orbital parameters of the exoplanetary system WASP-33b from orbital dynamics. *Mon. Not. R. Astron. Soc.* **2016**, in press.

23.   Iorio, L. Classical and relativistic node precessional effects in WASP-33b and perspectives for detecting them. *Astrophys. Space Sci.* **2011**, *331*, 485–496. [CrossRef]

24.   Goenner, H.F.M. On the History of Unified Field Theories. *Living Rev. Relativ.* **2004**, *7*, 1830–1923. [CrossRef]

25.   Sciama, D.W. The Physical Structure of General Relativity. *Rev. Mod. Phys.* **1964**, *36*, 463–469. [CrossRef]

26.   Kibble, T.W.B. Lorentz invariance and the gravitational field. *J. Math. Phys.* **1961**, *2*, 212–221. [CrossRef]

27.   Mielke, E.W. Is Einstein-Cartan Theory Coupled to Light Fermions Asymptotically Safe? *J. Grav.* **2013**, *2013*, 5. [CrossRef]

28.   Hehl, F.W. Gauge theory of gravity and spacetime. In Proceedings of the Workshop Towards a Theory of Spacetime Theories, Wuppertal, Germany, 21–23 July 2010.

29.   Aldrovandi, R.; Pereira, J.G. *Teleparallel Gravity*; Springer: Berlin, Germany, 2013.

30.   Hehl, F.W.; von der Heyde, P.; Kerlick, G.D.; Nester, J.M. General Relativity with Spin and Torsion: Foundations and Prospects. *Rev. Mod. Phys.* **1976**, *48*, 393–415. [CrossRef]

31.   Ni, W.T. Searches for the role of spin and polarization in gravity. *Rep. Prog. Phys.* **2010**, *73*, 056901. [CrossRef]

32. Popławski, N.J. Cosmology with torsion: An alternative to cosmic inflation. *Phys. Lett. B* **2010**, *694*, 181–185. [CrossRef]

33. Popławski, N.J. Non-singular, big-bounce cosmology from spinor-torsion coupling. *Phys. Rev. D* **2012**, *85*, 107502. [CrossRef]

34. Obukhov, Y.N. *Poincaré Gauge Gravity: Selected Topics*; Cornell University: Ithaca, NY, USA, 2006.

35. Hammond, R.T. Torsion gravity. *Rep. Prog. Phys.* **2002**, *65*, 599–649. [CrossRef]

36. Hojman, S.; Rosenbaum, M.; Ryan, M.P.; Shepley, L. Gauge invariance, minimal coupling and torsion. *Phys. Rev. D* **1978**, *17*, 3141–3146. [CrossRef]

37. Hojman, S.; Rosenbaum, M.; Ryan, M.P. Propagating torsion and gravitation. *Phys. Rev. D* **1979**, *19*, 430–437. [CrossRef]

38. Hayashi, K.; Shirafuji, T. New General Relativity. *Phys. Rev. D* **1979**, *19*, 3524–3553. [CrossRef]

39. Weitzenböck, R. *Invariantentheorie*; Popko Noordhoff: Groningen, The Netherlands, 1923.

40. Bel, L. *Connecting Connections. A Bricklayer View of General Relativity*; Cornell University: Ithaca, NY, USA, 2008.

41. Mao, Y.; Tegmark, M.; Guth, A.H.; Cabi, S. Constraining torsion with Gravity Probe B. *Phys. Rev. D* **2007**, *76*, 104029. [CrossRef]

42. March, R.; Belletini, G.; Tauraso, R.; Dell'Agnello, S. Constraining spacetime torsion with the Moon and Mercury. *Phys. Rev. D* **2011**, *83*, 104008. [CrossRef]

43. March, R.; Belletini, G.; Tauraso, R.; Dell'Agnello, S. Constraining spacetime torsion with LAGEOS. *Gen. Relativ. Gravit.* **2011**, *43*, 3099–3126. [CrossRef]

44. Paolozzi, A.; Ciufolini, I. LARES successfully launched in orbit: Satellite and mission description. *Acta Astronaut.* **2013**, *91*, 313–321. [CrossRef]

45. Iorio, L. The impact of the new Earth gravity models on the measurement of the Lense-Thirring effect with a new satellite. *New Astron.* **2005**, *10*, 616–635. [CrossRef]

46. Renzetti, G. On Monte Carlo simulations of the LAser RElativity Satellite experiment. *Acta Astronaut.* **2015**, *113*, 164–168. [CrossRef]

47. Ciufolini, I.; Moreno Monge, B.; Paolozzi, A.; Koenig, R.; Sindoni, G.; Michalak, G.; Pavlis, E.C. Monte Carlo simulations of the LARES space experiment to test General Relativity and fundamental physics. *Class. Quantum Grav.* **2013**, *30*, 235009. [CrossRef]

48. Iorio, L. The impact of the orbital decay of the LAGEOS satellites on the frame-dragging tests. *Adv. Space Res.* **2016**, in press.

49. Hehl, F.W.; Obukhov, Y.N. Élie Cartan torsion in geometry and in field theory, an essay. *Annal. Found. Louis Broglie* **2007**, *32*, 157–194.

50. Hehl, F.W.; Obukhov, Y.N.; Puetzfeld, D. On Poincaré gauge theory of gravity, its equations of motion, and Gravity Probe B. *Phys. Lett. A* **2013**, *377*, 1775–1781. [CrossRef]

51. Kleinert, H.; Pelster, A. Autoparallels from a new action principle. *Gen. Relativ. Grav.* **1999**, *31*, 1439–1447. [CrossRef]

52. Kleinert, H.; Shabanov, S.V. Spaces with Torsion from Embedding and the Special Role of Autoparallel Trajectories. *Phys. Lett. B* **1998**, *428*, 315–321. [CrossRef]

53. Kleinert, H. Nonholonomic Mapping Principle for Classical and Quantum Mechanics in Spaces with Curvature and Torsion. *Gen. Rel. Grav.* **2000**, *32*, 769–839. [CrossRef]

54. Anderson, J.D.; Nieto, M.M. *Relativity in Fundamental Astronomy: Dynamics, Reference Frames, and Data Analysis*; Klioner, S.A., Seidelmann, P.K., Soffel, M.H., Eds.; Proceedings IAU Symposium No. 261; Cambridge University Press: Cambridge, UK, 2010; pp. 189–197.

55. Lämmerzahl, C.; Preuss, O.; Dittus, H. Is the physics of the Solar System really understood? *Lasers, Clocks Drag-Free Control* **2008**, *349*, 75–101.

56. Iorio, L. Gravitational anomalies in the Solar System? *Int. J. Mod. Phys. D* **2015**, *24*, 1530015. [CrossRef]

57. Laskos-Grabowski, P. The Einstein-Cartan Theory: The Meaning and Consequences of Torsion. Master's Thesis, University of Wrocław, Wrocław, Poland, 2009.

58. Puetzfeld, D.; Obukhov, Y.N. Probing non-Riemannian spacetime geometry. *Phys. Lett. A* **2008**, *372*, 6711–6716. [CrossRef]

59. Landau, L.D.; Lifshitz, E.M. *The Classical Theory of Fields. Course of Theoretical Physics*, 4th ed.; Butterworth-Heinemann: Oxford, UK, 1987; Volume 2.

60.   Kleinert, H. *Path Integrals in Quantum Mechanics, Statistics, Polymer Physics and Financial Markets*, 15th ed.; World Scientific Publish. Co.: Singapore, Singapore, 2009.

61.   Kleinert, H. *Multivalued Fields: in Condensed Matter, Electromagnetism, and Gravitation*; World Scientific Publ. Co.: Singapore, Singapore, 2008.

62.   Kleinert, H.; Pelster, A. Novel Geometric Gauge Invariance of Autoparallels. *Acta Phys. Pol.* **1998**, *29*, 1015–1023.

63.   Pollard, H. *Mathematical Introduction to Celestial Mechanics*; Prentice-Hall Inc.: Englewood Cliffs, NY, USA, 1966.

64.   Danby, J.M.A. *Fundamentals of Celestial Mechanics*, 2nd ed.; Willmann-Bell, Inc.: Richmond, Virginia, USA, 1988.

65.   Burns, J.A. Elementary derivation of the perturbation equations of celestial mechanics. *Am. J. Phys.* **1976**, *44*, 944–949. [CrossRef]

66.   Krasinsky, G.A.; Brumberg, V.A. Secular increase of astronomical unit from analysis of the major planet motions, and its interpretation. *Celest. Mech. Dyn. Astron.* **2004**, *90*, 267–288. [CrossRef]

67.   Standish, E.M. The astronomical unit now. In *Transit of Venus: New Views of the Solar System and Galaxy*; Kurtz, D.W., Ed.; Cambridge University Press: Cambridge, UK, 2005; p. 163.

68.   Vinti, J.P. Classical solution of the two-body problem if the gravitational constant diminishes inversely with the age of the Universe. *Mon. Not. R. Astron. Soc.* **1974**, *169*, 417–427. [CrossRef]

69.   Iorio, L. An empirical explanation of the anomalous increases in the astronomical unit and the lunar eccentricity. *Astron. J.* **2011**, *142*, 68. [CrossRef]

70.   Acedo, L. Anomalous post-Newtonian terms and the secular increase of the astronomical unit. *Adv. Space Res.* **2013**, *52*, 1297–1303. [CrossRef]

71.   Li, X.; Chang, Z. Kinematics in Randers-Finsler geometry and secular increase of the astronomical unit. *Chin. Phys. C* **2011**, *35*, 914–919. [CrossRef]

72.   Miura, T.; Arakida, H.; Kasai, M.; Kuramata, S. Secular increase of the astronomical unit: A possible explanation in terms of the total angular momentum conservation law. *Publ. Astron. Soc. Jpn.* **2009**, *61*, 1247–1250. [CrossRef]

73.   Iorio, L. Secular increase of the astronomical unit and perihelion precessions as tests of the Dvali Gabadadze Porrati multi-dimensional braneworld scenario. *J. Cosmol. Astropart. Phys.* **2005**, *2005*, 6. [CrossRef]

74.   NASA Planetary Fact Sheet. Available online: http://nssdc.gsfc.nasa.gov/planetary/factsheet/ (accessed on 14 May 2015).

75.   Acedo, L. Constraints on non-standard gravitomagnetism by the anomalous perihelion precession of the planets. *Galaxies* **2014**, *2*, 466–481. [CrossRef]

76.   Iorio, L. A Critical Analysis of a Recent Test of the Lense-Thirring Effect with the LAGEOS Satellites. *J. Geod.* **2006**, *80*, 128–136. [CrossRef]

77.   Iorio, L.; Ruggiero, M.L.; Corda, C. Novel considerations about the error budget of the LAGEOS-based tests of frame-dragging with GRACE geopotential models. *Acta Astronaut.* **2013**, *91*, 141–148. [CrossRef]

78.   Renzetti, G. Some reflections on the LAGEOS frame-dragging experiment in view of recent data analyses. *New Astron.* **2014**, *29*, 25–27.

79.   Giles, P. Time-Distance Measurements of Large-Scale Flows in the Solar Convection Zone. Ph.D. Thesis, Stanford University, Stanford, CA, USA, 1999.

80.   Stark, D.; Wöhl, H. On the solar rotation elements as determined from sunspot observations. *Astron. Astrophys.* **1981**, *93*, 241–244.

7

Warm Inflation

Øyvind Grøn

Art and Design, Faculty of Technology, Oslo and Akershus University College of Applied Sciences,
P.O. Box 4 St., Olavs Plass, NO-0130 Oslo, Norway; Oyvind.Gron@hioa.no

Academic Editors: Elias C. Vagenas and Lorenzo Iorio

Abstract: I show here that there are some interesting differences between the predictions of warm and cold inflation models focusing in particular upon the scalar spectral index n_s and the tensor-to-scalar ratio r. The first thing to be noted is that the warm inflation models in general predict a vanishingly small value of r. Cold inflationary models with the potential $V = M^4 \left(\phi / M_P \right)^p$ and a number of e-folds $N = 60$ predict $\delta_{nsC} \equiv 1 - n_s \approx \left(p + 2 \right) / 120$, where n_s is the scalar spectral index, while the corresponding warm inflation models with constant value of the dissipation parameter Γ predict $\delta_{nsW} = \left[\left(20 + p \right) / \left(4 + p \right) \right] / 120$. For example, for $p = 2$ this gives $\delta_{nsW} = 1.1 \delta_{nsC}$. The warm polynomial model with $\Gamma = V$ seems to be in conflict with the Planck data. However, the warm natural inflation model can be adjusted to be in agreement with the Planck data. It has, however, more adjustable parameters in the expressions for the spectral parameters than the corresponding cold inflation model, and is hence a weaker model with less predictive force. However, it should be noted that the warm inflation models take into account physical processes such as dissipation of inflaton energy to radiation energy, which is neglected in the cold inflationary models.

Keywords: General relativity; Cosmology; The inflationary era

1. Introduction

In the usual (cold) inflationary models, dissipative effects with decay of inflaton energy into radiation energy are neglected. However, during the evolution of warm inflation dissipative effects are important, and inflaton field energy is transformed to radiation energy. This produces heat and viscosity, which make the inflationary phase last longer. Warm inflation models were introduced and developed by Berera and coworkers [1–14]. However, even earlier inflation models with dissipation of inflaton energy to radiation and particles had been considered [15–22]. Introductions to warm inflation models and references to works prior to 2009 on warm inflation are found in [8] and [23]. For later works, see [9] and [24] and references in these articles. Further developments are found in the articles [25–43].

In this scenario, there is no need for a reheating at the end of the inflationary era. The universe heats up and becomes radiation dominated during the inflationary era, so there is a smooth transition to a radiation dominated phase (Figure 1).

In the present work, I will review the foundations of warm inflation and some of the most recent phenomenological models of this type, focusing in particular on the comparison with the experimental measurements of the scalar spectral index n_s and the tensor to scalar ratio r by the Planck observatory.

The article is organized as follows. In Section 2, the definition and current measurements of these quantities are given. Then, the optical parameters in the warm inflation scenario are considered. We go on and study some phenomenological models in the subsequent sections: monomial-, natural- and viscous inflation. The models are compared in Section 7, and the results are summarized in the final section.

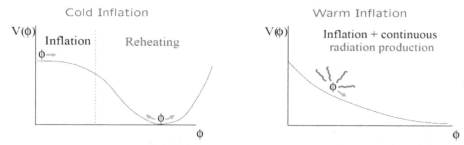

Figure 1. Illustration of the difference between cold inflation and warm inflation (Berera et al. (2009)).

2. Definition and Measured Values of the Optical Parameters

We shall here briefly review a few of the mathematical quantities that are used to describe the temperature fluctuations in the CMB. The power spectra of scalar and tensor fluctuations are represented by [44]

$$P_S = A_S\left(k_*\right)\left(\frac{k}{k_*}\right)^{n_S-1+(1/2)\alpha_S\ln(k/k_*)+\cdots}, \quad P_T = A_T\left(k_*\right)\left(\frac{k}{k_*}\right)^{n_T+(1/2)\alpha_T\ln(k/k_*)+\cdots},$$
$$A_S = \frac{V}{24\pi^2\varepsilon M_P^4} = \left(\frac{H^2}{2\pi\dot\phi}\right)^2, \qquad A_T = \frac{2V}{3\pi^2 M_P^4} = \varepsilon\left(\frac{2H^2}{\pi\dot\phi}\right)^2. \tag{2.1}$$

Here, k is the wave number of the perturbation which is a measure of the average spatial extension for a perturbation with a given power, and k_* is the value of k at a reference scale usually chosen as the scale at horizon crossing, called the pivot scale. One often writes $k = \dot a = aH$, where a is the scale factor representing the ratio of the physical distance between reference particles in the universe relative to their present distance. The quantities A_S and A_T are amplitudes at the pivot scale of the scalar- and tensor fluctuations, and n_S and n_T are the *spectral indices* of the corresponding fluctuations. We shall represent the scalar spectral index by the quantity $\delta_{ns} \equiv 1 - n_S$. The quantities n_S and n_T are called the *tilt* of the power spectrum of curvature perturbations and tensor modes, respectively, because they represent the deviation of the values $\delta_{ns} = n_t = 0$ that represent a scale invariant spectrum.

The quantities α_S and α_T are factors representing the k-dependence of the spectral indices. They are called the *running of the spectral indices* and are defined by

$$\alpha_S = \frac{dn_S}{d\ln k}, \quad \alpha_T = \frac{dn_T}{d\ln k} \tag{2.2}$$

They will, however, not be further considered in this article.

As mentioned above, if $n_S = 1$ the spectrum of the scalar fluctuations is said to be *scale invariant*. An invariant mass-density power spectrum is called a *Harrison-Zel'dovich spectrum*. One of the predictions of the inflationary universe models is that the cosmic mass distribution has a spectrum that is *nearly* scale invariant, but not exactly. The observations and analysis of the Planck team [45] have given the result $n_S = 0.968 \pm 0.006$. Hence, we shall use $n_S = 0.968$ as the preferred value of n_S. Different inflationary models will be evaluated against the Planck 2015 value of the tilt of the scalar curvature fluctuations, $\delta_{ns} = 0.032$.

The tensor-to-scalar ratio r is defined by

$$r \equiv \frac{P_T\left(k_*\right)}{P_S\left(k_*\right)} = \frac{A_T}{A_S} \tag{2.3}$$

As noted by [46], the tensor-to-scalar ratio is a measure of the energy scale of inflation, $V^{1/4} = (100r)^{1/4}\,10^{16}GeV$. From Equations (2.1) and (2.3), we have

$$r = 16\varepsilon \tag{2.4}$$

The Planck observational data have given $r < 0.11$.

3. Optical Parameters in Warm Inflation

During the warm inflation era, both the inflaton field energy with density ρ_ϕ and the electromagnetic radiation with energy density ρ_r are important for the evolution of the universe. The first Friedmann equation takes the form

$$H^2 = \frac{\kappa}{3}\left(\rho_\phi + \rho_r\right) \tag{3.1}$$

We shall here use units so that $\kappa = 1/M_P^2$ where M_P is the reduced Planck mass. In these models, the continuity equations for the inflaton field and the radiation take the form

$$\dot{\rho}_\phi + 3H\left(\rho_\phi + p_\phi\right) = -\Gamma\dot{\phi}^2 \quad , \quad \dot{\rho}_r + 4H\rho_r = \Gamma\dot{\phi}^2 \tag{3.2}$$

respectively, where the dot denotes differentiation with respect to cosmic time, and Γ is a dissipation coefficient of a process which transforms inflaton energy into radiation. In general, Γ is temperature dependent. The density and pressure of the inflaton field are given in terms of the kinetic and potential energy of the inflaton field as

$$\rho_\phi = \frac{\dot{\phi}^2}{2} + V \quad , \quad p_\phi = \frac{\dot{\phi}^2}{2} - V \tag{3.3}$$

During warm inflation, the dark energy predominates over radiation, i.e., $\rho_\phi >> \rho_r$, and H, ϕ and Γ are slowly varying so that the production of radiation is quasi-static, $\ddot{\phi} << H\dot{\phi}, \dot{\rho}_r << 4H\rho_r$ and $\dot{\rho}_r << \Gamma\dot{\phi}^2$. Note that in the slow roll era the kinetic energy of the inflaton field energy can be neglected compared to its potential energy. Then, the inflaton field obeys the equation of state $p_\phi \approx -\rho_\phi$. Also, in this era, the second of Equation (3.2) gives $\rho_r = 0$ in the case of vanishing dissipation, $\Gamma = 0$, i.e., in the warm inflation model all of the radiation is produced by dissipation of the inflaton energy. Then, the first Friedmann equation and the equation for the evolution of the inflaton field take the form

$$3H^2 = \kappa\rho_\phi = \kappa V \quad , \quad \left(3H + \Gamma\right)\dot{\phi} = -V' \tag{3.4}$$

respectively. Here, a prime denotes differentiation with respect to the inflaton field ϕ.

Defining the so-called dissipative ratio by

$$Q \equiv \Gamma/3H \tag{3.5}$$

the last of Equation (3.4) may be written as

$$3H\left(1 + Q\right)\dot{\phi} = -V' \tag{3.6}$$

The quantity Q represents the effectiveness at which inflaton energy is transformed to radiation energy. If $Q >> 1$ one says that there is a strong, dissipative regime, and if $Q << 1$ there is a weak dissipative regime.

During warm inflation, the second of the Equation (3.2) reduces to

$$\rho_r = (3/4)\,Q\dot{\phi}^2 \tag{3.7}$$

In the warm inflation scenario, a thermalized radiation component is present with $T > H$, where both T and H are expressed in units of energy. Then, the tensor-to-scalar ratio defined in Equation (2.3), is modified with respect to standard cold inflation, so that [12]

$$r_W = \frac{H/T}{(1+Q)^{5/2}}r \tag{3.8}$$

Hence, the tensor-to-scalar ratio is suppressed by the factor $(T/H)(1+Q)^{5/2}$ compared with the standard cold inflation.

Hall, Moss and Berera [9] have calculated the spectral index in warm inflation for the strong dissipative regime with $Q >> 1$ or $\Gamma >> 3H$. We shall here follow Visinelli [47] and permit arbitrary values of Q. Differentiating the first of the Equation (3.4) and using Equation (3.6) gives

$$\dot{H} = -(\kappa/2)(1+Q)\dot{\phi}^2 \tag{3.9}$$

Hence $\dot{H} < 0$.

We define the potential slow roll parameters ε and η by

$$\varepsilon \equiv \frac{1}{2\kappa}\left(\frac{V'}{V}\right)^2 \quad , \quad \eta \equiv \frac{1}{\kappa}\frac{V''}{V} \tag{3.10}$$

These expressions are to be evaluated at the beginning of the slow roll era. Using Equations (3.4), (3.6) and (3.9) and the first of Equation (3.10) we get

$$\varepsilon = -(1+Q)\frac{\dot{H}}{H^2} \tag{3.11}$$

Differentiation of Equation (3.6) and using that $\left(\dot{\phi}\right)' = \ddot{\phi}/\dot{\phi}$ gives

$$V'' = \frac{\Gamma'V'}{\Gamma+3H} - 3H(1+Q)\frac{\ddot{\phi}}{\dot{\phi}} - 3\dot{H} \tag{3.12}$$

Dividing by κV and using the first of Equation (3.4) in the two last terms leads to

$$\eta = \frac{Q}{1+Q}\frac{1}{\kappa}\frac{\Gamma'V'}{\Gamma V} - \frac{1+Q}{H}\frac{\ddot{\phi}}{\dot{\phi}} - \frac{\dot{H}}{H^2} \tag{3.13}$$

Defining

$$\beta \equiv \frac{1}{\kappa}\frac{\Gamma'V'}{\Gamma V} \tag{3.14}$$

and using Equation (3.12) we get

$$\frac{\ddot{\phi}}{H\dot{\phi}} = -\frac{1}{1+Q}\left(\eta - \beta + \frac{\beta - \eta}{1+Q}\right) \tag{3.15}$$

in agreement with Equation (3.14) of Visinelli [47] .

It follows from Equation (3.6) that

$$\frac{d}{d\phi} = -\frac{3H(1+Q)}{V'}\frac{d}{dt} \tag{3.16}$$

From Equation (3.5) and the first of Equation (3.4) we have

$$H\Gamma = \kappa V Q \tag{3.17}$$

Using Equations (3.14), (3.16) and (3.17) can be written as

$$\frac{\dot{\Gamma}}{H\Gamma} = -\frac{\beta}{1+Q} \tag{3.18}$$

During slow roll the second of the Equation (3.2) reduces to

$$4H\rho_r = \Gamma\dot{\phi}^2 \tag{3.19}$$

Differentiation gives

$$\frac{\dot{\rho}_r}{H\rho_r} = \frac{\dot{\Gamma}}{H\Gamma} + 2\frac{\ddot{\phi}}{H\dot{\phi}} - \frac{\dot{H}}{H^2} \tag{3.20}$$

Inserting Equations (3.11), (3.15) and (3.18) into Equation (3.20) gives

$$\frac{\dot{\rho}_r}{H\rho_r} = -\frac{1}{1+Q}\left(2\eta - \beta - \varepsilon + 2\frac{\beta - \varepsilon}{1+Q}\right) \tag{3.21}$$

We now define $\delta_{ns} \equiv 1 - n_s$, where n_s is the scalar spectral index. Visinelli [48] has deduced

$$\delta_{ns} = 4\frac{\dot{H}}{H^2} - 2\frac{\ddot{\phi}}{H\dot{\phi}} - \frac{\dot{\omega}}{H(1+\omega)} \tag{3.22}$$

where

$$\omega = \frac{T}{H}\frac{2\sqrt{3}\pi Q}{\sqrt{3+4\pi Q}} \tag{3.23}$$

Since $\rho_r \propto T^4$ we have that

$$\omega \propto \frac{\rho_r^{1/4}Q}{H\sqrt{3+4\pi Q}} \tag{3.24}$$

Differentiating this we get

$$\frac{\dot{\omega}}{H\omega} = \frac{1}{4}\frac{\dot{\rho}_r}{H\rho_r} - \frac{\dot{H}}{H^2} + \frac{3+2\pi Q}{3+4\pi Q}\frac{\dot{Q}}{HQ} \tag{3.25}$$

Differentiating Equation (3.5) gives

$$\frac{\dot{Q}}{HQ} = \frac{\dot{\Gamma}}{H\Gamma} - \frac{\dot{H}}{H^2} \tag{3.26}$$

Using Equations (3.11) and (3.18) then leads to

$$\frac{\dot{Q}}{HQ} = \frac{\varepsilon - \beta}{1+Q} \tag{3.27}$$

Inserting Equations (3.11), (3.21) and (3.27) into Equation (3.25) gives

$$\dot{\omega} = -\frac{H\omega}{1+Q}\left[\frac{2\eta - \beta - 5\varepsilon}{4} + \frac{1}{2}\frac{\beta - \varepsilon}{1+Q} + \frac{3+2\pi Q}{3+4\pi Q}(\beta - \varepsilon)\right] \tag{3.28}$$

Visinelli has rewritten this as follows

$$\dot{\omega} = -\frac{H\omega}{1+Q}\left[\frac{2\eta + \beta - 7\varepsilon}{4} + \frac{6+(3+4\pi)Q}{(1+Q)(3+4\pi Q)}(\beta - \varepsilon)\right] \tag{3.29}$$

Inserting the expressions (3.11), (3.15) and (3.29) into Equation (3.22) gives

$$\delta_{ns} = \frac{1}{1+Q}\left[4\varepsilon - 2\left(\eta - \beta + \frac{\beta - \varepsilon}{1+Q}\right) + \frac{\omega}{1+\omega}\left(\frac{2\eta + \beta - 7\varepsilon}{4} + \frac{6+(3+4\pi)Q}{(1+Q)(3+4\pi Q)}(\beta - \varepsilon)\right)\right] \tag{3.30}$$

The usual cold inflation is found in the limit $Q \to 0$ and $T << H$, i.e., $\omega \to 0$. Then,

$$\delta_{ns} \to 2\,(3\varepsilon - \eta) \tag{3.31}$$

In the strong regime of warm inflation, $Q >> 1$, $\omega >> 1$ we get

$$\delta_{ns} = \frac{3}{2Q}\left[\frac{3}{2}\,(\varepsilon + \beta) - \eta\right] \tag{3.32}$$

In the weak regime, $Q << 1$, Equation (3.16) leads to

$$\delta_{ns} = 2\,(3\varepsilon - \eta) - \frac{\omega/4}{1+\omega}\,(15\varepsilon - 2\eta - 9\beta) \tag{3.33}$$

It may be noted that in warm inflation the condition for slow roll is that the absolute values of ε, η and β are much smaller than $1 + Q$.

Visinelli has found that the tensor-to-scalar ratio in warm inflation is

$$r = \frac{16\varepsilon}{(1+Q)^2\,(1+\omega)} \tag{3.34}$$

In the cold inflation limit, this reduces to

$$r \to 16\varepsilon \tag{3.35}$$

In the strong dissipation regime warm inflation gives in general

$$r \to \frac{16}{Q^2\omega}\varepsilon << \varepsilon \tag{3.36}$$

Hence, all the warm inflation models predict an extremely small tensor-to-scalar-ratio in the strong dissipation regime with $Q >> 1$ and $\omega >> 1$.

4. Warm Monomial Inflation

Visinelli [48] has investigated warm inflation with a polynomial potential which we write in the form

$$V = M^4\,(\phi/M_P)^p \tag{4.1}$$

since the potential and the inflaton field have dimensions equal to the fourth and first power of energy, respectively. Here, M represents the energy scale of the potential when the inflaton field has Planck mass. Furthermore he assumes that the dissipative term is also monomial

$$\Gamma = \Gamma_0\,(\phi/M_P)^{q/2} \tag{4.2}$$

He considered models with $p > 0$ and $q > p$. However, in the present article, we shall also consider polynomial models with $p < 0$. From Equations (3.3) and (3.4) we have

$$Q = Q_0\left(\frac{\phi}{M_P}\right)^{\frac{q-p}{2}}, \quad Q_0 = \frac{\Gamma_0 M_P}{\sqrt{3}M^2} \tag{4.3}$$

The constant Q_0 represents the strength of the dissipation. For $q = p$ the dissipative ratio is constant, $Q = Q_0$. We shall here consider the strong dissipative regime where $Q >> 1$. Then, the second of Equation (3.3) reduces to

$$\dot{\phi} = -\frac{V'}{\Gamma} \tag{4.4}$$

Inserting Equations (4.1) and (4.2) gives

$$\dot{\phi} = -\frac{pM^4}{\Gamma_0 M_P}\left(\frac{\phi}{M_P}\right)^{p-\frac{q}{2}-1} \tag{4.5}$$

Integration leads to

$$\phi(t) = \left[\frac{4+q-2p}{2}\left(K - \frac{pM^4}{\Gamma_0 M^{p-\frac{q}{2}}}t\right)\right]^{\frac{2}{4+q-2p}} \quad , \quad q > 2(p-2) \tag{4.6}$$

where K is a constant of integration. The initial condition $\phi(0) = 0$ gives $K = 0$.

The special cases (i) $\Gamma = V/M_P^3$, i.e., $\Gamma_0 = M^4/M_P^3$, $q = 2p$ and (ii) $\Gamma = \Gamma_0$, i.e., $q = 0$, both with the initial condition $\phi(0) = 0$, i.e., $K = 0$, have been considered by Sharif and Saleem (2015). For these cases, the condition $\phi(t) > 0$ requires $p < 0$. In the first case, Equation (3.6) reduces to

$$\phi = M_P\sqrt{-2pM_Pt} \tag{4.7}$$

Note that the time has dimension inverse mass with the present units, so that M_Pt is dimensionless.

Visinelli, however, has considered polynomial models with $p > 0$. Then, we have to change the initial condition. The corresponding solution of Equation (4.5) with $q = 2p$ and the inflaton field equal to the Planck mass at the Planck time gives

$$\phi = M_P\sqrt{1 - 2pM_P(t - t_P)} \tag{4.8}$$

It may be noted that $q = 2(p-2)$ gives a different time evolution of the inflaton field. Then, Equation (3.5) with the boundary condition $\phi(t_P) = M_P$ has the solution

$$\phi = M_P\exp\left[-\frac{pM^4}{\Gamma_0 M_P^2}(t - t_P)\right] \tag{4.9}$$

In this case, the inflaton field decreases or increases exponentially, depending upon the sign of p. Inserting Equations (4.1) and (4.2) into Equations (3.9) and (3.13), the slow-roll parameters are

$$\varepsilon = \frac{p^2}{2}\left(\frac{M_P}{\phi}\right)^2 \quad , \quad \eta = \frac{2(p-1)}{p}\varepsilon \quad , \quad \beta = \frac{q}{p}\varepsilon \tag{4.10}$$

With these expressions Equation (3.32) valid in the regime of strong dissipation, $Q \gg 1$, gives

$$\delta_{ns} = \frac{3(4+3q-p)}{4p}\frac{\varepsilon}{Q} \tag{4.11}$$

The slow-roll regime ends when at least one of the parameters (4.10) is not much smaller than $1 + Q$. In the strong dissipative regime $Q \gg 1$ and $\varepsilon_f = Q_f$. Using Equations (4.3) and (4.10) we then get

$$\phi_f = M_P\left(\frac{p^2}{2Q_0}\right)^{\frac{2}{4+q-p}} \tag{4.12}$$

The number of e-folds, N, in the slow roll era for this model has been calculated by Visinelli [48] . It is defined by

$$N = \ln\frac{a_f}{a} = \int\limits_t^{t_f} Hdt = \int\limits_\phi^{\phi_f} \frac{H}{\dot{\phi}}d\phi \tag{4.13}$$

Using Equations (3.3) and (3.5) we get

$$N = \frac{1}{M_P^2} \int_{\phi_f}^{\phi} (1+Q) \frac{V}{V'} d\phi \tag{4.14}$$

Inserting the potential (4.1), performing the integration and considering the strong dissipative regime gives

$$N \approx \frac{2Q_0}{p(4+q-p)} \left[\left(\frac{\phi}{M_P} \right)^{\frac{4+q-p}{2}} - \left(\frac{\phi_f}{M_P} \right)^{\frac{4+q-p}{2}} \right] \tag{4.15}$$

The time dependence of the inflaton field is given by Equation (4.6) when $p < 0$ showing that $\phi_f > \phi$ in this case, and by Equation (4.8) when $p > 0$ implying $\phi_f < \phi$ in that case, showing that $N > 0$ in both cases (not dot here)

$$\frac{\phi}{M_P} \approx \left(\frac{p(4+q-p)N}{2Q_0} \right)^{\frac{2}{4+q-p}} \tag{4.16}$$

Inserting this into the first of Equations (4.10) and (4.3) gives

$$\varepsilon \approx \frac{p^2}{2} \left[\frac{2Q_0}{p(4+q-p)N} \right]^{\frac{4}{4+q-p}} \quad , \quad Q \approx Q_0 \left[\frac{p(4+q-p)N}{2Q_0} \right]^{\frac{q-p}{4+q-p}} \tag{4.17}$$

Inserting these expressions into Equation (4.11) gives

$$\delta_{ns} \approx \frac{3(4+3q-p)}{4(4+q-p)} \frac{1}{N} \tag{4.18}$$

Note that with $q = 0$, i.e., a constant value of the dissipation parameter Γ, Equation (4.18) reduces to

$$\delta_{ns} = \frac{3}{4N} \tag{4.19}$$

for all values of p. Then $N = 60$ gives $\delta_{ns} = 0.012$ which is smaller than the preferred value from the Planck data, $\delta_{ns} = 0.032$. Inserting $q = 2p$ in Equation (4.18) and solving the equation with respect to p gives,

$$p = \frac{4(4N\delta_{ns} - 3)}{15 - 4N\delta_{ns}} \tag{4.20}$$

The Planck values $\delta_{ns} = 0.032$, $N = 60$ give $p = 2.56$ and $q = 5.11$.

Panotopoulos and Videla [24] have investigated the tensor-to-scalar ratio in warm in inflation for inflationary models with an inflaton field given by the potential

$$V = (M/M_P)^4 \phi^4 \tag{4.21}$$

where M is the energy scale of the potential when the inflaton field has Planck mass, M_P. Let us choose $p = q = 4$ in the monomial models above. Inserting this in Equation (3.18) gives $\delta_{ns} = 9/4N$. With $\delta_{ns} = 0.032$ we get $N = 70$.

In this case $\delta_{ns} = 2/N$ for cold inflation. For $\delta_{ns} = 0.032$ this corresponds to $N \approx 62$ which is an acceptable number of e-folds. Then, the tensor-to-scalar ratio is $r = 0.32$, which is much larger than allowed by the Planck observations [45]. Panotopoulos and Videla found the corresponding $\delta_{ns}, r-$ relation in warm inflation with $\Gamma = aT$, where a is a dimensionless parameter. They considered two cases.

(A)   The weak dissipative regime. In this case $Q << 1$ and Equation (3.7) reduces to $r_W = (H/T)\, r$. They then found

$$r_W \approx \frac{0.01}{\sqrt{a}} \delta_{ns} \tag{4.22}$$

With the Planck values $\delta_{ns} = 0.032$ and $r_W < 0.12$ this requires $a > 7 \cdot 10^{-6}$. However, they also found that in this case $\delta_{ns} = 1/N$ giving $N = 31$ which is too small to be compatible with the standard inflationary scenario.

(B)   The strong dissipative regime. Then, $R >> 1$ and $r_W \approx \left(H/TR^{5/2} \right) r$. They then found

$$\delta_{ns} = \frac{45}{28N} \quad , \quad r_W = \frac{3.8 \cdot 10^{-7}}{a^4} \delta_{ns} \tag{4.23}$$

Then $N = 50$ and $a > 1.8 \cdot 10^{-2}$, so this is a promising model.

5. Warm Natural Inflation

Visinelli [47] has also investigated warm natural inflation with the potential

$$V(\phi) = V_0 \left(1 + \cos\widetilde{\phi} \right) = 2V_0 \cos^2 \left(\widetilde{\phi}/2 \right) \tag{5.1}$$

where $\widetilde{\phi} = \phi/M$, and M is the spontaneous symmetry breaking scale, and $M > M_P$ in order for inflation to occur. The constant V_0 is a characteristic energy scale for the model. The potential V has a minimum at $\widetilde{\phi} = \pi$. Inserting the potential (5.1) into the expressions (3.9) we get

$$\varepsilon = \frac{b}{2} \frac{1 - \cos\widetilde{\phi}_i}{1 + \cos\widetilde{\phi}_i} \quad , \quad \eta = \varepsilon - \frac{b}{2} \quad , \quad b = \left(\frac{M_P}{M} \right)^2 \tag{5.2}$$

From Equation (3.3) with the potential (5.1) we have

$$H = \sqrt{(\kappa/3)\, V_0 \left(1 + \cos\widetilde{\phi} \right)} \tag{5.3}$$

Equations (3.4) and (5.3) then give

$$Q = \frac{\Gamma M_P}{\sqrt{3V_0 \left(1 + \cos\widetilde{\phi} \right)}} \tag{5.4}$$

During the slow roll era we must have $\varepsilon << R$. Using the expressions (5.2) and (5.4) we find that this corresponds to

$$\frac{1 - \cos\widetilde{\phi}}{\sqrt{1 + \cos\widetilde{\phi}}} << 1/\beta \quad , \quad \beta = \frac{\sqrt{6V_0}}{\Gamma M_P} b \tag{5.5}$$

Inserting Equations (5.2) and (5.4) into Equation (3.31) with $\beta = 0$ gives in the strong dissipative regime

$$\delta_{ns} = \frac{3}{4\alpha} \frac{3 + \cos\widetilde{\phi}_i}{\sqrt{1 + \cos\widetilde{\phi}_i}} \tag{5.6}$$

We shall now express the δ_{ns} in terms of the number of e-folds of expansion during the slow roll era for this inflationary universe model, again following Visinelli. Assuming that the dissipation parameter Γ is independent of ϕ, i.e., that $\beta = 0$, the number of e-folds is given by

$$N = -\Gamma \int_{\phi_i}^{\phi_f} \frac{H(\phi)}{V'(\phi)} d\phi \tag{5.7}$$

Differentiating the potential (5.1) and inserting Equation (5.3) we get

$$N = \frac{\alpha}{2} \int_{\widetilde{\phi}_i}^{\widetilde{\phi}_f} \frac{\sqrt{1+\cos x}}{\sin x} dx = \frac{\alpha}{\sqrt{2}} \ln \frac{\tan\left(\widetilde{\phi}_f/4\right)}{\tan\left(\widetilde{\phi}_i/4\right)} \tag{5.8}$$

Hence,

$$\tan\frac{\widetilde{\phi}_i}{4} = \tan\frac{\widetilde{\phi}_f}{4} exp\left(-\frac{\beta N}{2}\right) \tag{5.9}$$

Visinelli has argued that

$$\widetilde{\phi}_f = \pi - \beta \tag{5.10}$$

giving

$$\tan\frac{\widetilde{\phi}_f}{4} = \frac{1 - \tan\left(\beta/4\right)}{1 + \tan\left(\beta/4\right)} \tag{5.11}$$

Inserting this into Equation (5.9) gives

$$\tan\frac{\widetilde{\phi}_i}{4} = \gamma exp\left(-\frac{\beta N}{2}\right) \quad , \quad \gamma = \frac{1 - \tan\left(\beta/4\right)}{1 + \tan\left(\beta/4\right)} \tag{5.12}$$

Applying the trigonometric identity

$$\sqrt{1+\cos\theta} = \sqrt{2}\frac{1 - \tan^2\left(\theta/4\right)}{1 + \tan^2\left(\theta/4\right)} \tag{5.13}$$

in the expression (5.12) and inserting the result into Equation (5.6) we finally arrive at

$$\delta_{ns} = \frac{3}{8}\beta\frac{exp\left(2\beta N\right) + \gamma^4}{exp\left(2\beta N\right) - \gamma^4} \tag{5.14}$$

Here, we must have $\beta \ll 1$ in order to give the Planck value $\delta_{ns} = 0.032$ for $N = 60$. Hence, Equation (5.12) gives $\gamma \approx 1$. A good approximation for δ_{ns} is therefore

$$\delta_{ns} \approx (3/8)\,\beta\coth\left(\beta N\right) \tag{5.15}$$

Inserting $\delta_{ns} = 0.032$ and $N = 60$ gives $\beta = 0.08$.
Visinelli (2011) further found that the tensor-to-scalar ratio for this inflationary model is

$$r = 128\kappa\sqrt{\frac{\pi}{\Gamma}}\frac{\dot{\phi}^2}{T\sqrt{H}} \tag{5.16}$$

Differentiating the expression (5.3) gives

$$\dot{H} = -\frac{\kappa V_0}{6M}\frac{s_\phi\dot{\phi}}{H} \quad , \quad s_\phi \equiv \sin\widetilde{\phi} \tag{5.17}$$

Combining this with Equation (3.8) in the strong dissipative regime and using Equation (3.4) gives

$$\dot{\phi} = \frac{3V_0 s_\phi}{M\Gamma} \tag{5.18}$$

The energy density of the radiation is

$$\rho_\gamma = aT^4 \tag{5.19}$$

where $a = 7.5657 \times 10^{-16}$ J·m^{-3}·K^{-4} $= 4.69 \times 10^{-6}$ GeV·m^{-3}·K^{-4} is the radiation constant. Combining with Equation (3.6) we get

$$T = \left(\frac{\Gamma}{4aH}\right)^{1/4} \dot{\phi}^{1/2} \tag{5.20}$$

Equations (5.15), (5.18) and (5.19) give

$$r = B\frac{s_\phi^{3/2}}{(1+\cos\widetilde{\phi})^{1/8}} \quad , \quad B = \frac{384 \cdot 3^{5/8}\kappa^{7/8}\sqrt{6\pi}V^{11/8}a^{1/4}}{M^{3/2}\Gamma^{9/4}} \tag{5.21}$$

Visinelli [47] has evaluated the constant B and concluded that for this type of inflationary universe model the expected value of r is extremely low. If observations give a value $r > 10^{-14}$ this model has to be abandoned. On the other hand, the predictions of this model are in accordance with the observations so far.

6. Warm Viscous Inflation

As noted by del Campo, Herrera and Pavón [29], it has been usual, for the sake of simplicity, to study warm inflation models containing an inflaton field and radiation, only, (comma here) ignoring the existence of particles with mass that will appear due to the decay of the inflaton field. However, these particles modify the fluid pressure in two ways: (i) The relationship between pressure and energy density is no longer $p = (1/3)\rho$ as it is for radiation. A simple generalization is to use the equation of state $p = w\rho$, where w is a constant with value $0 \leq w \leq 1$; (ii) Due to interactions between the particles and the radiation there will appear a bulk viscosity so that the effective pressure takes the form

$$p_{eff} = p - 3\varsigma H \tag{6.1}$$

where ς is a coefficient of bulk viscosity.

We shall now consider isotropic universe models corresponding to the anisotropic models considered by Sharif and Saleem [37]. Equation (3.8) can be written

$$\dot{\phi} = \pm M_P\sqrt{-2\dot{H}/(1+Q)} \tag{6.2}$$

For these models, the time dependence of the scale factor during the inflationary era may be written

$$a(t) = a_0\exp\left(\frac{t}{t_1}\right)^\beta \quad , \quad 0 < \beta \leq 1, \tag{6.3}$$

where a_0 is the value of the scale factor at $t = 0$ before the slow roll era has started, and t_1 is the Hubble time of the corresponding De Sitter model having $\beta = 1$. The Hubble parameter and its rate of change with time is

$$H = \frac{\beta}{t_1}\left(\frac{t}{t_1}\right)^{\beta-1} \quad , \quad \dot{H} = \frac{\beta(\beta-1)}{t_1^2}\left(\frac{t}{t_1}\right)^{\beta-2} \tag{6.4}$$

Note that $\dot{H} < 0$ for $\beta < 1$. Inserting the second expression into Equation (6.2) gives

$$\dot{\phi} = \pm\frac{M_P}{t_1}\sqrt{\frac{2\beta(1-\beta)}{1+Q}}\left(\frac{t}{t_1}\right)^{\frac{\beta}{2}-1} \tag{6.5}$$

Sharif and Saleem considered two cases. In the first one $\Gamma = \Gamma(\phi) = \kappa V(\phi)/M_P$. Equations (3.3) and (3.4) then gives $Q = H/M_P$. Furthermore, for several reasons, they restricted their analysis to the strong dissipative regime where $Q >> 1$. Equation (6.5) then reduces to

$$\dot{\phi} = \pm M_P\sqrt{2M_P(1-\beta)}t^{-1/2} \tag{6.6}$$

Integrating with the initial condition $\phi(0) = 0$ and assuming that $\phi(t) > 0$ we get

$$\phi(t) = 2M_P\sqrt{2M_P(1-\beta)t} \tag{6.7}$$

Hence, ϕ is an increasing function of time. Inserting the first of the expressions (6.4) into the first of the Equation (3.3) gives

$$V(t) = 3\left(\frac{\beta M_P}{t_1}\right)^2\left(\frac{t}{t_1}\right)^{2(\beta-1)} \tag{6.8}$$

Combining this with Equation (6.7) leads to

$$V(\phi) = 3\left(\frac{\beta M_P}{t_1}\right)^2\left(\frac{\phi}{2M_P\sqrt{2(1-\beta)}M_Pt_1}\right)^{4(\beta-1)} \tag{6.9}$$

Sharif and Saleem used the Hubble slow roll parameters,

$$\varepsilon_H \equiv -\frac{\dot{H}}{H^2} = \frac{1}{2(1+Q)}\left(\frac{V'}{V}\right)^2 \quad, \quad \eta_H \equiv -\frac{\ddot{H}}{2H\dot{H}} = \frac{1}{1+Q}\left[\frac{V''}{V} - \frac{1}{2}\left(\frac{V'}{V}\right)^2\right] \tag{6.10}$$

Note that $\varepsilon_H = 1+q$, where q is the deceleration parameter. In the present case and in the strong dissipative regime, we can replace $1+Q$ by $H = \sqrt{\kappa V/3}$. Then $\varepsilon_H = (1/Q)\varepsilon$ and $\eta_H = (1/Q)(\eta-\varepsilon)$. Differentiating the expression (6.9) then gives

$$\varepsilon_H = \frac{1-\beta}{\beta}\left(\frac{\phi}{2M_P\sqrt{2(1-\beta)}M_Pt_1}\right)^{-2\beta} \quad, \quad \eta_H = \frac{3-2\beta}{2\beta}\left(\frac{\phi}{2M_P\sqrt{2(1-\beta)}M_Pt_1}\right)^{-2\beta} = \frac{3-2\beta}{2(1-\beta)}\varepsilon_H \tag{6.11}$$

The slow roll era ends when the inflaton field has a value ϕ_f so that $\varepsilon_H(\phi_f) = 1$, corresponding to $\varepsilon(\phi_f) = Q$, which gives

$$\left(\frac{\phi_f}{2M_P\sqrt{2(1-\beta)}M_Pt_1}\right)^{2\beta} = \frac{1-\beta}{\beta} \tag{6.12}$$

The number of e-folds is given by Equation (4.15), which in the present case takes the form

$$N = \frac{1}{\sqrt{3}M_P}\int_{\phi_f}^{\phi}\frac{V^{3/2}}{V'}d\phi \tag{6.13}$$

Inserting the potential (6.9) and integrating gives

$$N = \left(\frac{\phi_f}{2M_P\sqrt{2(1-\beta)}M_Pt_1}\right)^{2\beta} - \left(\frac{\phi}{2M_P\sqrt{2(1-\beta)}M_Pt_1}\right)^{2\beta} = \frac{1-\beta}{\beta} - \left(\frac{\phi}{2M_P\sqrt{2(1-\beta)}M_Pt_1}\right)^{2\beta} \tag{6.14}$$

Hence

$$\left(\frac{\phi}{2M_P\sqrt{2(1-\beta)}M_Pt_1}\right)^{2\beta} = \frac{1-\beta}{\beta} - N \tag{6.15}$$

Since the left hand side is positive, this requires that $N < (1-\beta)/\beta$ or $\beta < 1/(N+1)$. For $N > 50$ this means that $0 < \beta < 0.02$.

Sharif and Saleem have calculated the scalar spectral index with the result

$$\delta_{ns} = \frac{3\beta-2}{\beta}\left(\frac{\phi}{2M_P\sqrt{2(1-\beta)}M_Pt_1}\right)^{-2\beta} \tag{6.16}$$

Using Equation (6.15) we get

$$\delta_{ns} = \frac{3\beta - 2}{1 - \beta - \beta N} \approx \frac{2 - 3\beta}{\beta} \frac{1}{N} \tag{6.17}$$

This equation can be written

$$\beta \approx \frac{2}{3 + N\delta_{ns}} \tag{6.18}$$

Inserting the Planck value $\delta_{ns} = 0.032$ and $N = 60$, give $\beta = 0.41$ corresponding to $p = -2.36$. This value of β is not allowed by Equation (6.15).

In the second case, Sharif and Saleem assumed that $\Gamma = \Gamma_0$. Equations (3.3) and (3.4) then give $Q = \Gamma_0/3H$. Using Equations (6.2) and (6.4) and integrating with the initial condition $\phi(0) = 0$, leads to

$$\phi(t) = \lambda \left(\frac{t}{t_1}\right)^{\beta - 1/2} \quad , \quad V(\phi) = 3\left(\frac{\beta M_P}{t_1}\right)^2 \left(\frac{\phi}{\lambda}\right)^{\frac{4(1-\beta)}{2\beta - 1}} \quad , \quad \lambda = \frac{2\beta M_P}{2\beta - 1}\sqrt{\frac{6(1-\beta)}{t_1 \Gamma_0}} \tag{6.19}$$

In this case ε_H and η_H becomes

$$\varepsilon_H = \frac{1 - \beta}{\beta}\left(\frac{\phi}{\lambda}\right)^{-\frac{2\beta}{2\beta - 1}} \quad , \quad \eta_H = \frac{2 - \beta}{\beta}\left(\frac{\phi}{\lambda}\right)^{-\frac{2\beta}{2\beta - 1}} = \frac{2 - \beta}{1 - \beta}\varepsilon_H \tag{6.20}$$

The final value of ϕ_f is given by

$$\left(\frac{\phi_f}{\lambda}\right)^{\frac{2\beta}{2\beta - 1}} = \frac{1 - \beta}{\beta} \tag{6.21}$$

The number of e-folds is

$$N = \left(\frac{\phi}{\lambda}\right)^{\frac{2\beta}{2\beta - 1}} - \left(\frac{\phi_f}{\lambda}\right)^{\frac{2\beta}{2\beta - 1}} = \left(\frac{\phi}{\lambda}\right)^{2\beta} - \frac{1 - \beta}{\beta} \tag{6.22}$$

Hence

$$\left(\frac{\phi}{\lambda}\right)^{\frac{2\beta}{2\beta - 1}} = N + \frac{1 - \beta}{\beta} \tag{6.23}$$

The scalar spectral index is

$$\delta_{ns} = \frac{4 + \beta}{2\beta}\left(\frac{\phi}{\lambda}\right)^{-\frac{2\beta}{2\beta - 1}} = \frac{4 + \beta}{2(\beta N + 1 - \beta)} \approx \frac{4 + \beta}{2\beta}\frac{1}{N} \tag{6.24}$$

which can be written

$$\beta = \frac{4}{2N\delta_{ns} - 1} \tag{6.25}$$

Inserting the Planck value $\delta_{ns} = 0.032$ and $N = 60$ gives $\beta = 1.4$ outside the range $\beta < 1$ which requires $N > 78$. However, in the anisotropic case considered by Sharif and Saleem, one may obtain agreement with the Planck data for $\beta < 1$. As noted above, the tensor to scalar ratio has a very small value in these models. The time evolution of the inflaton field is given by Equation (6.7).

7. Comparison of Models

The models of Sharif and Saleem are a class of the monomial models. Comparing Equations (4.1) and (6.9) we have $p = 4(\beta - 1)$ or $\beta = 1 + p/4$. Hence, for $\beta < 1$ we must have $p < 0$ while Visinelli considered models with $p > 0$. Furthermore, in the first case of Sharif and Saleem with $\Gamma = V$ we have

$q = 2p$ and in the case with $\Gamma = \Gamma_0$ we have $q = 0$. Also, it should be noted that Visinelly has deduced the expression for the spectral parameters from the potential slow roll parameters, while Sharif and Saleem have used the Hubble slow roll parameters, and they have got slightly different expressions.

Let us consider an isotropic monomial model with scale as given in Equation (6.3). Then, we have two formulae for the potential—Equations (4.1) and (6.9). Hence

$$t_1 = \left(\sqrt{3}\beta\right)^{\frac{1}{\beta}} \left[8\left(1-\beta\right)\right]^{\frac{1-\beta}{\beta}} \left(\frac{M_P}{M}\right)^{2/\beta} t_P \tag{7.1}$$

where $t_P = 1/M_P$ is the Planck time. As mentioned above in Sharif and Saleem's first case $\Gamma = \Gamma(\phi) = \kappa V(\phi)/M_P$. Combining this with the first Equation (3.3) we get $\Gamma = 3H^2/M_P$. Furthermore they considered the strong dissipative regime with $\Gamma >> 3H$. Hence $H >> M_P$. The slow roll era begins at a point of time, t_i, when the inflaton field is given by Equation (6.23). This leads to

$$t_i = \left(N + \frac{1-\beta}{\beta}\right)^{1/\beta} t_1 \tag{7.2}$$

The Hubble parameter is given by the first equation in (6.4) with a maximal value at the beginning of the inflationary era. Hence, the condition $H >> M_P$ requires that

$$t_i = \left(\frac{\beta}{M_P t_1}\right)^{\frac{1}{1-\beta}} t_1 \tag{7.3}$$

Inserting the expression (7.2) for t_1 we arrive at

$$t_i << \left(\beta N + 1 - \beta\right) t_P \tag{7.4}$$

Hence in this model with for example $\beta = 1/2$ and $N = 60$ the inflationary era begins much earlier than at around 30 Planck times. Inserting the inequality (7.4) into Equation (7.1) we get

$$M >> \sqrt{\sqrt{3}\left[8\left(1-\beta\right)\right]^{1-\beta}\left(\beta N + 1 - \beta\right)} M_P \tag{7.5}$$

Hence $M >> M_P$, so these models are large field inflation models.

V. Kamali and M. R. Setare [49] have considered warm viscous inflation models in the context of brane cosmology using the so-called chaotic potential (3.1) with $p = 2$, i.e., $\beta = 3/2$. We have considered the corresponding models in ordinary (not brane) spacetime which corresponds to taking the limit that the brane tension $\lambda \to \infty$ in their equations. They first considered the case $\Gamma = \Gamma_0$, i.e., $q = 0$. Then, the time evolution of the inflaton field is given by Equation (4.9) with $p = 2$. As noted above, in this case $\delta_{ns} = 0.012$ which is smaller than the preferred value from the Planck data. It may be noted that Kamali and M. R. Setare got a different result. Letting $\lambda \to \infty$ in their Equation (68) gives $\delta_{ns} = 0$, i.e., a scale invariant spectrum.

Next, they considered the case $\Gamma = \Gamma(\phi) = \alpha V(\phi)$. With $\alpha = 1$ this corresponds to the first case considered by Sharif and Saleem [37].

8. Conclusions

Warm inflation is a promising model of inflation, taking account of dissipative processes that are neglected in the usual, cold inflationary models. In warm inflation, radiation is produced by dissipation of the inflaton field, and reheating is not necessary. This type of inflationary model was introduced and developed initially by Berera and coworkers. Also, interactions between the inflaton field and the radiation provide a mechanism for producing viscosity.

In this article, I have given a review of some recent models with particular emphasis on their predictions of optical parameters, making it possible to evaluate the models against the observational data obtained by the Planck team. In particular, power law potential inflation, PI, and natural inflation, NI, in the warm inflation scenario have been considered.

I have emphasized that there are some interesting differences between the predictions of these models and the corresponding cold inflation models. The first thing to be noted is that the warm inflation models in general predict a vanishingly small value of the tensor-to-scalar ratio, r. I the present paper I have parametrized the scalar spectral index n_s by $\delta_{ns} = 1 - n_s$. The Planck data favor the value $\delta_{ns} = 0.032$, $r < 0.11$ and a number of e-folds $N = 60$.

Cold PI with the potential (4.1) predicts $\delta_{ns} = \frac{2(p+2)}{p+4N}$ and $r = \frac{16p}{p+4N}$. Inserting $\delta_{ns} = 0.032$ and $N = 60$ gives $p = 1.8$ and $r = 0.12$. The corresponding warm PI model with constant value of the dissipation parameter Γ predicts, according to Equation (6.24), $\delta_{ns} = \frac{20+p}{4+p}\frac{1}{2N}$ giving $p = 2.8$. The corresponding model with $\Gamma = \Gamma(\phi) = V(\phi)$ predicts $\delta_{ns} = -\frac{4+3p}{4+p}\frac{1}{N}$ giving $p = -2.36$. However, according to Equation (6.15), this model is only consistent for $-4 < p < -3.92$. Hence, this model is in conflict with the Planck data.

Cold natural inflation predicts

$$\delta_{ns} = b\frac{(2+b)\,e^{bN}+b}{(2+b)\,e^{bN}-b} \quad , \quad r = \frac{8b^2}{(2+b)\,e^{bN}-b} \quad , \quad b = \left(\frac{M_P}{M}\right)^2 \tag{8.1}$$

Inserting $\delta_{ns} = 0.032$ and $N = 60$ gives $b = 0.032$ or $M = 5.5M_P$, giving $r = 0.0006$. Since $M > M_P$ this is large field inflation according to the standard definition of this classification (Lyth [50], Dine and Pack [51]). The corresponding warm natural inflation model has two parameters, Γ and V_0, contained in β in the expression for δ_{ns}. Hence, some assumption concerning the relationship between Γ and V_0, is needed to make a prediction of the value of δ_{ns} in this model.

Acknowledgments: I would like to thank Luca Visinelli for useful correspondence concerning this work and the referees for valuable suggestions and for providing several references to old articles describing inflation models with dissipation of inflaton energy.

References

1. Berera, A. Warm inflation. *Phys. Rev. Lett.* **1995**, *75*, 3218–3221. [CrossRef] [PubMed]
2. Berera, A. Thermal properties of an inflationary universe. *Phys. Rev. D* **1996**, *54*, 2519–2534. [CrossRef]
3. Berera, A. Interpolating the stage of exponential expansion in the early universe: Possible alternative with no reheating. *Phys. Rev. D* **1997**, *55*, 3346–3357. [CrossRef]
4. Berera, A. Warm inflation in the adiabatic regime—A model, an existence proof for inflationary dynamics in quantum field theory. *Nucl. Phys. B* **2000**, *585*, 666–714. [CrossRef]
5. Berera, A. The warm inflationary universe. *Contemp. Phys.* **2006**, *47*, 33–49. [CrossRef]
6. Berera, A. Developments in inflationary cosmology. *Pramana* **2009**, *72*, 169–182. [CrossRef]
7. Berera, A.; Gleiser, M.; Ramos, R.O. Strong Dissipative Behavior in Quantum Field Theory. *Phys. Rev. D* **1998**, *58*, 123508. [CrossRef]
8. Hall, L.; Moss, I.G.; Berera, A. Constraining warm inflation with the cosmic microwave background. *Phys. Lett. B* **2004**, *589*, 1–6. [CrossRef]
9. Hall, L.; Moss, I.G.; Berera, A. Scalar perturbation spectra from warm inflation. *Phys. Rev. D* **2004**, *69*, 083525. [CrossRef]
10. Berera, A.; Moss, I.G.; Ramos, R.O. Warm Inflation and its Microphysical Basis. *Rep. Prog. Phys.* **2009**, *72*, 026901. [CrossRef]
11. Bartrum, S.; Berera, A.; Rosa, J.G. Warming up for Planck. *J. Cosmol. Astropart. Phys.* **2013**, *2013*, 025. [CrossRef]
12. Bastero-Gil, M.; Berera, A. Warm Inflation model building. *Int. J. Mod. Phys.* **2009**, *A24*, 2207–2240. [CrossRef]
13. Bastero-Gil, M.; Berera, A.; Ramos, R.O.; Rosa, J.G. General dissipation coefficient in low-temperature warm inflation. *J. Cosmol. Astropart. Phys.* **2013**, *2013*, 016. [CrossRef]

14. Bastero-Gil, M.; Berera, A.; Kronberg, N. Exploring parameter space of warm-inflation models. *J. Cosmol. Astropart. Phys.* **2015**, *2015*, 046. [CrossRef]

15. Abbott, L.F.; Farhi, E.; Wise, M.B. Particle production in the new inflationary cosmology. *Phys. Lett. B* **1982**, *117*, 29–33. [CrossRef]

16. Albrecht, A.; Steinhardt, P.J.; Turner, M.S.; Wilczek, F. Reheating an Inflationary Universe. *Phys. Rev. Lett.* **1982**, *48*, 1437–1440. [CrossRef]

17. Morikawa, M.; Sasaki, M. Entropy Production in the Inflationary Universe. *Prog. Theor. Phys.* **1984**, *72*, 782–798. [CrossRef]

18. Hosoya, A.; Sakagami, M. Time development of Higgs field at finite temperature. *Phys. Rev. D* **1984**, *29*, 2228–2239. [CrossRef]

19. Moss, I.G. Primordial inflation with spontaneous symmetry breaking. *Phys. Lett. B* **1985**, *154*, 120–124. [CrossRef]

20. Lonsdale, S.R.; Moss, I.G. A superstring cosmological model. *Phys. Lett. B* **1987**, *189*, 12–16. [CrossRef]

21. Yokoyama, J.; Maeda, K. On the Dynamics of the Power Law Inflation Due to an Exponential Potential. *Phys. Lett. B* **1988**, *207*, 31–35. [CrossRef]

22. Liddle, A.R. Power Law Inflation with Exponential Potentials. *Phys. Lett. B* **1989**, *220*, 502–508. [CrossRef]

23. Del Campo, S. Warm Inflationary Universe Models. In *Aspects of Today's Cosmology*; Alfonso-Faus, A., Ed.; InTech: Rijeka, Croatia, 2011.

24. Panotopoulos, G.; Videla, N. Warm $(\lambda/4)\,\phi^4$ inflationary universe model in light of Planck 2015 results. 2015, arXiv:1510.0698.

25. Bellini, M. Warm inflation and classicality conditions. *Phys. Lett. B* **1998**, *428*, 31–36. [CrossRef]

26. Lee, W.; Fang, L.-Z. Mass density perturbations from ination with thermal dissipation. *Phys. Rev. D* **1999**, *59*, 083503. [CrossRef]

27. Maia, J.M.F.; Lima, J.A.S. Extended warm inflation. *Phys. Rev. D* **1999**, *60*, 101301. [CrossRef]

28. Herrera, R.; del Campo, S.; Campuzano, C. Tachyon warm inflationary universe models. *J. Cosmol. Astropart. Phys.* **2006**, *2006*, 9. [CrossRef]

29. Del Campo, S.; Herrera, R.; Pavón, D. Cosmological perturbations in warm inflationary models with viscous pressure. *Phys. Rev. D* **2007**, *75*, 083518. [CrossRef]

30. Hall, L.M.H.; Peiris, H.V. Cosmological Constraints on Dissipative Models of Ination. *J. Cosmol. Astropart. Phys.* **2008**, *2008*, 027. [CrossRef]

31. Moss, I.G.; Xiong, C. On the consistency of warm inflation. 2008, arXiv:0808.0261. [CrossRef]

32. Deshamukhya, A.; Panda, S. Warm tachyonic inflation in warped background. *Int. J. Mod. Phys. D* **2009**, *18*, 2093–2106. [CrossRef]

33. Nozari, K.; Fazlpour, B. Non-Minimal Warm Ination and Perturbations on the Warped DGP Brane with Modified Induced Gravity. *Gen. Relativ. Gravit.* **2011**, *43*, 207–234. [CrossRef]

34. Cai, Y.F.; Dent, J.B.; Easson, D.A. Warm DBI Inflation. *Phys. Rev. D* **2011**, *83*, 101301. [CrossRef]

35. Cerezo, R.; Rosa, J.G. Warm inflection. *High Energy Phys.* **2013**, *2013*, 24. [CrossRef]

36. Sharif, M.; Saleem, R. Warm Anisotropic Inflationary Universe Model. *Eur. Phys. J. C* **2014**, *74*, 2738. [CrossRef]

37. Sharif, M.; Saleem, R. Warm anisotropic inflation with bulk viscous pressure in intermediate era. *Astropart. Phys.* **2015**, *62*, 241–248. [CrossRef]

38. Setare, M.R.; Kamali, V. Warm-intermediate inflationary model with viscous pressure in high dissipative regime. *Gen. Relativ. Gravit.* **2014**, *46*, 1698. [CrossRef]

39. Chimento, L.P.; Jacubi, A.S.; Zuccala, N.A.; Pavon, D. Synergistic warm ination. *Phys. Rev. D* **2002**, *65*, 083510. [CrossRef]

40. Kinney, W.H.; Kolb, E.W.; Melchiorri, A.; Riotto, A. Inflation model constraints from the Wilkinson Microwave Anisotropy Probe three-year data. *Phys. Rev. D* **2006**, *74*, 023502. [CrossRef]

41. Mishra, H.; Mohanty, S.; Nautiyal, A. Warm natural inflation. *Phys. Lett. B* **2012**, *710*, 245–250. [CrossRef]

42. Sánchez, J.C.; Bastero-Gill, B.M.; Berera, A.; Dimoupoulos, K. Warm hilltop inflation. *Phys. Rev. D* **2008**, *77*, 123527. [CrossRef]

43. Setare, M.R.; Sepehri, A.; Kamali, V. Constructing warm inflationary model in brane-antibrane system. *Phys. Lett. B* **2014**, *735*, 84–89. [CrossRef]

44. Kinney, W.H. Cosmology, inflation, and the physics of nothing. In *Techniques and Concepts of High-Energy Physics XII*; NATO Science Series; Springer: Berlin, Germany, 2003; Volume 123, pp. 189–243.

45. Ade, P.A.R.; Aghanim, N.; Arnaud, M.; Arroja, F.; Ashdown, M.; Aumont, J.; Baccigalupi, C.; Ballardini, M.; Banday, A.J.; Barreiro, R.B.; et al. Planck 2015 results. XIII. Constraints on inflation. 2015, arXiv:1502.01589.

46. Baumann, D. TASI Lectures on Inflation. 2012, arXiv:0907.5424.

47. Visinelli, L. Natural Warm Inflation. 2011, arXiv:1107.3523. [CrossRef]

48. Visinelli, L. Observational constraints on Monomial Warm Inflation. *JCAP07* **2016**, 054. [CrossRef]

49. Kamali, V.; Setare, M.R. Warm-viscous inflation model on the brane in light of Planck data. *Class. Quantum Gravity* **2015**, *32*, 235005. [CrossRef]

50. Lyth, D.H. Particle physics models of inflation. *Lect. Notes Phys.* **2008**, *738*, 81–118.

51. Dine, M.; Pack, L. Studies in small field inflation. *J. Cosmol. Astropart. Phys.* **2012**, *2012*, 033. [CrossRef]

8

A Brief History of Gravitational Waves

Jorge L. Cervantes-Cota [1], Salvador Galindo-Uribarri [1] and George F. Smoot [2,3,4,*]

[1] Department of Physics, National Institute for Nuclear Research, Km 36.5 Carretera Mexico-Toluca, Ocoyoacac, C.P. 52750 Mexico, Mexico; jorge.cervantes@inin.gob.mx (J.L.C.-C.); salvador.galindo@inin.gob.mx (S.G.-U.)
[2] Helmut and Ana Pao Sohmen Professor at Large, Institute for Advanced Study, Hong Kong University of Science and Technology, Clear Water Bay, Kowloon, 999077 Hong Kong, China
[3] Université Sorbonne Paris Cité, Laboratoire APC-PCCP, Université Paris Diderot, 10 rue Alice Domon et Leonie Duquet, 75205 Paris Cedex 13, France
[4] Department of Physics and LBNL, University of California; MS Bldg 50-5505 LBNL, 1 Cyclotron Road Berkeley, 94720 CA, USA
* Correspondence: gfsmoot@lbl.gov.

Academic Editors: Lorenzo Iorio and Elias C. Vagenas

Abstract: This review describes the discovery of gravitational waves. We recount the journey of predicting and finding those waves, since its beginning in the early twentieth century, their prediction by Einstein in 1916, theoretical and experimental blunders, efforts towards their detection, and finally the subsequent successful discovery.

Keywords: gravitational waves; General Relativity; LIGO; Einstein; strong-field gravity; binary black holes

1. Introduction

Einstein's General Theory of Relativity, published in November 1915, led to the prediction of the existence of gravitational waves that would be so faint and their interaction with matter so weak that Einstein himself wondered if they could ever be discovered. Even if they were detectable, Einstein also wondered if they would ever be useful enough for use in science. However, exactly 100 years after his theory was born, on 14 September 2015, these waves were finally detected and are going to provide scientific results.

In fact at 11:50:45 a.m. CET on 14 September 2015 Marco Drago—a postdoc—was seated in front of a computer monitor at the Max Planck Institute for Gravitational Physics in Hanover, Germany, when he received an e-mail, automatically generated three minutes before from the monitors of LIGO (for its acronym Laser Interferometer Gravitational wave Observatory). Marco opened the e-mail, which contained two links. He opened both links and each contained a graph of a signal similar to that recorded by ornithologists to register the songs of birds. One graph came from a LIGO station located at Hanford, in Washington State, and the other from Livingston Station in the state of Louisiana [1].

Marco is a member of a team of 30 physicists working in Hanover, analyzing data from Hanford and Livingston. Marco's duty is to be aware of and analyze the occurrence of an "event" that records the passage of a gravitational wave, in one of the four lines that automatically track the signals from the detectors in the two LIGO observatories on the other side of the Atlantic.

Marco noticed that the two graphs were almost identical, despite having been registered independently in sites separated by 1900 km (see Figure 1a); for comparison we include sonograms from animals (see Figure 1b,c). The time that elapsed between the two signals differed by about 7 milliseconds. These almost simultaneous records of signals coming from sites far away from each other, the similarity of their shapes and their large size, could not be anything but, either: a possible record of a gravitational wave traveling at the speed of light or, a "signal" artificially "injected" to the detectors by one of the four members of the LIGO program who are allowed to "inject" dummy signals. The reason why artificial signals are injected to the system is to the test whether the operation of the detectors is correct and if the duty observers are able to identify a real signal.

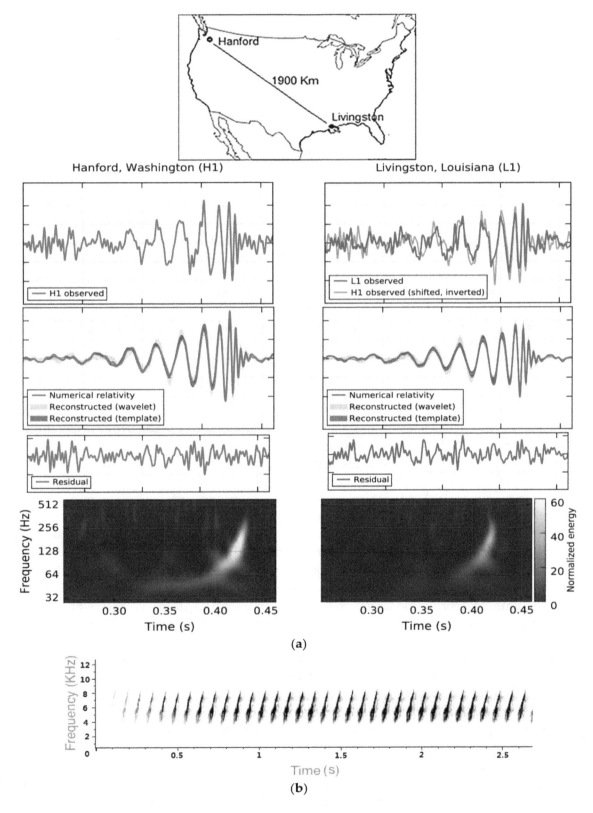

Figure 1. *Cont.*

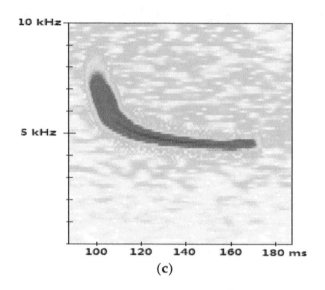

(c)

Figure 1. (a) Waveforms from LIGO sites [2] and their location and sonograms. Figure from https://losc.ligo.org/events/GW150914/. The gravitational-wave event GW150914 observed by the LIGO Hanford (H1, left column panels) and Livingston (L1, right column panels) detectors. Times are shown relative to 14 September 2015 at 09:50:45 UTC. For visualization, all time series are filtered with a 35–350 Hz band-pass filter to suppress large fluctuations outside the detectors' most sensitive frequency band, and band-reject filters to remove the strong instrumental spectral lines; (b) Chirping Sparrow (Spizella passerina) song: frequency versus time (in seconds), showing a song made up of a series of chirps; (c) A Pipistrelle bat call for echolocation. Bats use ultrasound for "seeing" and for social calls. This spectrogram is a graphic representation of frequencies against time. The color represents the loudness of each frequency. This spectrogram shows a falling call, which becomes a steady note. The yellow and green blotches are noise. This is nearly a time reverse of black hole merger sonograms.

Following the pre-established protocols, Marco tried to verify whether the signals were real or the "event" was just a dummy injected signal. Since aLIGO was still in engineering mode, there was no way to inject fake signals, i.e., hardware injections. Therefore, everyone was nearly 100% certain that this was a detection. However, it was necessary to go through the protocols of making sure that this was the case.

Marco asked Andrew Lundgren, another postdoc at Hanover, to find out if the latter was the case. Andrew found no evidence of a "dummy injection." On the other hand, the two signals detected were so clear, they did not need to be filtered to remove background noise. They were obvious. Marco and Andrew immediately phoned the control rooms at Livingston and Hanford. It was early morning in the United States and only someone from Livingston responded. There was nothing unusual to report. Finally, one hour after receiving the signal, Marco sent an e-mail to all collaborators of LIGO asking if anyone was aware of something that might cause a spurious signal. No one answered the e-mail.

Days later, LIGO leaders sent a report stating that there had not been any "artificial injection". By then the news had already been leaked to some other members of the world community of astrophysicists. Finally after several months, the official news of the detection of gravitational waves was given at a press conference on 11 February 2016, after the team had ascertained that the signals were not the result of some experimental failure, or any signal locally produced, earthquake, or electromagnetic fluctuation. This announcement is the most important scientific news so far this century. Gravitational waves were detected after 60 years of searching and 100 years since the prediction of their existence. The scientific paper was published in *Physical Review Letters* [2]. This discovery not only confirmed one of the most basic predictions of General Relativity but also opened a new window of observation of the universe, and we affirm without exaggeration that a new era in astronomy has been born.

In what follows we shall narrate the journey experienced in search of gravitational waves, including their conception in the early twentieth century, their prediction by Einstein in 1916, the theoretical controversy, efforts towards detection, and the recent discovery.

2. Lost and Found Gravitational Waves

On 5 July 1905 the *Comptes Rendus of the French Academy of Sciences* published an article written by Henri Poincare entitled "Sur la dynamique d' l'électron". This work summarized his theory of relativity [3]. The work proposed that gravity was transmitted through a wave that Poincaré called a gravitational wave (*onde gravifique*).

It would take some years for Albert Einstein to postulate in 1915 in final form the Theory of General Relativity [4]. His theory can be seen as an extension of the Special Theory of Relativity postulated by him 10 years earlier in 1905 [5]. The General Theory explains the phenomenon of gravity. In this theory, gravity is not a force—a difference from Newton's Law—but a manifestation of the curvature of space–time, this curvature being caused by the presence of mass (and also energy and momentum of an object). In other words, Einstein's equations match, on the one side, the curvature of local space–time with, on the other side of the equation, local energy and momentum within that space–time.

Einstein's equations are too complicated to be solved in full generality and only a few very specific solutions that describe space–times with very restrictive conditions of symmetry are known. Only with such restrictions it is possible to simplify Einstein's equations and so find exact analytical solutions. For other cases one must make some simplifications or approximations that allow a solution, or there are cases where equations can be solved numerically using computers, with advanced techniques in the field called Numerical Relativity.

Shortly after having finished his theory Einstein conjectured, just as Poincaré had done, that there could be gravitational waves similar to electromagnetic waves. The latter are produced by accelerations of electric charges. In the electromagnetic case, what is commonly found is dipolar radiation produced by swinging an electric dipole. An electric dipole is formed by two (positive and negative) charges that are separated by some distance. Oscillations of the dipole separation generate electromagnetic waves. However, in the gravitational case, the analogy breaks down because there is no equivalent to a negative electric charge. There are no negative masses. In principle, the expectation of theoretically emulate gravitational waves similar to electromagnetic ones faded in Einstein's view. This we know from a letter he wrote to his colleague, Karl Schwarzschild, on 19 February 1916. In this letter, Einstein mentioned in passing:

> "*Since then [November 14] I have handled Newton's case differently, of course, according to the final theory [the theory of General Relativity]. Thus there are no gravitational waves analogous to light waves. This probably is also related to the one-sidedness of the sign of the scalar T, incidentally [this implies the nonexistence of a "gravitational dipole"] [6]*

However, Einstein was not entirely convinced of the non-existence of gravitational waves; for a few months after having completed the General Theory, he refocused efforts to manipulate his equations to obtain an equation that looked like the wave equation of electrodynamics (Maxwell's wave equation), which predicts the existence of electromagnetic waves. However, as mentioned, these equations are complex and Einstein had to make several approximations and assumptions to transform them into something similar to Maxwell's equation. For some months his efforts were futile. The reason was that he used a coordinate system that hindered his calculations. When, at the suggestion of a colleague, he changed coordinate systems, he found a solution that predicted three different kinds of gravitational waves. These three kinds of waves were baptized by Hermann Weyl as longitudinal-longitudinal, transverse-longitudinal, and transverse-transverse [7].

These approaches made by Einstein were long open to criticism from several researchers and even Einstein had doubts. In this case Einstein had manipulated his field equations into a first approximation for wave-emitting bodies whose own gravitational field is negligible and with waves that propagate in empty and flat space.

Yet the question of the existence of these gravitational waves dogged Einstein and other notable figures in the field of relativity for decades to come. By 1922 Arthur Eddington wrote an article entitled "The propagation of gravitational waves" [8]. In this paper, Eddington showed that two of the three types of waves found by Einstein could travel at any speed and this speed depends on the coordinate system; therefore, they actually were spurious waves. The problem Eddington found in Einstein's original calculations is that the coordinate system he used was in itself a "wavy" system and therefore two of the three wave types were simply flat space seen from a wavy coordinates system; i.e., mathematical artifacts were produced by the coordinate system and were not really waves at all. So the existence of the third wave (the transverse-transverse), allegedly traveling at the speed of light, was also questioned. Importantly, Eddington did prove that this last wave type propagates at the speed of light in all coordinate systems, so he did not rule out its existence.

In 1933 Einstein emigrated to the United States, where he had a professorship at the Institute for Advanced Study in Princeton. Among other projects, he continued to work on gravitational waves with the young American student Nathan Rosen.

In 1936 Einstein wrote to his friend, renowned physicist Max Born, "Together with a young collaborator [Rosen], I arrive at the interesting result that *gravitational waves do not exist*, though they have been assumed a certainty to the first approximation" (emphasis added) [9] (p. 121, Letter 71).

That same year, Einstein and Rosen sent on 1 June an article entitled "Are there any gravitational waves?" to the prestigious journal *Physical Review*, whose editor was John T. Tate [10]. Although the original version of the manuscript does not exist today, it follows from the abovementioned letter to Max Born that the answer to the title of the article was "they do not exist".

The editor of the *Physical Review* sent the manuscript to Howard Percy Robertson, who carefully examined it and made several negative comments. John Tate in turn wrote to Einstein on 23 July, asking him to respond to the reviewer's comments. Einstein's reaction was anger and indignation; he sent the following note to Tate [10]:

July 27, 1936

Dear Sir.

 "We (Mr. Rosen and I) had sent you our manuscript for publication and had not authorized you to show it to specialists before it is printed. I see no reason to address the—in any case erroneous—comments of your anonymous expert. On the basis of this incident I prefer to publish the paper elsewhere."

Respectfully

Einstein

P.S. Mr. Rosen, who has left for the Soviet Union, has authorized me to represent him in this matter.

On July 30th, John Tate replied to Einstein that he very much regretted the withdrawal of the article, saying "I could not accept for publication in *The Physical Review* a paper which the author was unwilling I should show to our Editorial Board before publication" [10].

During the summer of 1936 a young physicist named Leopold Infeld replaced Nathan Rosen as the new assistant to Einstein. Rosen had departed a few days before for the Soviet Union. Once he arrived at Princeton, Infeld befriended Robertson (the referee of the Einstein–Rosen article). During one of their encounters the topic of gravitational waves arose. Robertson confessed to Infeld his skepticism about the results obtained by Einstein. Infeld and Robertson discussed the point and reviewed together the Einstein and Rosen manuscript, confirming the error. Infeld in turn informed Einstein about the conversation with Robertson.

An anecdote illustrating the confused situation prevailing at that time is given in Infeld's autobiography. Infeld refers to the day before a scheduled talk that Einstein was to give at Princeton on the "Nonexistence of gravitational waves". Einstein was already aware of the error in his manuscript, which was previously pointed out by Infeld. There was no time to cancel the talk. The next day Einstein gave his talk and concluded, "*If you ask me whether there are gravitational waves or not, I must answer that I don't know. But it is a highly interesting problem*" [10].

After having withdrawn the Einstein–Rosen paper from the *Physical Review*, Einstein had summited the very same manuscript to the *Journal of the Franklin Society* (Philadelphia). This journal accepted the paper for publication without modifications. However, after Einstein learned that the paper he had written with Rosen was wrong, he had to modify the galley proofs of the paper. Einstein sent a letter to the editor on 13 November 1936 explaining the reasons why he had to make fundamental changes to the galley proofs. Einstein also renamed the paper, entitling it "On gravitational waves", and modified it to include different conclusions [10]. It should be noted that this would not have happened if Einstein had accepted in the first instance Robertson's valid criticisms. Tellingly, the new conclusions of his rewritten article read [11]:

"Rigorous solution for Gravitational cylindrical waves is provided. For convenience of the reader the theory of gravitational waves and their production, known in principle, is presented in the first part of this article. After finding relationships that cast doubt on the existence of gravitational fields rigorous wavelike solutions, we have thoroughly investigated the case of cylindrical gravitational waves. As a result, there are strict solutions and the problem is reduced to conventional cylindrical waves in Euclidean space".

Furthermore, Einstein included this explanatory note at the end of his paper [11],

"*Note—The second part of this article was considerably altered by me after the departure to Russia of Mr. Rosen as we had misinterpreted the results of our formula. I want to thank my colleague Professor Robertson for their friendly help in clarifying the original error. I also thank Mr. Hoffmann your kind assistance in translation.*"

In the end, Einstein became convinced of the existence of gravitational waves, whereas Nathan Rosen always thought that they were just a formal mathematical construct with no real physical meaning.

3. Pirani's Trip to Poland; the Effect of a Gravitational Wave

To prove the existence of a gravitational wave it becomes necessary to detect its effects. One of the difficulties presented by the General Theory of Relativity resides in how to choose the appropriate coordinate system in which one observer may calculate an experimentally measurable quantity, which could, in turn, be compared to a real observation. Coordinate systems commonly used in past calculations were chosen for reasons of mathematical simplification and not for reasons of physical convenience. In practice, a real observer in each measurement uses a local Cartesian coordinate system relative to its state of motion and local time. To remedy this situation, in 1956 Felix A. E. Pirani published a work that became a classic article in the further development of the Theory of Relativity. The article title was "On the physical significance of the Riemann tensor" [12]. The intention of this work was to demonstrate a mathematical formalism for the deduction of physical observable quantities applicable to gravitational waves. Curiously, the work was published in a Polish magazine. The reason for this was that Pirani, who at that time worked in Ireland, went to Poland to visit his colleague Leopold Infeld, of whom we have already spoken; the latter had returned to his native Poland in 1950 to help boost devastated Polish postwar physics. Because Infeld went back to Poland, and because of the anti-communist climate of that era, Infeld was stripped of his Canadian citizenship. In solidarity Pirani visited Poland and sent his aforementioned manuscript to *Acta Physica Polonica*. The importance of Pirani's Polish paper is that he used a very practical approach that got around this whole problem of the coordinate system, and he showed that the waves would move particles back and forth as they pass by.

4. Back and Forth as Waves Pass by

One of the most famous of Einstein's collaborators, Peter Bergmann, wrote a well-known popular book *The Riddle of Gravitation*, which describes the effect a gravitational wave passing over a set of particles would have [13]. Following Bergmann, we shall explain this effect.

When a gravitational wave passes through a set of particles positioned in an imaginary circle and initially at rest, the passing wave will move these particles. This motion is perpendicular (transverse) to the direction in which the gravitational wave travels. For example, suppose a gravitational pulse passes in a direction perpendicular to this page, Figure 2 shows how a set of particles, initially arranged in a circle, would sequentially move (a, b, c, d).

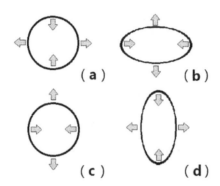

Figure 2. (**a,b,c,d**) Sequential effect of a gravitational wave on a ring of particles. In the image of Figure 2a is observed as the particles near horizontal move away from each other while those are near vertical move together to reach finally the next moment as shown in (Figure 2b). At that moment all the motions are reversed and so on. This is shown in Figure 2c,d. All these motions occur successively in the plane perpendicular (transverse) to the direction of wave propagation.

At first sight the detection of gravitational waves now seems very simple. One has to compare distances between perpendicularly placed pairs of particles and wait until a gravitational wave transits. However, one has to understand in detail how things happen. For instance, a ruler will not stretch, in response to a gravitational wave, in the same way as a free pair of particles, due to the elastic properties of the ruler, c.f. note 11, p. 19 in [14]. Later it was realized that changes produced in such disposition of particle pairs (or bodies) can be measured if, instead of the distances, we measure the time taken by light to traverse them, as the speed of light is constant and unaltered by a gravitational wave.

Anyway, Pirani's 1956 work remained unknown among most physicists because scientists were focusing their attention on whether or not gravitational waves carry energy. This misperception stems from the rather subtle matter of defining energy in General Relativity. Whereas in Special Relativity energy is conserved, in General Relativity energy conservation is not simple to visualize. In physics, a conservation law of any quantity is the result of an underlying symmetry. For example, linear momentum is conserved if there is spatial translational symmetry, that is, if the system under consideration is moved by a certain amount and nothing changes. In the same way, energy is conserved if the system is invariant under time. In General Relativity, time is part of the coordinate system, and normally it depends on the position. Therefore, globally, energy is not conserved. However, any curved space–time can be considered to be locally flat and, locally, energy is conserved.

During the mid-1950s the question of whether or not gravitational waves would transmit energy was still a hot issue. In addition, the controversy could not be solved since there were no experimental observations that would settle this matter. However, this situation was finally clarified thanks to the already mentioned work by Pirani [12], and the comments suggested by Richard Feynman together with a hypothetical experiment he proposed. The experiment was suggested and comments were delivered by Feynman during a milestone Congress held in 1957 in Chapel Hill, North Carolina.

We will come back to this experiment later, but first we shall speak about the genesis of the Chapel Hill meeting.

5. What Goes Up Must Come Down

The interest in the search for gravitational waves began at a meeting occurred in Chapel Hill, North Carolina in 1957. The meeting brought together many scientists interested in the study of gravity. What is unusual is that this meeting would not have been possible without the funding of an eccentric American millionaire named Roger W. Babson.

On 19 January 1949 Roger W. Babson founded the Gravity Research Foundation (GRF), which still exists today. Babson's motivation for establishing the foundation was a "debt" that he thought he owed to Newton's laws—which, according to his understanding, led him to become a millionaire [15]. Babson earned the greatest part of his fortune in the New York Stock Exchange by applying his own version of Newton's Gravity law, "What goes up must come down". Thus he bought cheap shares on their upward route and sold them before their price collapsed. His ability to apply the laws of Newton was surprising because he anticipated the 1929 Wall Street crash. "To every action there is a reaction", he used to preach.

Babson's interest in gravity arose when he was a child, following a family tragedy. Babson's older sister drowned when he was still an infant. In his version of the unfortunate accident, he recalls, "... she could not fight gravity." The story of this eccentric millionaire is detailed on the website of the GRF foundation [15]. Babson became obsessed with finding a way to control the force of gravity and therefore he established the aforementioned foundation, which had as its main activity arranging a yearly essay contest that dealt with "the chances of discovering a partial insulation, reflector, or absorber of gravity". An annual award of $1000 (a considerable amount at the time) was offered to the best essay. The essays submitted for the competition were limited to 2000 words. This annual award attracted several bizarre competitors and was awarded several times to risible submissions. However, in 1953 Bryce DeWitt, a young researcher at Lawrence Livermore Laboratory in California, decided to write an essay and enter the contest because he needed the money to pay his home's mortgage.

The essay presented by DeWitt in the 1953 competition was a devastating critique of the belief that it is possible to control gravity. In DeWitt's own words, his writing "essentially nagged [the organizers] for that stupid idea" [16]. To his surprise, his essay was the winner despite having been written in one night. DeWitt notes those were "the faster 1000 dollars earned in my whole life!" [16].

But DeWitt never imagined he would earn many thousands more dollars than he won with his essay. The reason for it might be found in the final paragraph of his essay,

"In the near future, external stimuli to induce young people to engage in gravitational physics research, despite its difficulties, are urgently needed" [16].

This final paragraph of DeWitt's essay echoed in Babson's mind. Perhaps, he thought, why not focus my philanthropy to support serious studies of gravitation? Perhaps he thought his GRF could refocus its activities onto the scientific study of gravitation.

Babson shared this new enthusiasm with a friend, Agnew Bahnson, also a millionaire and also interested in gravity. Bahnson was a little more practical than Babson and convinced him to found an independent institute separate from GRF. Thus arose the idea of founding a new Institute of Field Physics (IOFP), whose purpose would be pure research in the gravitational fields. The idea of founding the IOFP was clever because the old GRF was severely discredited among the scientific community. For example, one of the promotional brochures GRF mentioned as an example of the real possibility of gravity control the biblical episode where Jesus walks on water. Such was the ridicule and vilification of the GRF in scientific circles that a famous popularizer of mathematics, Martin Gardner, devoted an entire chapter of one of his books to ridiculing the GRF. In this work Gardner claims that the GRF "is perhaps the most useless project of the twentieth century" [17]. So Bahnson knew that to research on gravity in the discredited GRF had very little chance of attracting serious scientists. We must mention that today GRF enjoys good prestige and many well-known scientists have submitted their essays to its annual competition. That proves that it is worth trying for a thousand dollars.

In order to start the new IOFP institute off on the right foot, Bahnson contacted a famous Princeton physicist, John Archibald Wheeler, who supported the idea of hiring Bryce DeWitt to lead the new institute, whose headquarters would be established in Chapel Hill, NC, Bahnson's hometown and headquarters of the University of North Carolina. Wheeler, knowing the vast fortune of the couple of millionaires, hastened to send a telegram to DeWitt. In one of his lines the telegram said "Please do not give him a 'no' for answer from the start" [16]. That's how DeWitt won more than one thousand dollars, actually much more. In January 1957 the IOFP was formally inaugurated, holding a scientific conference on the theme "The role of gravitation in Physics". As we shall review below, the Chapel Hill conference rekindled the crestfallen and stagnant study of gravitation prevailing in those days.

6. The Chapel Hill Conference 1957

The 1957 Chapel Hill conference was an important event for the study of gravity. Attendance was substantial: around 40 speakers from institutions from 11 countries met for six days, from 18 to 23 January 1957 on the premises of the University of North Carolina at Chapel Hill. Participants who attended the meeting were predominantly young physicists of the new guard: Feynman, Schwinger, Wheeler, and others. During the six-day conference, discussions focused on various topics: classical gravitational fields, the possibility of unification of gravity with quantum theory, cosmology, measurements of radio astronomy, the dynamics of the universe, and gravitational waves [18].

The conference played a central role in the future development of classical and quantum gravity. It should be noted that the Chapel Hill 1957 conference today is known as the GR1 conference. That is, the first of a series of GR meetings that have been held regularly in order to discuss the state of the art in matters of Gravitation and General Relativity (GR = General Relativity). The conference has been held in many countries and possesses international prestige. The last was held in New York City in 2016.

In addition to the issues and debates on the cosmological models and the reality of gravitational waves, during the conference many questions were formulated, including ideas that are topical even today. To mention a few, we can say that one of the assistants, named Hugh Everett, briefly alluded to his parallel universes interpretation of quantum physics. On the other hand, DeWitt himself pointed out the possibility of solving gravitation equations through the use of electronic computers and warned of the difficulties that would be encountered in scheduling them for calculations, thus foreseeing the future development of the field of Numerical Relativity. However, what concerns us here is that gravitational waves were also discussed at the conference; chiefly, the question was whether gravitational waves carrying energy or not.

Hermann Bondi, a distinguished physicist at King's College London, presided over session III of Congress entitled "General Relativity not quantized". In his welcome address to the participants he warned "...still do not know if a transmitter transmits energy radiation ..." [18] (p. 95). With these words Bondi marked the theme that several of the speakers dealt with in their presentations and subsequent discussion. Some parts of the debate focused on a technical discussion to answer the question about the effect a gravitational pulse would have on a particle when passing by, i.e., whether or not the wave transmits energy to the particle.

During the discussions, Feynman came up with an argument that convinced most of the audience.

His reasoning is today known as the "sticky bead argument". Feynman's reasoning is based on a thought experiment that can be described briefly as follows: Imagine two rings of beads on a bar (see Figure 3, upper part). The bead rings can slide freely along the bar. If the bar is placed transversely to the propagation of a gravitational wave, the wave will generate tidal forces with respect to the midpoint of the bar. These forces in turn will produce longitudinal compressive stress on the bar. Meanwhile, and because the bead rings can slide on the bar and also in response to the tidal forces, they will slide toward the extreme ends first and then to the center of the bar (Figure 3, bottom). If contact between the beads and the bar is "sticky", then both pieces (beads and bar) will be heated by friction. This heating implies that energy was transmitted to the bar by the gravitational wave, showing that gravitational waves carry energy [18].

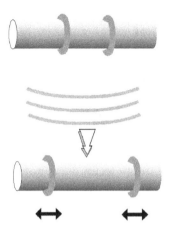

Figure 3. Sketch of the "sticky bead argument".

In a letter to Victor Weisskopf, Feynman recalls the 1957 conference in Chapel Hill and says, "I was surprised to find that a whole day of the conference was spent on this issue and that 'experts' were confused. That's what happens when one is considering energy conservation tensors, etc. instead of questioning, can waves do work?" [19].

Discussions on the effects of gravitational waves introduced at Chapel Hill and the "sticky bead argument" convinced many—including Hermann Bondi, who had, ironically, been among the skeptics on the existence of gravitational waves. Shortly after the Chapel Hill meeting Bondi issued a variant of the "sticky bead argument" [20].

Among the Chapel Hill audience, Joseph Weber was present. Weber was an engineer at the University of Maryland. He became fascinated by discussions about gravitational waves and decided to design a device that could detect them. Thus, while discussions among theoretical physicists continued in subsequent years, Weber went even further because, as discussed below, he soon began designing an instrument to make the discovery.

7. The First Gravitational Wave Detector

The year following the meeting at Chapel Hill, Joseph Weber began to speculate how he could detect gravitational waves. In 1960 he published a paper describing his ideas on this matter [21]. Basically he proposed the detection of gravitational waves by measuring vibrations induced in a mechanical system. For this purpose, Weber designed and built a large metal cylinder as a sort of "antenna" to observe resonant vibrations induced in this antenna that will eventually be produced by a transit of a gravitational wave pulse. This is something like waiting for someone to hit a bell with a hammer to hear its ring.

It took his team several years to build the "antenna", a task that ended by the mid-sixties. In 1966, Weber, in a paper published in *Physical Review*, released details of his detector and provided evidence of its performance [22]. His "antenna" was a big aluminum cylinder about 66 cm in diameter and 153 cm in length, weighing 3 tons. The cylinder was hanging by a steel wire from a support built to isolate vibrations of its environment (see Figure 4). In addition, the whole arrangement was placed inside a vacuum chamber. To complete his instrument, Weber placed a belt of detectors around the cylinder. The detectors were piezoelectric crystals to sense cylinder vibrations induced by gravitational waves. Piezoelectric sensors convert mechanical vibrations into electrical impulses.

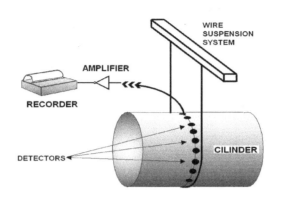

Figure 4. Sketch of Weber's cylinder detector and photo of Joseph Weber at the antenna.

Weber built two detectors. The first one was at the University of Maryland and the other was situated 950 km away, in Argonne National Laboratory near Chicago. Both detectors were connected to a registration center by a high-speed phone line. The idea of having two antennas separated by a large distance allowed Weber to eliminate spurious local signals, that is, signals produced by local disturbances such as thunderstorms, cosmic rays showers, power supply fluctuations, etc. In other words, if a detected signal was not recorded simultaneously in both laboratories, the signal should be discarded because it was a local signal and therefore spurious.

For several years, Weber made great efforts to isolate his cylinders from spurious vibrations, local earthquakes, and electromagnetic interference, and argued that the only significant source of background noise was random thermal motions of the atoms of the aluminum cylinder. This thermal agitation caused the cylinder length to vary erratically by about 10^{-16} meters, less than the diameter of a proton; however, the gravitational signal he anticipated was not likely to get much greater than the threshold stochastic noise caused by thermal agitation.

It took several years for Weber and his team to begin detecting what they claimed were gravitational wave signals. In 1969 he published results announcing the detection of waves [23]. A year later, Weber claimed that he had discovered many signals that seemed to emanate from the center of our galaxy [24]. This meant that in the center of the Milky Way a lot of stellar mass became energy ($E = mc^2$) in the form of gravitational waves, thus reducing the mass of our galaxy. This "fact" presented the problem that a mass conversion into energy as large as Weber's results implied involved a rapid decrease of the mass that gravitationally keeps our galaxy together. If that were the case, our galaxy would have already been dispersed long ago. Theoretical physicists Sciama, Field, and Rees calculated that the maximum conversion of mass into energy for the galaxy, so as not to expand more than what measurements allowed, corresponded to an upper limit of 200 solar masses per year [25]. However, Weber's measurements implied that a conversion of 1000 solar masses per year was taking place. Something did not fit. Discussions took place to determine what mechanisms could make Weber's measurements possible. Among others, Charles Misner, also from the University of Maryland, put forward the idea that signals, if stemming from the center of the Milky Way, could have originated by gravitational synchotron radiation in narrow angles, so as to avoid the above constraints considered for isotropic emission. Some others, like Peter Kafka of the Max Planck Institute in Munich, claimed in an essay for the Gravity Research Foundation's contest in 1972 (in which he won the second prize) that Weber's measurements, if they were isotropically emitted, and taking into account the inefficiency of bars, would imply a conversion of three million solar masses per year in the center of the Milky Way [26]. It soon became clear that Weber's alleged discoveries were not credible. Weber's frequent observations of gravitational waves related to very sporadic events and raised many suspicions among some scientists. It seemed that Weber was like those who have a hammer in hand and to them everything looks like a nail to hit.

Despite Weber's doubtful measurements, he began to acquire notoriety. In 1971, the famous magazine *Scientific American* invited him to write an article for their readers entitled "The detection of gravitational waves" [27].

Whether it was the amazing—for some—findings of Weber, or doubtful findings of others, or the remarks made by Sciama, Field, Rees, and Kafka, the fact is that many groups of scientists thought it was a good idea to build their own gravitational wave detectors to repeat and improve on Weber's measurements. These first-generation antennas were aluminum cylinders weighing about 1.5 tons and operating at room temperature [28]. Joseph Weber is considered a pioneer in experimental gravitation and therefore he is honored by the American Astronomical Society, which awards every year the Joseph Weber Award for Astronomical Instrumentation.

By the mid-seventies, several detectors were already operative and offered many improvements over Weber's original design; some cylinders were even cooled to reduce thermal noise. These experiments were operating in several places: at Bell Labs Rochester-Holmdel; at the University of Glasgow, Scotland; in an Italo-German joint program in Munich and Frascati; in Moscow; in Tokyo; and at the IBM labs in Yorktown Heights [28]. As soon as these new instruments were put into operation, a common pattern emerged: there were no signals. In the late seventies, everyone except Weber himself agreed that his proclaimed detections were spurious. However, the invalidation of Weber's results urged other researchers to redouble the search for gravitational waves or devise indirect methods of detection.

At that time great pessimism and disappointment reigned among the "seekers" of gravitational waves. However, in 1974 an event occurred that raised hopes. In that year Joseph Hooten Taylor and Alan Russell Hulse found an object in the sky (a binary pulsar) that revealed that an accelerated mass radiated gravitational energy. While this observation did not directly detect gravitational waves, it pointed to their existence. The announcement of the detection of gravitational radiation effects was made in 1979 [29].

This announcement sparked renewed interest in the future discovery of gravitational waves, and urged other researchers to redouble the search for the lost waves and devise other methods of detection. Some were already trying the interferometric method (see Figure 5).

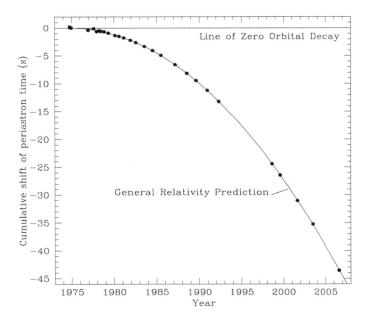

Figure 5. Binary Pulsar Advance of the Periastron (point of closest approach of the stars) versus Time. General Relativity predicts this change because of the energy radiated away by gravitational waves. Hulse and Taylor were awarded the Nobel Prize for this observation in 1993. Figure taken from (Living Rev.Rel.11:8, 2008).

To proceed with the description of the method that uses interferometers, it is necessary to know the magnitude of the expected effects of a gravitational wave on matter. This magnitude is properly quantified by the "h" parameter.

8. The Dimensionless Amplitude, h

The problem with gravitational waves, as recognized by Einstein ever since he deduced for the first time their existence, is that their effect on matter is almost negligible. Among other reasons, the value of the gravitational constant is very, very small, which makes a possible experimental observation extremely difficult.

Furthermore, not all waves are equal, as this depends on the phenomenon that generates them; nor is the effect of a wave on matter has the same intensity. To evaluate the intensity of the effect that a particular wave produces on matter, a dimensionless factor, denoted by the letter "h," has been defined. The dimensionless amplitude "h" describes the maximum displacement per unit length that would produce waves on an object. To illustrate this definition we refer to Figure 6. This figure shows two particles represented by gray circles. The pair is shown originally spaced by a distance "l" and locally at rest.

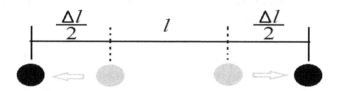

Figure 6. Definition of dimensionless amplitude h = Δl/L.

By impinging a gravitational wave perpendicularly on the sheet of paper, both particles are shifted respective to the positions marked by black circles. This shift is denoted by Δl/ 2, which means a relative shift between the pair of particles is now equal to Δl/L ≈ h, where Δl is the change in the spacing between particles due to gravitational wave, l is the initial distance between particles, and h is the dimensionless amplitude. In reality the factor h is more complex and depends upon the geometry of the measurement device, the arrival direction, and the frequency and polarization of the gravitational wave [14]. Nature sets a natural amplitude of $h \sim 10^{-21}$.

This factor h is important when considering the design of a realistic gravitational wave detector. We must mention that the value of h depends on the kind of wave to be detected and this in turn depends on how the wave was produced and how far its source is from an observer. Later we shall return to the subject and the reader shall see the practicality of the factor h.

To identify the sources that produce gravitational waves it is important to consider their temporal behavior. Gravitational waves are classified into three types: stochastic, periodic, and impulsive (bursts) [28]. Stochastic waves contribute to the gravitational background noise and possibly have their origin in the Big Bang. There are also expected stochastic backgrounds due to Black Hole-Black Hole coalescences. These types of waves fluctuate randomly and would be difficult to identify and separate due to the background noise caused by the instruments themselves. However, their identification could be achieved by correlating data from different detectors; this technique applies to other wave types too [14]. The second type of wave, periodic, corresponds to those whose frequency is more or less constant for long periods of time. Their frequency can vary up to a limit (quasiperiodic). For example, these waves may have their origin in binary neutron stars rotating around their center of mass, or from a neutron star that is close to absorb material from another star (accreting neutron star). The intensity of the generated waves depends on the distance from the binary source to the observer. The third type of wave comes from impulsive sources such as bursts that emit pulses of intense gravitational radiation. These may be produced during the creation of Black Holes in a supernova explosion or through the merging of two black holes. The greater the mass, the more intense the signal. They radiate at a frequency inversely proportional to their mass. Such sources are more intense and are expected to have higher amplitudes. Figure 7 shows various examples of possible sources of gravitational waves in which the three different wave types appear in different parts of the spectrum.

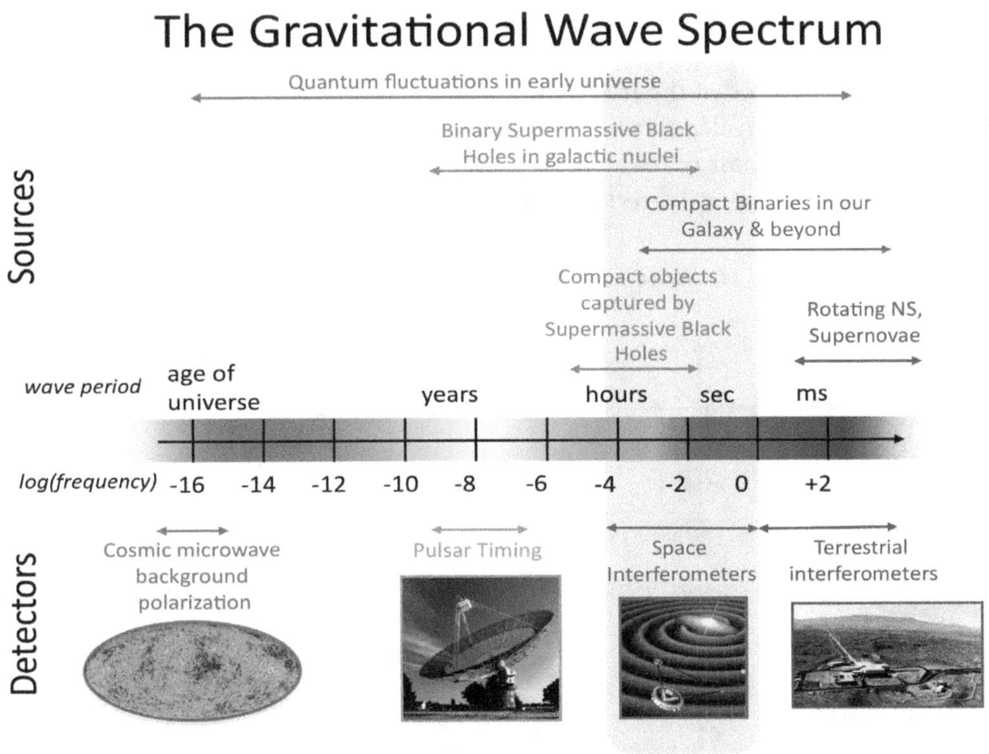

The Gravitational Wave Spectrum

Figure 7. Gravitational wave spectrum showing wavelength and frequency along with some anticipated sources and the kind of detectors one might use. Figure credit: NASA Goddard Space Flight Center.

Different gravitational phenomena give rise to different gravitational wave emissions. We expect primordial gravitational waves stemming from the inflationary era of the very early universe. Primordial quantum fields fluctuate and yield space–time ripples at a wide range of frequencies. These could in principle be detected as B-mode polarization patterns in the Cosmic Microwave Background radiation, at large angles in the sky. Unsuccessful efforts have been reported in recent years, due to the difficulty of disentangling the *noisy* dust emission contribution of our own galaxy, the BICEP2 and PLANCK projects. On the other hand, waves of higher frequencies but still very long wavelengths arising from the slow inspiral of massive black holes in the centers of merged galaxies will cause a modified pulse arrival timing, if very stable pulsars are monitored. Pulsar timing also places the best limits on potential gravitational radiation from cosmic string residuals from early universe phase transitions. Other facilities are planned as space interferometers, such as the Laser Interferometer Space Antenna (LISA), which is planned to measure frequencies between 0.03 mHz and 0.1 Hz. LISA plans to detect gravitational waves by measuring separation changes between fiducial masses in three spacecrafts that are supposed to be 5 million kilometers apart! The expected sources are merging of very massive Black Holes at high redshifts, which corresponds to waves emitted when the universe was 20 times smaller than it is today. It should also detect waves from tens of stellar-mass compact objects spiraling into central massive Black Holes that were emitted when the universe was one half of its present size. Last but not least, Figure 7 shows terrestrial interferometers that are planned to detect waves in the frequency from Hertz to 10,000 Hertz. The most prominent facilities are those of LIGO in the USA, VIRGO in Italy, GEO600 in Germany, and KAGRA in Japan, which are all running or expected to run soon. They are just beginning to detect Black Hole mergings, as was the case of the 14 September event [2]. We will go into detail about the interferometric technique later.

9. The Origin of the Interferometric Method

It is not known for sure who invented the interferometer method to detect gravitational waves, possibly because the method had several precursors. After all, an idea can arise at the same time among various individuals and this indeed seems to be the case here. However, before going into detail about the historical origins of the method, we shall briefly discuss the basics of this technique.

Figure 8 shows a very simplified interferometer. It consists of a light source (a laser), a pair of reflective mirrors attached to a pair of test masses (not shown in the figure), a beam splitter (which can be a semi-reflecting mirror or half-silvered mirror), and a light detector or photodetector.

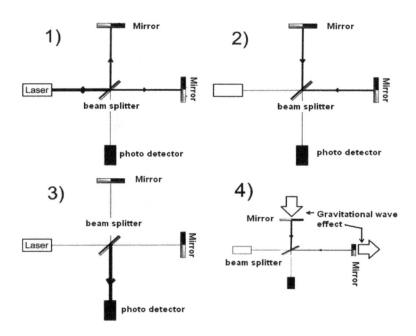

Figure 8. Schematic of an interferometer for detecting gravitational waves.

The laser source emits a beam of monochromatic light (i.e., at a single frequency) that hits the beam splitter surface. This surface is partially reflective, so part of the light is transmitted through to the mirror at the right side of the diagram while some is reflected to the mirror at the upper side of the sketch (Figure 8(1)). Then, as seen in Figure 8(2), both beams recombine when they meet at the splitter and the resulting beam is reflected toward the detector (Figure 8(3)). Finally, the photodetector measures the light intensity of the recombined beam. This intensity is proportional to the square of the height of the recombined wave.

Initially, both reflecting mirrors are positioned at nearly the same distance from the beam splitter. In reality what is needed is that the interferometer is locked on a dark fringe. Deviations from a dark fringe are then measured with the passage of a gravitational wave.

If the distance between one of the mirrors to the light splitter varies by an amount Δl with respect to distance to the same splitter of the second mirror, then the recombined beam will change its intensity. From measuring the intensity change of the recombined light beam, it is possible to obtain Δl.

When a gravitational wave passes through the interferometer at a certain direction, for example perpendicular to the plane where the pair of mirrors lies, both mirrors shift positions. One of the mirrors slightly reduces its distance to the beam splitter, while the second mirror slightly increases its distance to the splitter (see Figure 8(4)). The sum of the two displacements is equal to Δl. The photodetector records a variation in the intensity of the recombined light, thereby detecting the effect of gravitational waves.

A very important specific feature of the interferometer effectiveness is given by the length of its arms. This is the distance "l" between the wave splitter and its mirrors. On the other hand, the wavelength of the gravitational wave sets the size of the detector L needed. The optimal size of the arms turns out to be one-fourth of the wavelength. For a typical gravitational wave frequency of 100 Hz, this implies L = 750 km, which is actually too long to make except by folding the beams back and forth via the Fabry–Pérot technique, which helps to achieve the desired optimal size. In practice it is of the utmost importance to have an interferometer with very long arms matching the frequencies one plans to observe.

The importance of a long arm "l" is easy to explain if we remember the definition of the dimensionless amplitude h = Δl/L, which we already discussed above (see Figure 6). If a gravitational wave produces a displacement Δl for the distance between the mirrors, according to the definition of h, the resulting change Δl will be greater the longer the interferometer arm l is, since they are directly proportional (Δl = L × h). Therefore, the reader can notice that interferometric arm lengths must be tailored depending on what type of gravitational source is intended for observation. The explanation just given corresponds to a very basic interferometer, but in actuality these instruments are more complex. We will come back to this.

10. Genesis of the Interferometer Method (Or, Who Deserves the Credit?)

Let us now turn our attention to the origin of the interferometer method. The first explicit suggestion of a laser interferometer detector was outlined in the former USSR by Gertsenshtein and Pustovoid in 1962 [30]. The idea was not carried out and eventually was resurrected in 1966 behind the "Iron Curtain" by Vladimir B. Braginskiĭ, but then again fell into oblivion [31].

Some year before, Joseph Weber returned to his laboratory at the University of Maryland after having attended the 1957 meeting at Chapel Hill. Weber came back bringing loads of ideas. Back at his university, Weber outlined several schemes on how to detect gravitational waves. As mentioned, one of those was the use of a resonant "antenna" (or cylinder), which, in the end, he finally built. However, among other various projects, he conceived the use of interferometer detectors. He did not pursue this conception, though, and the notion was only documented in the pages of his laboratory notebook [28] (p. 414). One can only say that history produces ironies.

By the end of 1959 Weber began the assembly of his first "antenna" with the help of his students Robert L. Forward and David M. Zipoy [32]. Forward would later (in 1978) turn out to be the first scientist to build an interferometric detector [33].

In the early seventies Robert L. Forward, a former student of Joseph Weber at that time working for Hughes Research Laboratory in Malibu, California, decided, with the encouragement of Rainer Weiss, to build a laboratory interferometer with Hughes' funds. Forward's interest in interferometer detectors had evolved some years before when he worked for Joseph Weber at his laboratory in the University of Maryland on the development and construction of Weber's antennas.

By 1971 Forward reported the design of the first interferometer prototype (which he called a "Transducer Laser"). In his publication Forward explained, "The idea of detecting gravitational radiation by using a laser to measure the differential motion of two isolated masses has often been suggested in past[5]" The footnote reads, "To our knowledge, the first suggestion [of the interferometer device] was made by J. Weber in a telephone conversation with one of us (RLF) [Forward] on 14 September 1964" [34].

After 150 hours of observation with his 8.5-m arms interferometer, Robert Forward reported "an absence of significant correlation between the interferometer and several Weber bars detectors, operating at Maryland, Argonne, Glasgow and Frascati". In short, Robert Forward did not observe gravitational waves. Interestingly, in the acknowledgments of his article, Forward recognizes the advice of Philip Chapman and Weiss [33].

Also in the 1970s, Weiss independently conceived the idea of building a Laser Interferometer, inspired by an article written by Felix Pirani, the theoretical physicist who, as we have already mentioned, developed in 1956 the necessary theory to grow the conceptual framework of the method [12]. In this case it was Weiss who developed the method. However, it was not only Pirani's paper that influenced Weiss; he also held talks with Phillip Chapman, who had glimpsed, independently, the same scheme [35]. Chapman had been a member of staff at MIT, where he worked on electro-optical systems and gravitational theory. He left MIT to join NASA, where he served from 1967 to 1972 as a scientist–astronaut (he never went to space). After leaving NASA, Chapman was employed as a researcher in laser propulsion systems at Avco Everett Research Laboratory in Malibu, California. It was at this time that he exchanged views with Weiss. Chapman subsequently lost interest in topics related to gravitation and devoted himself to other activities.

In addition, Weiss also held discussions with a group of his students during a seminar on General Relativity he was running at MIT. Weiss gives credit to all these sources in one of his first publications on the topic: "The notion is not new; It has appeared as a gedanken experiment in F.A.E. Pirani's studies of the measurable properties of the Riemann tensor. However, the realization that with the advent of lasers it is feasible to detect gravitational waves by using this technique [interferometry], grew out of an undergraduate seminar that I ran at MIT several years ago, and has been independently discovered by Dr. Phillip Chapman of the the National Aeronautics and Space Administration, Houston" [35].

Weiss recalled, in a recent interview, that the idea was incubated in 1967 when he was asked by the head of the teaching program in physics at MIT to give a course of General Relativity. At that time Weiss's students were very interested in knowing about the "discoveries" made by Weber in the late sixties. However, Weiss recalls that "I couldn't for the life of me understand the thing he was doing" and "I couldn't explain it to the students". He confesses "that was my quandary at the time" [36].

A year later (in 1968) Weiss began to suspect the validity of Weber's observations because other groups could not verify them. He thought something was wrong. In view of this he decided to spend a summer in a small cubicle and worked the whole season on one idea that had occurred to him during discussions with his students at the seminar he ran at MIT [36].

After a while, Weiss started building a 1.5-m long interferometer prototype, in the RLE (Research Laboratory of Electronics) at MIT using military funds. Some time later, a law was enacted in the United States (the "Mansfield amendment"), which prohibited Armed Forces financing projects that were not of strictly military utility. Funding was suddenly suspended. This forced Weiss to seek financing from other U.S. government and private agencies [36].

11. Wave Hunters on a Merry-Go-Round (GEO)

In 1974, NSF asked Peter Kafka of the Max Planck Institute in Munich to review a project. The project was submitted by Weiss, who requested $53,000 in funds for enlarging the construction of a prototype interferometer with arms nine meters in length [37]. Kafka agreed to review the proposal. Being a theoretician himself, Kafka showed the proposal documents to some experimental physicists at his institute for advice. To Kafka's embarrassment, the local group currently working on Weber bars became very enthusiastic about Weiss's project and decided to build their own prototype [38], headed by Heinz Billing.

This German group had already worked in collaboration with an Italian group in the construction of "Weber antennas". The Italian–German collaboration found that Weber was wrong [36]. The Weiss proposal fell handily to the Germans as they were in the process of designing a novel Weber antenna that was to be cooled to temperatures near absolute zero to reduce thermal noise in the new system. However, learning of Weiss's proposal caused a shift in the research plans of the Garching group. They made the decision to try the interferometer idea. Germans contacted Weiss for advice and they also offered a job to one of his students on the condition that he be trained on the Weiss 1.5-m prototype. Eventually Weiss sent David Shoemaker, who had worked on the MIT prototype, to join the Garching group. Shoemaker later helped to build a German 3-m prototype and later a 30-m interferometer [39]. This interferometer in Garching served for the development of noise suppression methods that would later be used by the LIGO project.

It is interesting to mention that Weiss's proposal may seem modest nowadays (9-m arms), but he already had in mind large-scale interferometers. His prototype was meant to lead to a next stage featuring a one-kilometer arm length device, which his document claimed it would be capable of detecting waves from the Crab pulsar (PSR B0531 + 21) if its periodic signal were integrated over a period of months. Furthermore, the project envisaged a future development stage where long baseline interferometers in outer space could eventually integrate Crab pulsar gravitational waves in a matter of few hours.

NSF then supported Weiss's project and funds were granted in May 1975 [37,38].

At that time Ronald Drever, then at the University of Glasgow, attended the International School of Cosmology and Gravitation in Erice, *La Città della Scienza*, Sicily in March 1975. There, a lecture entitled "Optimal detection of signals through linear devices with thermal source noises, an application to the Munich–Frascati Weber-type gravitational wave detectors" was delivered by the same Peter Kafka of Munich [40]. His lecture was again very critical of Weber's results and went on to showing that the current state of the art of Weber bars including the Munich–Frascati experiment, was far from the optimal sensitivity required for detection. In fact, the conclusion of his notes reads: "It seems obvious that only a combination of extremely high quality and extremely low temperature will bring resonance detectors [Weber bars] near the range where astronomical work is possible. Another way which seems worth exploring is *Laser interferometry with long free mass antennas*" (emphasis ours) [40].

Ronald Drever was part of Kafka's lecture audience. The lecture probably impressed him as he started developing interferometric techniques on his return to Glasgow. He began with simple tasks, a result of not having enough money. One of them was measuring the separation between two massive bars with an interferometer monitoring the vibrations of aluminum bar detectors. The bars were given to him by the group at the University of Reading, United Kingdom. The bars were two halves of the Reading group's split bar antenna experiment [41]. By the end of the 1970s he was leading a team at Glasgow that had completed a 10-meter interferometer. Then in 1979 Drever was invited to head up the team at Caltech, where he accepted a part-time post. James Hough took his place in Glasgow.

Likewise, in 1975 the German group at Munich (Winkler, Rüdiger, Schilling, Schnupp, and Maischberger), under the leadership of Heinz Billing, built a prototype with an arm length of 3 m [42]. This first prototype displayed unwanted effects such as laser frequency instabilities, lack of power, a shaky suspension system, etc. The group worked hard to reduce all these unwanted effects by developing innovative technologies that modern-day gravitational interferometers embraced. In 1983 the same group, now at the Max Planck Institute of Quantum Optics (MPQ) in Garching, improved their first prototype by building a 30 m arm length instrument [43]. To "virtually" increase the optical arm length of their apparatus, they "folded" the laser beam path by reflecting the beam backwards and forwards between the mirrors many times, a procedure that is known as "delay line." In Weiss's words, "the Max Planck group actually did most of the very early interesting development. They came up with a lot of what I would call the practical ideas to make this thing [gravitational interferometers] better and better" [44].

After a couple of years of operating the 30-m model, the Garching group was prepared to go for Big-Science. In effect, in June 1985, they presented a document "Plans for a large Gravitational wave antenna in Germany" at the Marcel Grossmann Meeting in Rome [45]. This document contains the first detailed proposal for a full-sized interferometer (3 km). The project was submitted for funding to the German authorities but there was not sufficient interest in Germany at that time, so it was not approved.

In the meantime, similar research was undertaken by the group at Glasgow, now under Jim Hough after Drever's exodus to Caltech. Following the construction of their 10-m interferometer, the Scots decided in 1986 to take a further step by designing a Long Baseline Gravitational Wave Observatory [46]. Funds were asked for, but their call fell on deaf ears.

Nevertheless, similar fates bring people together, but it is still up to them to make it happen. So, three years later, the Glasgow and the Garching groups decided to unite efforts to collaborate in a plan to build a large detector. It did not take long for both groups to jointly submit a plan for an underground 3-km installation to be constructed in the Harz Mountains in Germany, but again their proposal was not funded [47]. Although reviewed positively, a shortage of funds on both ends (the British Science and Engineering Research Council (SERC) and the Federal Ministry of Research and Technology (Bundesministerium für Forschung und Technologie BMFT)) prevented the approval. The reason for the lack of funds for science in Germany was a consequence of the German re-unification (1989–1990), as there was a need to boost the Eastern German economy; since private, Western funds lagged, public funds were funneled to the former German Democratic Republic. Incidentally the Harz Mountains are the land of German fairy tales.

In spite of this disheartening ruling, the new partners decided to try for a shorter detector and compensate by employing more advanced and clever techniques [48]. A step forward was finally taken in 1994 when the University of Hanover and the State of Lower Saxony donated ground to build a 600-m instrument in Ruthe, 20 km south of Hanover. Funding was provided by several German and British agencies. The construction of GEO 600 started on 4 September 1995.

The following years of continuous hard work by the British and Germans brought results. Since 2002, the detector has been operated by the Center for Gravitational Physics, of which the Max Planck Institute is a member, together with Leibniz Universität in Hanover and Glasgow and Cardiff Universities.

The first stable operation of the Power Recycled interferometer was achieved in December 2001, immediately followed by a short coincidence test run with the LIGO detectors, testing the stability of the system and getting acquainted with data storage and exchange procedures. The first scientific data run, again together with the LIGO detectors, was performed in August and September of 2002. In November 2005, it was announced that the LIGO and GEO instruments began an extended joint science run [49]. In addition to being an excellent observatory, the GEO 600 facility has served as a development and test laboratory for technologies that have been incorporated in other detectors all over the world.

12. The (Nearly) . . . Very Improbable Radio Gravitational Observatory—VIRGO

In the late 70s, when Allain Brillet was attracted to the detection of gravitational waves, the field was ignored by a good number of his colleagues after the incorrect claims of Joe Weber. However, Weiss's pioneering work on laser interferometers in the early 1970s seemed to offer more chances of detection beyond those of Weber bars.

Brillet's interest in the field started during a postdoc stay at the University of Colorado, Boulder under Peter L. Bender of the Laboratory of Astrophysics, who, together with Jim Faller, first proposed the basic concept behind LISA (the Laser Interferometer Space Antenna). Brillet also visited Weiss at MIT in 1980 and 1981 where he established good links with him that produced, as we shall see, a fruitful collaboration in the years to come.

Upon Brillet's return to France in 1982 he joined a group at Orsay UPMC that shared the same interests. This small group (Allain Brillet, Jean Yves Vinet, Nary Man, and two engineers) experienced difficulties and had to find refuge in the nuclear physics department of the Laboratoire de Physique de l'Institute Henri Poincaré, led by Philippe Tourrenc [50].

On 14 November 1983 a meeting on Relativity and Gravitation was organized by the Direction des Etudes Recherches et Techniques de la Délégation Générale pour l'Armement. One of the objectives of the meeting was the development of a French project of gravitational wave detectors [51]. There, Brillet gave a lecture that advocated for the use of interferometers as the best possible detection method [52]. His lecture raised some interest, but French agencies and academic departments were not willing to invest money or personnel in this area. The technology was not yet available, mainly in terms of power laser stability, high-quality optical components, and seismic and thermal noise isolation. In addition, at that point in time there was no significant experimental research on gravitation in France.

However, interest in gravitational wave detection began to change when Hulse and Taylor demonstrated the existence of gravitational waves. It was at the Marcel Grossman meeting of 1985 in Rome that Brillet met Adalberto Giazotto, an Italian scientist working at the Universita di Pisa on the development of suspension systems. At that meeting Giazotto put forward his ideas and the first results of his super-attenuators, devices that serve as seismic isolators to which interferometer mirrors could be attached. During the same meeting Jean-Yves Vinet (Brillet's colleague) gave a talk about his theory of recycling, a technique invented by Ronald Drever to reduce by a large factor the laser power required by gravitational interferometers. Conditions for a partnership were given.

Both scientists then approached the research leaders of a German project (the Max Planck Institute of Quantum Optics in Garching), hoping to collaborate on a big European detector, but they were told that their project was "close to being financed" and the team at Garching did not accept the idea of establishing this international collaboration, because it "would delay project approval" [50].

So they decided to start their own parallel project, the VIRGO Interferometer, named for the cluster of about 1500 galaxies in the Virgo constellation about 50 million light-years from Earth. As no terrestrial source of gravitational wave is powerful enough to produce a detectable signal, VIRGO must observe far enough out into the universe to see many of the potential source sites; the Virgo Cluster is the nearest large cluster.

At that time Brillet was told that CNRS (Le Centre National de la Recherche Scientifique) would not be able to finance VIRGO's construction on the grounds that priority was given to the Very Large Telescope in Chile. Even so, both groups (Orsay and Pisa) did not give in to dismay and continued their collaboration; in 1989 they were joined by the groups of Frascati and Naples. This time they decided to submit the VIRGO project to the CNRS (France) and the INFN (Istituto Nazionale di Fisica Nucleare, Italy) [50].

The VIRGO project was approved in 1993 by the French CNRS and in 1994 by the Italian INFN. The place chosen for VIRGO was the alluvial plain of Cascina near Pisa. The first problem INFN encountered was persuading the nearly 50 land title holders to cooperate and sell their parcels to the government. Gathering the titles took a long time. The construction of the premises started in 1996. To complicate matters further, VIRGO's main building was constructed on a very flat alluvial plain, so it was vulnerable to flooding. That took additional time to remedy [53].

From the beginning it was decided to use the VIRGO interferometer as its own prototype, in contrast to LIGO, which used MIT and Caltech and the German–British (Geo 600) installations to test previous designs before integrating them into the main instrument. This strategy was decided on the grounds that it would be faster to solve problems in actual size directly, rather than spend years on a smaller prototype and only then face the real difficulties.

Between 1996 and 1999, VIRGO had management problems as the construction was handled by an association of separate laboratories without a unified leadership, so it was difficult to ensure proper coordination [42]. As a result, in December 2000 the French CNRS and the Italian INFN created the European Gravitational Observatory (EGO consortium), responsible for the VIRGO site, the construction, maintenance, and operation of the detector, and its upgrades.

The construction of the initial VIRGO detector was completed in June 2003 [54]. It was not until 2007 that VIRGO and LIGO agreed to join in a collaborative search for gravitational waves. This formal agreement between VIRGO and LIGO comprises the exchange of data, and joint analysis and co-authorship of all publications concerned. Several joint data-taking periods followed between 2007 and 2011.

Even though a formal cooperation has been established, continued informal cooperation has been running for years ever since Alain Brillet visited the MIT laboratories back in 1980–1981. As a matter of fact, VIRGO and LIGO have exchanged a good number of students and postdocs. Just to name one, David Shoemaker, the current MIT LIGO Laboratory Director, received his PhD on the Nd-YAG lasers and recycling at Orsay before joining LIGO. Also, in 1990 Jean Yves Vinet provided

LIGO with a computer simulation program necessary to specify its optical system. LIGO adapted this computer code.

The year 2016 will be an important milestone for the construction of the advanced VIRGO detector. After a months-long commissioning period, the advanced VIRGO detector will join the two advanced LIGO detectors ("aLIGO") for a first common data-taking period that should include on the order of one gravitational wave event per month. With all three detectors operating, data can be further correlated and the direction of the gravity waves' source should be much more localized.

13. The Origin of the LIGO Project

In the summer of 1975 Weiss went to Dulles Airport in Washington, D.C. to pick up Kip Thorne, a renowned theoretical physicist from Caltech. The reason for visiting Washington was attending a NASA meeting on uses of space research in the field of cosmology and relativity. (One of us, GFS, attended this meeting as a young postdoc and remembers a presentation by a tired Rai Weiss on the concept of a "Laser Interferometer Space Array" for detecting gravitational waves. It was a very naïve and ambitious space project presentation and only in about 2035 (60 years later) does it appear likely to be a working realization. The meeting did open up to me the idea of doing science in space.) Weiss recalls, "I picked Kip up at the airport on a hot summer night when Washington, D.C., was filled with tourists. He did not have a hotel reservation so we shared a room for the night" [37].

They did not sleep that night because both spent the night discussing many topics, among them how to search for gravitational waves.

Weiss remembers that night "We made a huge map on a piece of paper of all the different areas in gravity. Where was there a future? Or what was the future, or the thing to do?" [55]. Thorne decided that night that the thing they ought to do at Caltech was interferometric gravitational wave detection. However, he would need help from an experimental physicist.

Thorne first thought of bringing to the United States his friend Vladimir B. Braginskiĭ, a Russian scientist who had closely worked with him and, moreover, had already acquired experience in the search for gravitational waves [56]. However, the Cold War prevented his transfer. Meanwhile, Weiss suggested another name, Ronald Drever. Weiss had only known Drever from his papers, not in person. Drever was famous for the Hughes–Drever Experiments, spectroscopic tests of the isotropy of mass and space confirming the Lorentz invariance aspects of the theory of Relativity, and had also been the leader of the group that built a "Weber cylinder" at the University of Glasgow [57]. At that time Drever was planning the construction of an interferometer.

In 1978 Thorne offered a job to Drever at Caltech for the construction of an interferometer. Drever accepted the offer in 1979, dividing his time between the Scottish university and Caltech. Hiring Drever half-time soon paid dividends because in 1983 he had already built his first instrument at Caltech, an interferometer whose "arms" measured 40 m. The instrument was noisier than expected and new ingenious solutions ranging from improving seismic isolation and laser power increase to stabilization were attempted. In 1983 Drever began full-time work at Caltech with the idea of gradually improving and increasing the size of the prototype as it was built and run. In contrast to the Caltech apparatus, as already mentioned, at MIT Weiss had built a modest 1.5-m prototype with a much smaller budget than the Californian instrument. In late 1979 the NSF granted modest funds to the Caltech interferometer group and gave a much smaller amount of money to the MIT team. Soon Drever and Weiss began to compete to build more sensitive and sophisticated interferometers.

The sensitivity of the interferometers can be enhanced by boosting the power of the lasers and increasing the optical path of the light beam as it travels through the interferometer arms.

To increase the sensitivity of the interferometer, Weiss put forward the use of an optical delay line. In the optical delay method, the laser light passes through a small hole in an adjacent wave divider mirror and the beam is reflected several times before emerging through the inlet port. Figure 9 shows a simplified diagram of the method. In this figure only a couple of light "bounces" are shown between the mirrors to maintain clarity of the scheme, but in reality the beam is reflected multiple times. This method effectively increases the length of the interferometer.

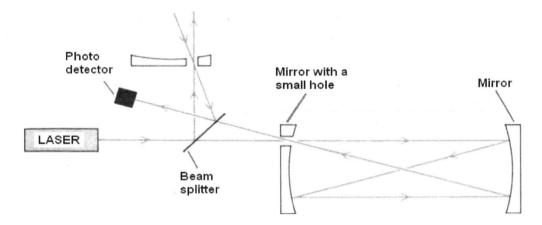

Figure 9. Optical delay method.

In the meantime Drever developed an arrangement that utilized "Fabry–Pérot cavities." In this method the light passes through a partially transmitting mirror to enter a resonant cavity flanked at the opposite end by a fully reflecting mirror. Subsequently, the light escapes through the first mirror, as shown in Figure 10.

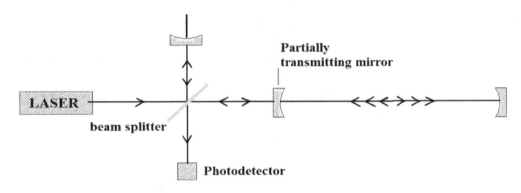

Figure 10. Fabry–Pérot method.

As already mentioned, Weiss experimented with an interferometer whose two L-shaped "arms" were 1.5 m long. Drever, meanwhile, had already built and operated a 40-m interferometer. With the Caltech group appearing to be taking the lead, Weiss decided in 1979 to "do something dramatic". That year Weiss held talks with Richard Isaacson, who at the time served as program director of the NSF Gravitational Physics division. Isaacson had a very strong professional interest in the search for gravitational waves as he had developed a mathematical formalism to approximate gravitational waves solutions from Einstein's equations of General Relativity in situations where the gravitational fields are very strong [58]. Weiss offered to conduct a study in collaboration with industry partners to determine the feasibility and cost of an interferometer whose arms should measure in kilometers. In turn Isaacson receive a document to substantiate a device on a scale of kilometers, with possible increased funding from the NSF. The study would be funded by NSF. The study by Weiss and colleagues took three years to complete. The produced document was entitled "A study of a long Baseline Gravitational Wave Antenna System," co-authored by Peter Saulson and Paul Linsay [59]. This fundamental document is nowadays known as the "The Blue Book" and covers very many important issues in the construction and operation of such a large interferometer. The Blue Book was submitted to the NSF in October 1983. The proposed budget was just under $100 million to build two instruments located in the United States.

Before "The Blue Book" was submitted for consideration to the NSF, Weiss met with Thorne and Drever at a Relativity congress in Italy. There they discussed how they could work together—this was mandatory, because the NSF would not fund two megaprojects on the same subject and with the same objective. However, from the very beginning it was clear that Drever did not want to collaborate with Weiss, and Thorne had to act as mediator. In fact, the NSF settled matters by integrating the MIT and Caltech groups together in a "shotgun wedding" so the "Caltech–MIT" project could be jointly submitted to the NSF [55].

14. The LIGO Project

The Caltech–MIT project was funded by NSF and named the "Laser Interferometer Gravitational-Wave Observatory", known by its acronym LIGO. The project would be led by a triumvirate of Thorne, Weiss, and Drever. Soon interactions between Drever and Weiss became difficult because, besides the strenuous nature of their interaction, both had differing opinions on technical issues.

During the years 1984 and 1985 the LIGO project suffered many delays due to multiple discussions between Drever and Weiss, mediated when possible by Thorne. In 1986 the NSF called for the dissolution of the triumvirate of Thorne, Drever, and Weiss. Instead Rochus E. Vogt was appointed as a single project manager [60].

In 1988 the project was finally funded by the NSF. From that date until the early 1990s, project progress was slow and underwent a restructuring in 1992. As a result, Drever stopped belonging to the project and in 1994 Vogt was replaced by a new director, Barry Clark Barish, an experimental physicist who was an expert in high-energy physics. Barish had experience in managing big projects in physics. His first activity was to review and substantially amend the original five-year old NSF proposal. With its new administrative leadership, the project received good financial support. Barish's plan was to build the LIGO as an evolutionary laboratory where the first stage, "initial LIGO" (or iLIGO), would aim to test the concept and offer the possibility of detecting gravitational waves. In the second stage ("aLIGO" or advanced LIGO), wave detection would be very likely (see Figure 11).

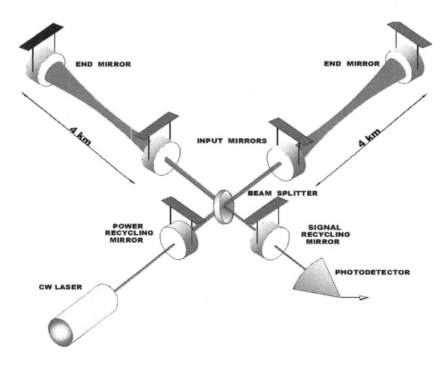

Figure 11. Advanced LIGO interferometer design concept. Figure made after T.F. Carruthers and D.H. Reitze, "LIGO", *Optics & Photonics News*, March 2015.

Two observatories, one in Hanford in Washington State and one in Livingston, Louisiana, would be built. Construction began in late 1994 and early 1995, respectively, and ended in 1997. Once the construction of the two observatories was complete, Barish suggested two organizations to be funded: the laboratory LIGO and scientific collaboration LIGO (LIGO Scientific Collaboration (LSC)). The first of these organizations would be responsible for the administration of laboratories. The second organization would be a scientific forum headed by Weiss, responsible for scientific and technological research. LSC would be in charge of establishing alliances and scientific collaborations with Virgo and GEO600.

Barish's idea to make LIGO an evolutionary apparatus proved in the end to pay dividends. The idea was to produce an installation whose parts (vacuum system, optics, suspension systems, etc.) could always be readily improved and buildings that could house those ever-improving interferometer components.

In effect, LIGO was incrementally improved by advances made in its own laboratories and those due to associations with other laboratories (VIRGO and GEO600). To name some: Signal-recycling mirrors were first used in the GEO600 detector, as well as the monolithic fiber-optic suspension system that was introduced into advanced LIGO. In brief, LIGO detection was the result of a worldwide collaboration that helped LIGO evolve into its present remarkably sensitive state.

15. Looking Back Over the Trek

The initial LIGO operated between 2002 and 2010 and did not detect gravitational waves. The upgrade of LIGO (advanced LIGO) began in 2010 to replace the detection and noise suppression and improve stability operations at both facilities. This upgrade took five years and had contribution from many sources. For example, the seismic suspension used in aLIGO is essentially the design that has been used in VIRGO since the beginning. While advanced VIRGO is not up and running yet, There have been many technological contributions from both LSC scientists on the European side and VIRGO.

aLIGO began in February 2015 [61]. The team operated in "engineering mode"—that is, in test mode—and in late September began scientific observation [62]. It did not take many days for LIGO to detect gravitational waves [2]. Indeed, LIGO detected the collision of two black holes of about 30 solar masses collapsed to 1300 million light years from Earth.

Even the latest search for gravitational waves was long and storied. The upgrade to aLIGO cost $200 million, and preparing it took longer than expected, so the new and improved instrument's start date was pushed back to 18 September 2015.

16. The Event: 14 September 2015

On Sunday September 13th the LIGO team performed a battery of last-minute tests. "We yelled, we vibrated things with shakers, we tapped on things, we introduced magnetic radiation, we did all kinds of things", one of the LIGO members said. "And, of course, everything was taking longer than it was supposed to". At four in the morning, with one test still left to do—a simulation of a truck driver hitting his brakes nearby—we stopped for the night. We went home, leaving the instrument to gather data in peace and quiet. The signal arrived not long after, at 4:50 a.m. local time, passing through the two detectors within seven milliseconds of each other. It was four days before the start of Advanced LIGO's first official run. It was still during the time meant for engineering tests, but nature did not wait.

The signal had been traveling for over a billion years, coming from a pair of 30-solar mass black holes orbiting around their common center of mass and slowly drawing into a tighter and tighter orbit from the energy being lost by gravitational radiation. The event was the end of this process —formed from final inspiral, the merger of the black holes to form a larger one, and the ring down of that new massive black hole. All of this lasted mere thousandths of a second, making the beautiful signals seen by both aLIGO detectors which immediately answered several questions: Are there gravitational waves and can we detect them? Is General Relativity likely right for strong fields? Are large black holes common?

The waveform detected by both LIGO observatories matched the predictions of General Relativity for a gravitational wave emanating from the inward spiral and merger of a pair of black holes of around 36 and 29 solar masses and the subsequent "ringdown" of the single resulting black hole. See Figures 12 and 13 for the signals and GR theoretical predictions.

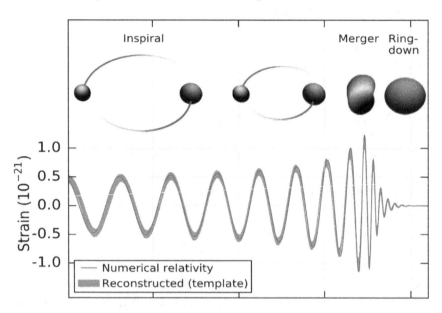

Figure 12. Showing the phases: binary orbits inspiraling, black holes merging, and final black hole ringing down to spherical or ellipsoidal shape [2].

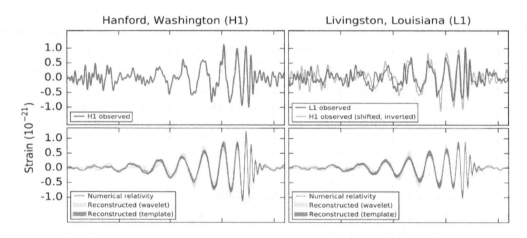

Figure 13. LIGO measurement of the gravitational waves at the Livingston (**right**) and Hanford (**left**) detectors, compared with the theoretical predicted values from General Relativity [2].

The signal was named GW150914 (from "Gravitational Wave" and the date of observation). It appeared 14 September 2015 and lasted about 0.2 s. The estimated distance to the merged black holes is $410 ^{+160} _{-180}$ Mpc or 1.3 billion light years, which corresponds to a redshift of about z = 0.09. The interesting thing is that the estimated energy output of the event in gravitational waves is about $(3.0 +/- 0.5) M_\odot \times c^2$. That is three times the rest mass of our sun converted into energy.

More recently, the aLIGO team announced the detection and analysis of another binary black hole merger event, GW151226. This event is apparently the merger of a 14 solar mass black hole with a 7.5 solar mass black hole, again at a distance of about 440 Mpc (about 1.4 billion light years). An interesting feature is that one of the black holes is measured to have significant spin, s = 0.2. There is also a report of a candidate event that occurred between the two confirmed events and had black holes and energy intermediate to the two confirmed events. These additional reports also show the beginning of measuring rates and the distribution of black hole binary systems. It has long been expected that as aLIGO improves its design sensitivity (about three times better than this first run), that a stable set of events would include the merger of binary neutron stars, for whose inventory we have better estimates. These early events predict that there will be a whole distribution of gravitational wave events observed in the future, from neutron star mergers, black hole mergers, and even neutron star–black hole mergers, see Figure 14.

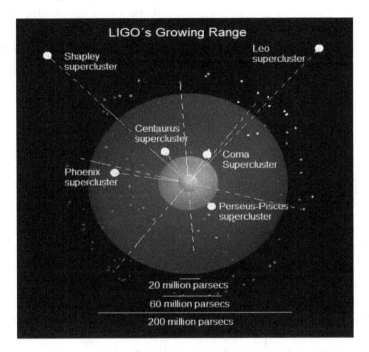

Figure 14. Showing the initial range of LIGO and the anticipated range of aLIGO. The volume is much greater and the anticipated rate of events and detections are expected to scale up with the volume. Note the large number of galaxies included in the observational volume. The anticipated factor of 3 in sensitivity should correspondingly increase the event rate by up to nine times.

Albert Einstein originally predicted the existence of gravitational waves in 1916, based upon General Relativity, but wrote that it was unlikely that anyone would ever find a system whose behavior would be measurably influenced by gravitational waves. He was pointing out that the waves from a typical binary star system would carry away so little energy that we would never even notice that the system had changed—and that is true. The reason we can see it from the two black holes is that they are closer together than two stars could ever be. The black holes are so tiny and yet so massive that they can be close enough together to move around each other very, very rapidly. Still, to get such a clear signal required a very large amount of energy and the development of extraordinarily sensitive instruments. This clearly settles the argument about whether gravitational waves really exist; one major early argument was about whether they carried any energy. They do! That was proved strongly and clearly.

Some analyses have been carried out to establish whether or not GW150914 matches with a binary black hole configuration in General Relativity [63]. An initial consistency test encompasses the mass and spin of the end product of the coalescence. In General Relativity, the final black hole product of a binary coalescence is a Kerr black hole, which is completely described by its mass and spin. It has been verified that the remnant mass and spin from the late-stage coalescence deduced by numerical relativity simulations, inferred independently from the early stage, are consistent with each other, with no evidence for disagreement from General Relativity. There is even some data on the ring down phase, but we can hope for a better event to provide quality observations to test this phase of General Relativity.

GW150914 demonstrates the existence of black holes more massive than $\simeq 25M_\odot$, and establishes that such binary black holes can form in nature and merge within a Hubble time. This is of some surprise to stellar theorists, who predicted smaller mass black holes would be much more common.

17. Conclusions

This observation confirms the last remaining unproven prediction of General Relativity (GR)—gravitational waves—and validates its predictions of space–time distortion in the context of large-scale cosmic events (known as strong field tests). It also inaugurates the new era of gravitational-wave astronomy, which promises many more observations of interesting and energetic objects as well as more precise tests of General Relativity and astrophysics. While it is true that we can never rule out deviations from GR at the 100% level, all three detections so far agree with GR to an extremely high level (>96%). This will put constraints on some non-GR theories and their predictions.

With such a spectacular early result, others seem sure to follow. In the four-month run, 47 days' worth of coincident data was useful for scientific analysis, i.e., this is data taken when both LIGOs were in scientific observation mode. The official statement is that these 47 days' worth of data have been fully analyzed and no further signals lie within them. We can expect many more events once the detectors are running again.

For gravitational astronomy, this is just the beginning. Soon, aLIGO will not be alone. By the end of the year VIRGO, a gravitational-wave observatory in Italy, should be operating to join observations and advanced modes. Another detector is under construction in Japan and talks are underway to create a fourth in India. Most ambitiously, a fifth, orbiting, observatory, the Evolved Laser Interferometer Space Antenna, or e-LISA, is on the cards. The first pieces of apparatus designed to test the idea of e-LISA are already in space and the first LISA pathfinder results are very encouraging.

Together, by jointly forming a telescope that will permit astronomers to pinpoint whence the waves come, these devices will open a new vista onto the universe. (On the science side, the data is analyzed jointly by members of both LIGO and VIRGO, even though these data only come from LIGO. This is due to the analysis teams now being fully integrated. As this is not widely known, people do not realize that there is a large contribution from VIRGO scientists to the observations and to the future.) As technology improves, waves of lower frequency—corresponding to events involving larger masses—will become detectable. Eventually, astronomers should be able to peer at the first 380,000 years after the Big Bang, an epoch of history that remains inaccessible to every other kind of telescope yet designed.

The real prize, though, lies in proving Einstein wrong. For all its prescience, the theory of relativity is known to be incomplete because it is inconsistent with the other great 20th-century theory of physics, quantum mechanics. Many physicists suspect that it is in places where conditions are most extreme—the very places that launch gravitational waves—that the first chinks in relativity's armor will be found, and with them we will get a glimpse of a more all-embracing theory.

Gravitational waves, of which Einstein remained so uncertain, have provided direct evidence for black holes, about which he was long uncomfortable, and may yet yield a peek at the Big Bang, an event he knew his theory was inadequate to describe. They may now lead to his theory's unseating. If so, its epitaph will be that in predicting gravitational waves, it predicted the means of its own demise.

Acknowledgments: GFS acknowledges Laboratoire APC-PCCP, Université Paris Diderot, Sorbonne Paris Cité (DXCACHEXGS), and also the financial support of the UnivEarthS Labex program at Sorbonne Paris Cité (ANR-10-LABX-0023 and ANR-11-IDEX-0005-02). JLCC acknowledges financial support from conacyt Project 269652 and Fronteras Project 281.

Author Contributions: J.L.C.-C., S.G.-U., and G.F.S. conceived the idea and equally contributed.

References

1. Davide, C. LIGO's path to victory. *Nature* **2016**, *530*, 261–262.
2. Abbot, B.P.; Abbott, R.; Abbott, T.D.; Abernathy, M.R.; Acernese, F.; Ackley, K.; Adams, C.; Adams, T.; Addesso, P.; Adhikari, P.X.; et al. Observation of Gravitational Waves from a Binary Black Hole Merger. *Phys. Rev. Lett.* **2016**, *116*, 061102. [CrossRef] [PubMed]
3. Henri, P. "Sur la Dynamique de l'électron". *Proc. Acad. Sci.* **1905**, *140*, 1504–1508. (In French)
4. Albert, E. Integration Neherunsweise Feldgleichungen der der Gravitation. *Sitzungsber. Preuss. Akad. Wiss. Berlin (Math.Phys.)* **1916**, *22*, 688–696. (In Germany)
5. Albert, E. Zur Elektrodynamik bewegter Körper. *Annalen Phys.* **1905**, *17*, 891–921. (In Germany)
6. Albert, E. The Collected Papers of Albert Einstein, Volume 8: The Berlin Years: Correspondence, 1914–1918 (English Translation Supplement, Translated by Ann M. Hentschel) Page 196, Doc 194. Available online: http://einsteinpapers.press.princetonedu/VOL8-trans/224 (accessed on 22 June 2016).
7. Hermann, W. Space, Time, Matter. Methuen & Co. Ltd.: London, 1922; Chapter 4; p. 376, Translated from German by Henry l. Brose. Available online: http://www.gutenberg.org/ebooks/43006 (accessed on 22 June 2016).
8. Stanley, E.A. The propagation of Gravitational waves. *Proc. R. Soc. Lond.* **1922**, *102*, 268–282.
9. Albert, E.; Max, B. *The Born-Einstein Letters. Friendship, Politics and Physics in Uncertain Times*; Macmillan: London, UK, 2005; p. 121.
10. Daniel, K. Einstein versus the Physical Review. *Phys. Today* **2005**, *58*, 43–48.
11. Albert, E.; Rosen, N. On gravitational waves. *J. Frank. Inst.* **1937**, *223*, 43–54.
12. Pirani Felix, A.E. On the Physical Significance of the Riemann Tensor. *Gen. Relativ. Gravit.* **2009**, *41*, 1215–1232. [CrossRef]
13. Bergmann Peter, G. *The Riddle of Gravitation*; Charles Scribner's Sons: New York, NY, USA, 1968.
14. Maggiore, M. *Gravitational Waves*; Oxford University Press: Oxford, UK, 2008.
15. Gravity Research Foundation. Available online: http://www.gravityresearchfoundation.org/origins.html (accessed on 21 June 2016).
16. Interview of Bryce DeWitt and Cecile DeWitt-Morette by Kenneth W. In *Ford on 1995 February 28, Niels Bohr Library & Archives*; American Institute of Physics: College Park, MD, USA. Available online: http://www.aip.org/history-programs/niels-bohr-library/oral-histories/23199 (accessed on 21 June 2016).
17. Martin, G. Chapter 8 "Sir Isaac Babson". In *Fads and Fallacies in the Name of Science*; Dover Publications: New York, NY, USA, 1957; p. 93.
18. DeWitt, M.C.; Rickles, D. The role of Gravitation in Physics. Report from the 1957 Chapel Hill Conference. Max Planck Research Library for the History and Development of Knowledge. Open Access Edition (2011). In particular see Chapter 27 "An Expanded Version of the Remarks by R.P. Feynman on the Reality of Gravitational Waves" also see comments Feymann on page 252. Available online: http://www.edition-open-sources.org/sources/5/index.html (accessed on 21 June 2016).
19. Feynman, R.P. Unpublished letter to Victor F. Weisskopf, January 4–February 11, 1961; in Box 3 File 8 of The Papers of Richard P. Feynman, the Archives, California Institute of Technology. Available online: http://www.oac.cdlib.org/findaid/ark:/13030/kt5n39p6k0/dsc/?query=lettertoWeisskopf1961#c02--1.7.7.3.2 (accessed on 21 June 2016).
20. Bondi, H. Plane gravitational waves in the general relativity. *Nature* **1957**, *179*, 1072–1073. [CrossRef]
21. Weber, J. Detection and Generation of Gravitational Waves. *Phys. Rev.* **1960**, *117*, 306–313. [CrossRef]
22. Weber, J. Observation of the Thermal Fluctuations of a Gravitational-Wave Detector. *Phys. Rev. Lett.* **1966**, *17*, 1228–1230. [CrossRef]
23. Weber, J. Evidence for Discovery of Gravitational Radiation. *Phys. Rev. Lett.* **1969**, *22*, 1320–1324. [CrossRef]
24. Weber, J. Anisotropy and Polarization in the Gravitational-Radiation Experiments. *Phys. Rev. Lett.* **1970**, *25*, 180–184. [CrossRef]
25. Sciama, D.; Field, G.; Rees, M. Upper Limit to Radiation of Mass Energy Derived from Expansion of Galaxy. *Phys. Rev. Lett.* **1969**, *23*, 1514–1515. [CrossRef]
26. Kafka, P. *Are Weber's Pulses Illegal?*; Gravity Research Foundation: Babson Park, MA, USA, 1972.
27. Weber, J. The Detection of Gravitational Waves. *Sci. Am.* **1971**, *224*, 22–29. [CrossRef]

28. Thorne Kip, S. Gravitational Radiation. In *Three Hundred Years of Gravitation*; Hawking, S., Israel, W.W., Eds.; Cambridge University Press: Cambridge, UK, 1987; pp. 330–458.

29. Taylor, J.H.; Fowler, L.A.; McCulloch, P.M. Overall measurements of relativistic effects in the binary pulsar PSR 1913 + 16. *Nature* **1979**, *277*, 437–440. [CrossRef]

30. Gertsenshtein, M.E.; Pustovoit, V.I. On the detection of low frequency gravitational waves. *Sov. J. Exp. Theor. Phys.* **1962**, *43*, 605–607.

31. Braginskiĭ Vladimir, B. Gravitational radiation and the prospect of its experimental discovery. *Sov. Phys. Usp.* **1966**, *8*, 513–521. [CrossRef]

32. Weber, J. How I discovered Gravitational Waves. *Pop. Sci.* **1962**, *5*, 106–107.

33. Forward, R.L. Wide band Laser-Interferometer Gravitational-radiation experiment. *Phys. Rev. D* **1978**, *17*, 379–390. [CrossRef]

34. Moss, G.E.; Miller, R.L.; Forward, R.L. Photon-noise-Limited Laser Transducer for Gravitational Antenna. *Appl. Opt.* **1971**, *10*, 2495–2498. [CrossRef] [PubMed]

35. Weiss, R. Quarterly Progress Report 1972, No 105, 54-76. Research Laboratory of Electronics, MIT. Available online: http://dspace.mit.edu/bitstream/handle/1721.1/56271/RLE_QPR_105_V.pdf?sequence= 1 (accessed on 21 June 2016).

36. Chu Jennifer "Rainer Weiss on LIGO's origins" MIT Q & A NEWS February 11, 2016. Available online: http://news.mit.edu/2016/rainer-weiss-ligo-origins-0211 (accessed on 23 June 2016).

37. Weiss, R. Interferometric Broad Band Gravitational Antenna. Grant identification number MPS75–04033. Available online: https://dspace.mit.edu/bitstream/handle/1721.1/56655/RLE_PR_119_ XV.pdf?sequence=1 (accessed on 23 June 2016).

38. Collins Harry, interview with Peter Kafka. See WEBQUOTE under "Kafka Referees Weiss's Proposal". Available online: http://sites.cardiff.ac.uk/harrycollins/webquote/ (accessed on 21 June 2016).

39. Shoemaker, D.; Schilling, R.; Schnupp, L.; Winkler, W.; Maischberger, K.; Rüdiger, A. Noise behavior of the Garching 30-meter gravitational wave detector prototype. *Phys. Rev. D* **1988**, *38*, 423–432. [CrossRef]

40. De Sabbata, V.; Weber, J. Topics in Theoretical and Experimental Gravitation Physics. In Proceedings of the International School of Cosmology and Gravitation, Sicily, Italy, 13–25 March 1975.

41. Allen, W.D.; Christodoulides, C. Gravitational radiation experiments at the University of Reading and the Rutherford Laboratory. *J. Phys. A Math. Gen.* **1975**, *8*, 1726–1733. [CrossRef]

42. Collins, H. *Gravity's Shadow*; University of Chicago Press: Chicago, IL, USA, 2004.

43. Shoemaker, D.; Winkler, W.; Maischberger, K.; Ruediger, A.; Schilling, R.; Schnupp, L. Progress with the Garching (West Germany) 30 Meter Prototype for a Gravitational Wave Detector. In Proceedings of the 4th Marcel Grossmann Meeting, Rome, Italy, 17–21 June 1985.

44. Cohen, S.K. Rainer Weiss Interviewed. 10 May 2000. Available online: http://resolver.caltech.edu/ CaltechOH:OH_Weiss_R (accessed on 21 June 2016).

45. Winkler, W.; Maischberger, K.; Rudiger, A.; Schilling, R.; Schnupp, L.; Shoemaker, D. Plans for a large gravitational wave antenna in Germany. In Proceedings of the 4th Marcel Grossmann Meeting, Rome, Italy, 17–21 June 1985.

46. Hough, J.; Meers, B.J.; Newton, G.P.; Robertson, N.A.; Ward, H.; Schutz, B.F.; Drever, R.W.P. *A British Long Baseline Gravitational Wave Observatory: Design Study Report*; Rutherford Appleton Laboratory Gravitational Wave Observatory: Oxon, UK, 1986.

47. Leuchs, G.; Maischberger, K.; Rudiger, A.; Roland, S.; Lise, S.; Walter, W. Vorschlag zum Bau eines groBen Laser-Interferometers zur Messung von Gravitationswellen –erweiterte Fasssung-. Available online: http://www.mpq.mpg.de/4464435/mpq_reports#1987 (accessed on 23 June 2016).

48. Hough, J.; Meers, B.J.; Newton, G.P.; Robertson, N.A.; Ward, H.; Leuchs, G.; Niebauer, T.M.; Rudiger, A.; Schilling, R.; Schnupp, L.; et al. Proposal for a Joint German-British Interferometric Gravitational Wave Detector. Available online: http://eprints.gla.ac.uk/114852/7/114852.pdf (accessed on 23 June 2016).

49. Biennial Reports 2004/05 Max Planck Institute for Gravitational Physics (PDF). Available online: https://www.aei.mpg.de/148353/biennial2004_05.pdf (accessed on 23 June 2016).

50. Brillet, A. Avant VIRGO, ARTEMIS UMR 7250, jeudi 26 mai 2016. Available online: https://artemis.oca.eu/ spip.php?article463 (accessed on 23 June 2016).

51. Bordé, Ch.J. Méthodes optiques de détection des ondes gravitationnelles—Préface. *Ann. Phys. Fr.* **1985**, *10*, R1–R2. (In French) [CrossRef]

52. Brillet, A. Interferometric gravitational wave antennae. *Ann. Phys. Fr.* **1985**, *10*, 219–226. [CrossRef]

53. Accadia, T.; Acernese, F.; Alshourbagy, M.; Amico, P.; Antonucci, F.; Aoudia, S.; Arnaud, N.; Arnault, C.; Arun, K.G.; Astone, P.; et al. VIRGO: A laser interferometer to detect gravitational waves. *J. Instrum.* **2012**, *7*, P03012. [CrossRef]

54. Ondes Gravitationnelles Inauguration du Détecteur Franco-Italien VIRGO. Available online: http://www2.cnrs.fr/presse/communique/206.htm?&debut=40 (accessed on 21 June 2016). (In French).

55. Janna, L. *Black Hole Blues and Other Songs from Outer Space*; Knopf: New York, NY, USA, 2016.

56. Braginski, V.B.; Manukin, A.B.; Popov, E.I.; Rudenko, V.N.; Korev, A.A. Search for Gravitational Radiation of Extraterrestrial Origin. *Sov. Phys. Usp.* **1937**, *15*, 831–832. [CrossRef]

57. Drever, R.W.P. A search for anisotropy of inertial mass using a free precession technique. *Philos. Mag.* **1961**, *6*, 683–687. [CrossRef]

58. Isaacson Richard, A. Gravitational Radiation in the Limit of High Frequency. I. The Linear Approximation and Geometrical Optics. *Phys. Rev.* **1968**, *166*, 1263–1271. [CrossRef]

59. Linsay, P.; Saulson, P.; Weiss, R. A Study of a Long Baseline Gravitational Wave Antenna System. Available online: https://dcc.ligo.org/public/0028/T830001/000/NSF_bluebook_1983.pdf (accessed on 23 June 2016).

60. Russell, R. Catching the wave. *Sci. Am.* **1992**, *266*, 90–99.

61. Davide, C. Hunt for gravitational waves to resume after massive upgrade: LIGO experiment now has better chance of detecting ripples in space-time. *Nature* **2015**, *525*, 301–302.

62. Jonathan, A. Advanced Ligo: Labs Their Ears Open to the Cosmos. Available online: http://www.bbc.com/news/science-environment-34298363 (accessed on 23 June 2016).

63. Abbott, B.P. Tests of General Relativity with GW150914. *Phys. Rev. Lett.* **2016**, *116*, 221101. [CrossRef] [PubMed]

9

On the Effect of the Cosmological Expansion on the Gravitational Lensing by a Point Mass

Oliver F. Piattella

Physics Department, Universidade Federal do Espírito Santo, Vitória 29075-910, Brazil; oliver.piattella@pq.cnpq.br

Academic Editors: Lorenzo Iorio and Elias C. Vagenas

Abstract: We analyse the effect of the cosmological expansion on the deflection of light caused by a point mass, adopting the McVittie metric as the geometrical description of a point-like lens embedded in an expanding universe. In the case of a generic, non-constant Hubble parameter, H, we derive and approximately solve the null geodesic equations, finding an expression for the bending angle δ, which we expand in powers of the mass-to-closest approach distance ratio and of the impact parameter-to-lens distance ratio. It turns out that the leading order of the aforementioned expansion is the same as the one calculated for the Schwarzschild metric and that cosmological corrections contribute to δ only at sub-dominant orders. We explicitly calculate these cosmological corrections for the case of the H constant and find that they provide a correction of order 10^{-11} on the lens mass estimate.

Keywords: McVittie metric; gravitational lensing; cosmology

PACS: 95.30.Sf; 04.70.Bw; 95.36.+x

1. Introduction

The effect of the cosmological constant Λ (and thus, by extension, of cosmology) on the bending of light is an issue which has raised interest since a pioneering work by Rindler and Ishak in 2007 [1]. The common intuition is that Λ cannot have any local effect on the deflection of light because it is homogeneously distributed in the universe, thus not forming lumps which may act as lenses. Moreover, General Relativity (GR) is assumed as the fundamental theory of gravity, and the Kottler metric is considered [2] (We use throughout this paper $G = c = 1$ units):

$$ds^2 = f(r)dt^2 - f^{-1}(r)dr^2 - r^2 d\Omega^2 ,\tag{1}$$

where $d\Omega^2 = d\theta^2 + \sin^2\theta d\phi^2$ and:

$$f(r) \equiv 1 - \frac{2M}{r} - \frac{\Lambda r^2}{3} ,\tag{2}$$

as the description of a point mass M embedded in a de Sitter space. It turns out that Λ does not appear in the null geodesic trajectory equation, cf. e.g., Equation (17) of Ref. [3]. Indeed:

$$\frac{d^2u}{d\phi^2} + u = 3Mu^2 ,\tag{3}$$

where $u = 1/r$ is the inverse of the radial distance from the lens. Therefore, one may conclude that Λ does not affect the bending of light, which is then entirely due to the presence of the point mass M.

On the other hand, in Ref. [1], the authors point out that the bending angle cannot be calculated as the angle between the asymptotic directions of the light ray, since these do not exist. Indeed, from Equation (2), one sees that $r \leq \sqrt{3/\Lambda}$, i.e., a cosmological horizon exists. In other words, the effect of Λ enters into the boundary conditions that we choose when solving Equation (3).

Thus, the authors of Ref. [1] find that Λ enters the definition of the deflection angle in the following way (cf. their Equation (17)):

$$\psi_0 \approx \frac{2M}{R} \left(1 - \frac{2M^2}{R^2} - \frac{\Lambda R^4}{24M^2} \right) , \tag{4}$$

where R is the closest approach distance. Twice ψ_0 is the deflection angle. Therefore, one identifies the well-known Schwarzschild contribution $4M/R$, weighed by Λ, which tends to thwart the deflection.

After Ref. [1], many authors confirmed with their calculations that Λ does enter the formula for the deflection angle, although sometimes in a way different from the one in Equation (4). See e.g., Refs. [4–12].

On the other hand, there are a few works that do not agree with the above-mentioned results (see e.g., [13–16]). The main criticism is that the **Hubble flow** is not properly taken into account, i.e., the relative motion among source, lens and observer is neglected. In particular, the authors of Ref. [15] argue that the Λ contribution in Equation (4) is cancelled by the aberration effect due to the cosmological relative motion. Another interesting remark made in Ref. [15] is that the contribution of Λ to the deflection angle does not vanish for $M \to 0$ in Equation (4). In this respect, consider also e.g., Equation (25) of Ref. [10]:

$$\delta = \frac{4M}{b} - Mb \left(\frac{1}{r_S^2} + \frac{1}{r_O^2} \right) + \frac{2Mb\Lambda}{3} - \frac{b\Lambda}{6}(r_S + r_O) - \frac{b^3\Lambda}{12} \left(\frac{1}{r_S} + \frac{1}{r_O} \right) + \frac{Mb^3\Lambda}{6} \left(\frac{1}{r_S^2} + \frac{1}{r_O^2} \right) + \cdots . \tag{5}$$

Here, b is the impact parameter and r_S and r_O are the radial distances from the lens to the source and to the observer, respectively. Taking the limit $M \to 0$ in the above equation does not imply $\delta \to 0$.

This seems to be odd since we do not expect lensing without a lens. However, this is the result that one obtains when the **Hubble flow** is not taken into account. Indeed, the author of Ref. [16] constructs "by hand" cosmological observers in the Kottler metric and finds that Λ has no observable effect on the deflection of light.

Therefore, according to the results of Refs. [13–16], the standard approach to gravitational lensing does not need modifications. For the sake of clarity, the standard approach to gravitational lensing consists of using the result on the deflection angle obtained from the Schwarzschild metric (which models the lens) together with the cosmological angular diameter distances calculated from the Friedmann–Lemaître–Robertson–Walker (FLRW) metric. See e.g., Ref. [17].

In Ref. [18], we also tackled the investigation of whether a cosmological constant might affect the gravitational lensing by adopting the McVittie metric [19] as the description of the lens. The McVittie metric is an exact **spherically symmetric solution** of Einstein equations in presence of a point mass and a cosmic perfect fluid. See e.g., Refs. [20–28] for mathematical investigations of the geometrical properties of McVittie metric. In Ref. [18], we considered a constant Hubble factor, thus the geometry involved is the very Kottler one considered by most of the authors cited in this paper, but written in a different reference frame. Our results corroborate those of Refs. [13–16].

In the present paper, we generalise the results of Ref. [18] to the case of a generic time-dependent H. See also Ref. [29]. This is necessary in order to make contact with the current standard model of cosmology, the ΛCDM model, in which pressure-less matter prevents H to be a constant. In particular, Friedmann equation for the **spatially-flat** ΛCDM model reads:

$$\frac{H^2}{H_0^2} = \Omega_m a^{-3} + \Omega_\Lambda , \tag{6}$$

where H_0 is the Hubble constant, a is the scale factor, Ω_m is the present density parameter of pressure-less matter and $\Omega_\Lambda = 1 - \Omega_m$ is the **present** density parameter of the cosmological constant. The time-derivative of H can be easily computed as:

$$\frac{\dot{H}}{H_0^2} = -\frac{3}{2}\Omega_{\mathrm{m}}a^{-3} \, . \tag{7}$$

Since $\Omega_{\mathrm{m}} \approx 0.3$, one can see that $\dot{H}_0$ and H_0^2 are of the same order at present time. Moreover, $|\dot{H}| \geq H^2$ for $1 + z \geq \sqrt[3]{2\Omega_\Lambda/\Omega_{\mathrm{m}}}$. This gives a redshift $z > 1.67$. Many of the observed sources and lenses have redshifts larger than this limit (see e.g., the CASTLES survey, https://www.cfa.harvard.edu/castles/); therefore, the above calculation shows that assuming H constant is a very bad approximation and if one plans to make contact with observation, it is necessary to go beyond the static case of the Kottler metric. The McVittie metric with a generic, non-constant H provides an opportunity to do this.

Very recently, the deflection of light in a cosmological context with a generic $H(t)$ has been considered in Ref. [30]. The main result is the following, cf. Equation (11) of Ref. [30]:

$$\Delta\tilde{\varphi} = \frac{4\tilde{M}_{\mathrm{MSH}}}{\tilde{R}} - 2H^2\tilde{R}^2 \, , \tag{8}$$

where $\tilde{M}_{\mathrm{MSH}}$ is the Misner–Sharp–Hernandez mass [31,32] contained in a radius $\tilde{R} = a(t)r$. Hence, it turns out that the effect of cosmology on the gravitational lensing depends on whether one takes into account the total contribution (local plus cosmological) to the mass or just the local one. However, in both cases, $\tilde{M}_{\mathrm{MSH}}$ can be decomposed in the local m contribution plus the cosmological one, which is cancelled by the $-2H^2\tilde{R}^2$ in the above Equation (8). Thus, it appears that the net result is that the cosmic fluid does not contribute directly to the gravitational lensing.

As a final remark, we must stress that the McVittie metric is an oversimplified model of an actual lens, which has a more complicated structure than that of a point. However, the results of our investigation may shed an important light and give valuable insights for future research that takes into account a more complex structure of the lens. As far as we know, lenses with structures different from a point have been considered only by Ref. [10].

The present paper is structured as follows. In Section 2, we present the McVittie metric and its principal features. In Section 3, we obtain the null geodesic equations and calculate the deflection angle. In Section 4, we focus on the case of a constant H and calculate exactly the subdominant contribution to the deflection angle, estimating a relative correction on the mass determination of about 10^{-11}, due to the **Hubble flow**. In Section 5, we present our conclusions. Throughout the paper, we use $G = c = 1$ units.

2. The McVittie Metric

The McVittie metric [19] can be written in the following form:

$$ds^2 = -\left(\frac{1-\mu}{1+\mu}\right)^2 dt^2 + (1+\mu)^4 a(t)^2(d\rho^2 + \rho^2 d\Omega^2) \, , \tag{9}$$

where $a(t)$ is the scale factor, $d\Omega^2 = d\theta^2 + \sin^2\theta d\phi^2$ and

$$\mu \equiv \frac{M}{2a(t)\rho} \, , \tag{10}$$

where M is the mass of the point. One can check that for $a = $ constant, the Schwarzschild metric in isotropic coordinates is recovered, whereas for $M = 0$, the FLRW metric is recovered.

When $\mu \ll 1$, the McVittie metric (9) can be approximated by:

$$ds^2 = -(1-4\mu)dt^2 + (1+4\mu)a(t)^2(d\rho^2 + \rho^2 d\Omega^2) \, , \tag{11}$$

i.e., it takes the form of a perturbed FLRW metric in the Newtonian gauge with gravitational potential 2μ. Since there is only a single gravitational potential, then no anisotropic pressure is present [17,33].

Since null geodesics are conformally invariant, the scale factor $a(t)$ can be written as a conformal factor in front of Equation (11), replacing the cosmic time with the conformal time. The time dependence of the photon trajectory appears then only in μ which, from Equation (10), represents an effective mass decreasing inversely proportional to the cosmic expansion. We are going to show that most of the deflection takes place at the closest approach distance to the lens. Therefore, we can already estimate that at this moment the effective mass of the lens would be M/a_L, where a_L is the scale factor evaluated at the moment of closest approach. Therefore, the bending angle shall be proportional to M/a_L, and this is indeed our result from Equation (47).

Note also that the gravitational potential $2\mu = M/(a\rho)$ resembles the Newtonian one but with an effective mass that reduces with time or a gravitational radius that increases with time. On the other hand, we know from cosmological linear perturbation theory that, in the matter-dominated epoch, the gravitational potential is time-independent. Therefore, the McVittie metric cannot offer a realistic model of a gravitational lens.

Calculating the Einstein tensor from the McVittie line element (9), one gets:

$$G^t{}_t = 3H^2, \quad G^r{}_r = G^\theta{}_\theta = G^\phi{}_\phi = 3H^2 + \frac{2\dot{H}(1+\mu)}{1-\mu}, \tag{12}$$

from which one deduces that the pressure of the cosmological medium has the following form:

$$P = -\frac{1}{8\pi}\left[3H^2 + \frac{2\dot{H}(1+\mu)}{1-\mu}\right], \tag{13}$$

i.e., it is not homogeneous and diverging when $\mu = 1$. If $H = $ constant, then there is no divergence and the pressure is also a constant. This is the case of the Schwarzschild–de Sitter space, described by the Kottler metric [2]. When $\dot{H} \neq 0$, **far away** from the point mass, i.e., for $\mu \ll 1$, one gets the usual result of cosmology:

$$P = -\frac{1}{8\pi}\left(3H^2 + 2\dot{H}\right) + \mathcal{O}(\mu) = -\rho - \frac{\dot{H}}{4\pi} + \mathcal{O}(\mu), \tag{14}$$

i.e., the acceleration equation. Isotropy is preserved since $G^r{}_r = G^\theta{}_\theta = G^\phi{}_\phi$, i.e., there is no anisotropic pressure, as we already mentioned.

Following Faraoni [28], but also Park [13], the McVittie metric (9) can be reformulated in terms of the areal radius

$$R = a\rho(1+\mu)^2, \tag{15}$$

and gets the following form:

$$ds^2 = -\left(1 - \frac{2M}{R} - H^2R^2\right)dt^2 + \frac{dR^2}{1-\frac{2M}{R}} - \frac{2HR}{\sqrt{1-\frac{2M}{R}}}dtdR + R^2d\Omega^2. \tag{16}$$

Changing the time coordinate to

$$F(r,t)dT = dt + \frac{HR}{\sqrt{1-\frac{2M}{R}}\left(1-\frac{2M}{R}-H^2R^2\right)}dR, \tag{17}$$

the above line element (16) can be finally cast as

$$ds^2 = -\left(1 - \frac{2M}{R} - H^2R^2\right)F^2dT^2 + \frac{dR^2}{1 - \frac{2M}{R} - H^2R^2} + R^2d\Omega^2 \ . \tag{18}$$

If H is constant, then F can be set to unity and we recover the Kottler metric. Using Equation (18) and calculating the Misner–Sharp–Hernandez mass [31,32] of a sphere of proper radius R, one finds [28]:

$$m_{\text{MSH}} = M + \frac{H^2R^3}{2} = M + \frac{4\pi}{3}\rho R^3 \ , \tag{19}$$

which contains the time-independent contribution M from the point mass plus the mass of the cosmic fluid contained in the sphere. Therefore, M has indeed the physical meaning of the mass of the point. In the Kottler case, one can also define a Komar integral, or Komar mass, and verify that it is indeed equal to M [34].

3. The Bending of Light in the McVittie Metric

We now revisit the calculation for the bending angle performed in Ref. [18], but take into account a general non-constant Hubble factor $H(t)$. We perform the calculations in two different ways: the one in this section is also used in Ref. [18] and is based on the approach usually adopted to study weak lensing (see e.g., Ref. [33,35]). In this approach, the origin of the coordinate system is occupied by the observer. The second way is the one in which the lens is put at the origin of the coordinate system and it is employed in Appendix A.

We adopt μ as perturbative parameter and work at the first order approximation in μ. The observed angle of a lensed source is of the order of the arc second, which corresponds to $\theta_O \approx 10^{-6}$ radians. See e.g., the CASTLES survey lens database (https://www.cfa.harvard.edu/castles/). At least for Einstein ring systems, the bending angle is of order $\delta \approx 10^{-6}$, i.e., it is of the same order as the observed deflection angle. However, at the same time, the bending angle is of the same order of μ. Therefore, we draw the conclusion that $\mu \approx 10^{-6}$ and the truncation error, when working at first order in μ, is $\mathcal{O}(10^{-12})$.

The geometry of the lensing process is depicted in Figure 1. In this scheme, the observer stays at the origin of the spatial coordinate system and x is the comoving coordinate along the observer-lens axis.

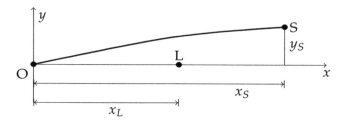

Figure 1. Scheme of lensing.

The observer has spatial position $(0,0,0)$ and the lens has spatial position $(x_L,0,0)$. The McVittie metric (9) written in Cartesian spatial coordinates is the following:

$$ds^2 = -\left(\frac{1-\mu}{1+\mu}\right)^2 dt^2 + (1+\mu)^4 a(t)^2 \delta_{ij}dx^i dx^j \ . \tag{20}$$

Note that the spherical symmetry of the McVittie metric implies rotational symmetry about the observer-lens axis. Therefore, we set $z = 0$ without losing generality and the source has thus spatial position $(x_S, y_S, 0)$.

Since in metric (9) the lens lays at the origin of the coordinate system, we have to perform a translation along the x axis in the Cartesian coordinates of metric (20), so that μ gets the following form:

$$\mu = \frac{M}{2a(t)\sqrt{(x - x_L)^2 + y^2 + z^2}} . \tag{21}$$

Introducing an affine parameter λ and the four-momentum $P^\mu = dx^\mu/d\lambda$, we can derive from metric (20) the following relation:

$$g_{\mu\nu}P^\mu P^\nu = 0 \quad \Rightarrow \quad P^0 = \frac{1 + \mu}{1 - \mu}p \sim (1 + 2\mu)p , \tag{22}$$

where $p^2 = g_{ij}P^i P^j$ is the proper momentum. The above equation represents the usual gravitational redshift experienced by a photon passing through the potential well generated by the point mass. Note that this potential well is not static since μ is time-dependent.

Now, we calculate the geodesic equations for the photon propagating in the McVittie metric (20):

$$\frac{d^2 x^\nu}{d\lambda^2} + \Gamma^\nu_{\alpha\beta}\frac{dx^\alpha}{d\lambda}\frac{dx^\beta}{d\lambda} = 0 . \tag{23}$$

The geodesic equation for $\nu = 0$ has the following form:

$$\frac{dP^0}{d\lambda} = 2\dot{\mu}(P^0)^2 + 4\mu_{,i}P^0 P^i - p^2[H(1 + 4\mu) + 2\dot{\mu}] . \tag{24}$$

Using Equation (22) and the fact that, from Equation (10), $\dot{\mu} = -H\mu$, one finds:

$$p\frac{dp}{dt} = -Hp^2 + 2H\mu p^2 + 4\mu_{,i}P^i p . \tag{25}$$

The zeroth-order term Hp^2 represents the usual cosmological redshift term. The spatial geodesics equations have the form:

$$\frac{dP^i}{d\lambda} = \frac{4\delta^{il}\mu_{,l}p^2}{a^2} - 2HpP^i - 4P^i P^k \mu_{,k} . \tag{26}$$

We now look for an equation for the quantity $dy/dx \equiv \tan\theta$, which represents the slope of the line tangent to the photon trajectory. Note that θ is a physical angle because of the isotropic form of metric (20).

Since we can invert $x(\lambda)$ to $\lambda(x)$, being it a monotonic function, we can rewrite Equation (26) for y and change the variable to x:

$$P^x \frac{d}{dx}\left(P^x \frac{dy}{dx}\right) = 4\mu_{,y}[(P^x)^2 + (P^y)^2] - 2Ha(1 + 2\mu)\sqrt{(P^x)^2 + (P^y)^2}P^x \frac{dy}{dx} - 4\frac{dx}{dz}P^x[P^y \mu_{,y} + P^x \mu_{,x}] , \tag{27}$$

where we used the fact that

$$p^2 = a^2(1 + 4\mu)[(P^x)^2 + (P^y)^2] . \tag{28}$$

Expanding the left-hand side and using $P^y/P^x = dy/dx$, we obtain:

$$\frac{d^2 y}{dx^2} + \frac{1}{P^x}\frac{dP^x}{dx}\frac{dy}{dx} = 4\mu_{,y}\left[1 + \left(\frac{dy}{dx}\right)^2\right] - 2Ha(1 + 2\mu)\sqrt{1 + \left(\frac{dy}{dx}\right)^2}\frac{dy}{dx} - 4\frac{dy}{dx}\left(\frac{dy}{dx}\mu_{,y} + \mu_{,z}\right) . \tag{29}$$

In order to determine the second term on the left-hand side, we use Equation (26) for x:

$$\frac{1}{P^x}\frac{dP^x}{dx} = 4\mu_{,x}\left[1 + \left(\frac{dy}{dx}\right)^2\right] - 2Ha(1 + 2\mu)\sqrt{1 + \left(\frac{dy}{dx}\right)^2} - 4\left(\frac{dy}{dx}\mu_{,y} + \mu_{,z}\right) . \tag{30}$$

Combining the two Equations (29) and (30), we finally find:

$$\frac{d^2y}{dx^2} = 4\mu_{,y}\left[1 + \left(\frac{dy}{dx}\right)^2\right] - 4\mu_{,x}\left[1 + \left(\frac{dy}{dx}\right)^2\right]\frac{dy}{dx}. \tag{31}$$

Let's discuss a little about the **spatial derivative** of μ. From Equation (21), we get:

$$\mu_{,y} = -\mu\frac{y}{\sqrt{(x - x_L)^2 + y^2}}, \tag{32}$$

and

$$\mu_{,x} = -\mu\frac{(x - x_L)}{\sqrt{(x - x_L)^2 + y^2}}. \tag{33}$$

Notice that, when $\mu = 0$, Equation (31) becomes:

$$\frac{d^2y}{dx^2} = 0, \tag{34}$$

i.e., the zeroth-order trajectory is, as expected, a straight line in comoving coordinates.

We now make a second approximation: we assume $dy/dx = \tan\theta$ to be small. From the CASTLES survey we know that the observed angle θ_O is of the order of the arc second, which corresponds to $\theta_O \approx 10^{-6}$ radians. The latter is larger that the actual angular position of the source, say θ_S, because of the lensing geometry, see e.g., Figure 2. For this reason, we can assume dy/dx to be small along all the trajectory. Since $dy/dx = \tan\theta$, then $dy/dx = \theta + \theta^3/3 + \cdots$. The truncation error is then of order $\mathcal{O}(\theta^3) \sim 10^{-18}$.

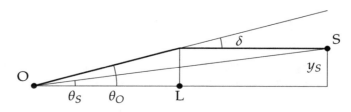

Figure 2. The thin-lens approximation.

We consider Equation (31) up to the lowest order term, i.e.,

$$\frac{d^2y}{dx^2} = 4\mu_{,y} + \mathcal{O}(\mu\theta), \tag{35}$$

where $\mathcal{O}(\mu\theta) \sim 10^{-12}$. Using Equation (32) for the derivative $\mu_{,y}$, the above equation becomes:

$$\frac{d^2y}{dx^2} = -\frac{2My}{a(x)\left[(x - x_L)^2 + y^2\right]^{3/2}}. \tag{36}$$

Note that a is a function of time, but inverting $x(t)$, we can write a as a function of x. For simplicity, we normalise x, y and $2M$ to x_L, thus obtaining:

$$\frac{d^2Y}{dX^2} = -\alpha\frac{Y}{a(X)\left[(X - 1)^2 + Y^2\right]^{3/2}}, \tag{37}$$

where $Y \equiv y/x_L$, $X \equiv x/x_L$ and $\alpha \equiv 2M/x_L$. The above equation was already found in Ref. [18]. Since $dY/dX = \tan\theta$ and $\tan\theta \sim \theta$, we can cast the above equation in the following form:

$$\frac{d\theta}{dX} = -\alpha \frac{Y}{a(X)\left[(X-1)^2 + Y^2\right]^{3/2}} \cdot \tag{38}$$

We *define* the bending angle as follows:

$$\delta \equiv \int_{\theta_S}^{\theta_O} d\theta = -\alpha \int_{X_S}^{0} \frac{Y(X)dX}{a(X)\left[(X-1)^2 + Y(X)^2\right]^{3/2}} , \tag{39}$$

i.e., as the variation of the slope of the trajectory between the source and the observer.

We shall solve the above equation keeping the first order in α. The order of magnitude of α can be estimated as follows:

$$\alpha \equiv \frac{2M}{x_L} \approx 2MH_0/z_L , \tag{40}$$

where we assumed a small redshift z_L. The above approximation becomes an exact result in the case of a constant Hubble parameter (see e.g., (51)).

Therefore, α is proportional to the ratio between the Schwarzschild radius of the lens and the Hubble radius. This is $H_0M \sim 10^{-12}$ for a galaxy of 10^{10} $M_\odot$ and $H_0M \sim 10^{-9}$ for a cluster of 10^3 galaxies each of mass 10^{10} $M_\odot$.

We now devote a small paragraph to the zeroth-order solution.

3.1. The Zeroth-Order Solution

The zeroth-order solution (i.e., the one for $\alpha = 0$) of Equation (37) is a straight line in comoving coordinates:

$$y = \theta_S(x - x_S) + y_S , \tag{41}$$

where $\theta_S \ll 1$ is the slope of the trajectory and (x_S, y_S) are the comoving coordinates of the source. See Figure 3.

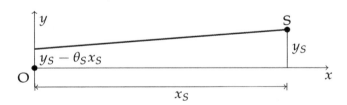

Figure 3. Zeroth-order solution.

When we pass from comoving to proper distances by multiplying by the scale factor $a(x)$ we obtain:

$$y_p = \theta_S x_p + \frac{a(x_p)}{a_S}(y_{pS} - \theta_S x_{pS}) , \tag{42}$$

where we used a subscript p to indicate the proper distance. The above is not a straight line trajectory, as also noticed by the authors of Ref. [15]. It is bent because of the $a(x_p)$ factor on the right-hand side, whose effect vanishes only for $y_{pS} = \theta_S x_{pS}$. The latter condition, when substituted in Equation (41), represents the ray which gets to $y = 0$ when $x = 0$, i.e., the observed ray. See Figure 4.

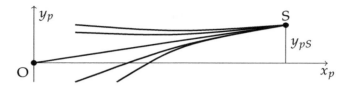

Figure 4. Zeroth-order solution, using proper distances.

The **Hubble flow** seems to bend away the trajectories such that we cannot detect any light. This happens isotropically, i.e., no observer could ever detect a bent ray but just the straight one coming directly from the source.

On the other hand, let's speculate about the following. If a cosmologically bent ray passes sufficiently close to a lens, then its trajectory could be bent back by the gravitational field of the lens, possibly allowing us to detect it. See Figure 5.

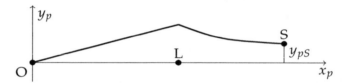

Figure 5. A cosmologically bent ray, bent back. The "back-bending".

This "back-bending" seems to suggest that the bending angle must increase and therefore cosmology must somehow enter the gravitational lensing phenomenon.

3.2. Calculation of the Bending Angle

We now integrate Equation (39) retaining the first order only in α. For this reason, the $Y(X)$ entering the integral is the zeroth-order solution, which we discussed in the previous subsection.

Since we are working at the first order in α, we can assume without losing generality that the zeroth-order trajectory is horizontal, i.e., $Y^{(0)} = Y_S \equiv y_S/x_L$.

The equation for the slope, i.e., Equation (38), becomes

$$\frac{d\theta}{dX} = -\frac{\alpha Y_S}{a(X)\left[(X-1)^2 + Y_S^2\right]^{3/2}}.$$

(43)

In order to determine $a(X)$, we take advantage of metric (20) and write:

$$dx^2 = \frac{1-8\mu}{1+(dy/dx)^2}\frac{dt^2}{a^2},$$

(44)

which is a very complicated integration to perform, since it includes the very trajectory we want to determine. On the other hand, we are staying at the lowest possible order of approximation; therefore:

$$dx = -\frac{dt}{a},$$

(45)

i.e., all the contributions coming from μ and θ of Equation (44) are of negligible order in Equation (43).

Now, write Equation (43) as follows:

$$d\theta = -\frac{\alpha dX}{a(X)Y_S^2\left[1 + \frac{(X-1)^2}{Y_S^2}\right]^{3/2}}.$$

(46)

When $(X - 1)^2 \gg Y_S^2$, the above integration, whatever function a might be of X, is $\mathcal{O}(Y_S)$. On the other hand, when $(X - 1)^2 \ll Y_S^2$, the above integration is $\mathcal{O}(1/Y_S^2)$.

Therefore, the main contribution comes from $X = 1$ and spans the interval $1 - Y_S < X < 1 + Y_S$. That is, most of the deflection takes place very close to the lens, as it happens for the case of the Schwarzschild metric. For this reason, we also approximate $a(X)$ with a_L, which is the scale factor when $x = x_L$.

Therefore, we end up with the following bending angle:

$$\delta = -\frac{\alpha}{a_L Y_S^2}(-2Y_S) = \frac{2\alpha}{a_L Y_S} = \frac{4M}{a_L y_S} . \tag{47}$$

The above formula is general, valid for any kind of **Hubble flow**. We derive it using another method in Appendix A and prove its validity in the case of a dust-dominated universe, for which an exact calculation is possible, in Appendix B.

Now we apply Equation (47) in the lens equation. Let us refer to Figure 2. The geometry of this figure is justified by the fact that, as we showed earlier, the bending happens predominantly at the closest approach distance to the lens.

In Figure 2, θ_S is the angular position of the source, so that $\theta_S D_S$ is the proper transversal position of the source, where D_S is the angular-diameter distance from the observer to the source. The angle θ_O is the angular apparent position of the source, so that $\theta_O D_S$ is the transversal apparent position of the source.

Therefore, the lens equation in the thin-lens approximation can be written as:

$$\theta_O D_S = \theta_S D_S + \delta D_{LS} , \tag{48}$$

where D_{LS} is the angular-diameter distance between lens and source. Using the result of Equation (47), we get

$$\theta_O - \theta_S = \frac{4M}{a_L y_S} \frac{D_{LS}}{D_S} . \tag{49}$$

In the standard lens equation, one has the closest approach distance to the lens, let's call it R, in place of $a_L y_S$. One then writes $R = \theta_O D_L$ and thus finds the usual formula, see e.g., [17].

Now, since we found that the deflection occurs almost completely at the closest position to the lens, we can approximate $y_S \approx y_L$. Moreover, one also has $y_L = \theta_O x_L$ and $D_L = a_L x_L$, from the definition of the angular-diameter distance to the lens. Thus, $a_L y_S \approx a_L \theta_O x_L = \theta_O D_L$, and we recover the usual well-known formula:

$$\theta_O(\theta_O - \theta_S) = \frac{4M}{D_L} \frac{D_{LS}}{D_S} . \tag{50}$$

Therefore, we can conclude that cosmology does not modify the bending angle at the leading order of the expansion in powers of μ and θ. The cosmological "drift" discussed earlier for the zeroth-order solution is already taken into account when using angular-diameter distances so that the final result does not change.

However, sub-dominant terms do carry a cosmological signature, as we show in the next section. Here, we address the simple case of a cosmological constant-dominated universe, where analytical calculations are possible.

4. Next-to-Leading Order Contributions to the Bending Angle in the Case of a Cosmological Constant-Dominated Universe

As we saw in Equation (47), the leading contribution in the expansion for the bending angle calculated in the McVittie metric is the same as the one calculated for the Schwarzschild metric. Therefore, it is interesting to check if next-to-leading orders do carry a signature of the cosmological embedding of the point lens. We tackle this issue here in the case of a cosmological constant-dominated universe, for which exact calculations are possible, and leave a more general treatment as a future work.

When $H = H_0 = \text{constant}$, one can find an analytic expression for $a(x)$:

$$x = \int_a^1 \frac{da'}{H_0 a^2} = \frac{1}{H_0}\left(\frac{1}{a}-1\right) = \frac{z}{H_0}\,,\tag{51}$$

where in the last equality we introduced the redshift. The scale factor as function of the comoving distance is thus:

$$\frac{1}{a(x)} = H_0 x + 1 = H_0 X x_L + 1 = z_L X + 1\,,\tag{52}$$

and Equation (43) becomes:

$$\frac{d\theta}{dX} = -\frac{\alpha Y_S(z_L X + 1)}{[(X-1)^2 + Y_S^2]^{3/2}}\,.\tag{53}$$

As we anticipated, this equation can be solved exactly and the bending angle, as we defined it in Equation (39), is the following:

$$\delta = \frac{\alpha}{Y_S}\left[\frac{1 + z_L + z_L Y_S^2}{\sqrt{1 + Y_S^2}} + \frac{(X_S - 1)(1 + z_L) - z_L Y_S^2}{\sqrt{(X_S - 1)^2 + Y_S^2}}\right]\,.\tag{54}$$

Expanding this solution for a small impact parameter Y_S, one gets:

$$\delta = \frac{2\alpha(1 + z_L)}{Y_S}\left[1 + Y_S^2 \frac{2(z_L - 1) + X_S[2 + X_S(z_L - 1) - 4z_L]}{4(z_L + 1)(X_S - 1)^2}\right]\,,\tag{55}$$

where we have already truncated $\mathcal{O}(Y_S^4)$ terms and put in evidence the leading order contribution $2\alpha(1 + z_L)/Y_S$ (see Equation (47)).

Recovering the physical quantities $Y_S = y_S/x_L$, $X_S = x_S/x_L$, $\alpha = 2M/x_L$ and using Equation (51) in order to express x as the redshift, we get:

$$\delta = \frac{4M(1 + z_L)}{y_S}\left[1 + \frac{y_S^2}{x_L^2}\frac{2z_L^2(z_L - 1) + z_S[2z_L + z_S(z_L - 1) - 4z_L^2]}{4(z_L + 1)(z_S - z_L)^2}\right]\,.\tag{56}$$

We already showed in the discussion leading to Equation (50) that $y_S \approx \theta_O x_L$, so that:

$$\delta = \frac{4M}{\theta_O D_L}\left[1 + \theta_O^2\frac{2z_L^2(z_L - 1) + z_S[2z_L + z_S(z_L - 1) - 4z_L^2]}{4(z_L + 1)(z_S - z_L)^2}\right]\,,\tag{57}$$

and in the lens equation:

$$\theta_O(\theta_O - \theta_S) = \frac{4M D_{LS}}{D_L D_S}\left[1 + \theta_O^2\frac{2z_L^2(z_L - 1) + z_S[2z_L + z_S(z_L - 1) - 4z_L^2]}{4(z_L + 1)(z_S - z_L)^2}\right]\,.\tag{58}$$

Let's focus on Einstein ring systems, i.e., $\theta_S = 0$. We have in this case the mass estimate (it is actually an estimate on the product $H_0 M$, due to the presence of the angular-diameter distances, see Ref. [17])):

$$\frac{4M D_{LS}}{D_L D_S} = \theta_O^2\left[1 - \theta_O^2\frac{2z_L^2(z_L - 1) + z_S[2z_L + z_S(z_L - 1) - 4z_L^2]}{4(z_L + 1)(z_S - z_L)^2}\right]\,.\tag{59}$$

The next-to-leading order correction is $\mathcal{O}(\theta_O^4)$ and depends on the redshifts of the lens and of the source.

Consider, for example, the Einstein ring Q0047-2808 of the CASTLES survey, for which $\theta_O = 2.7''$, $z_S = 3.60$ and $z_L = 0.48$. Substituting these numbers in Equation (59), the correction on the mass estimate is therefore

$$\frac{4MD_{LS}}{\theta_O^2 D_L D_S} = 1 + 0.12\,\theta_O^2 = 1 + 2.03 \cdot 10^{-11}\,. \tag{60}$$

This is an extremely small correction which nonetheless depends on cosmology. Note that it is only one order of magnitude larger than the terms $\mathcal{O}(\mu^2)$ that we have neglected in our calculations.

5. Conclusions

We investigated whether cosmology affects the gravitational lensing caused by a point mass. To this purpose, we used McVittie metric as the description of the point-like lens embedded in an expanding universe. The reason for this choice is to use a metric which properly takes into account the **Hubble flow** to which source, lens and observer are subject. We considered the general case in which the Hubble factor is a generic function of time and find that no contribution coming from cosmology enters the bending angle at the leading order (see Equation (47)), thus strengthening the results obtained by [13–16].

We addressed the sub-dominant contributions to the bending angle in the special case of a constant Hubble factor $H = H_0$, for which exact calculations are possible. We found that in this case cosmology does affect the bending of light, through a combination of the lens and source redshifts, given in Equation (58). This correction is of order 10^{-11} for the Einstein ring Q0047-2808.

We conclude that the standard approach to gravitational lensing on cosmological distances, which consists of patching together the results coming from the Schwarzschild metric (which models the lens) and Friedmann–Lemaître-Robertson–Walker (FLRW) metric (which serves to calculate the cosmological angular diameter distances), does not require modifications.

Future developments of this investigation should address the entity subdominant orders of the expansion for the bending angle in a model-independent way. We expect the latter to depend on H_L, i.e., the Hubble parameter evaluated at the lens redshift. If these corrections were measurable, they might provide a new cosmological probe for determining the value of the Hubble parameter at different redshifts.

Finally, we must stress that McVittie metric is a particular and oversimplified description of the geometry of a lens, this being a galaxy or a cluster of galaxies. Therefore, another improvement would be that of tackling the analysis of the gravitational lensing and of the bending angle by constructing for the lens a density profile more realistic than a Dirac delta (i.e., the one used here for a point mass). We expect that different lens density profiles would lead to different results in the mass estimates also from the point of view of the cosmological corrections, as shown in Ref. [10] for the case of the Kottler metric.

Acknowledgments: The author thanks CNPq (Brasilia, Brazil) for partial financial support. He is also indebted to D. Bacon, V. Marra, H. Velten and the anonymous referees for stimulating discussions and suggestions.

Appendix A. Standard Approach to Gravitational Lensing

We now place the lens at the origin of the reference frame and use polar coordinates, as in Figure A1.

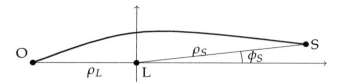

Figure A1. Scheme of lensing, with the lens at the origin of the coordinate system.

Again, we work at first order in μ. For a photon, metric (11) with $\theta = \pi/2$ gives:

$$0 = -(1 - 4\mu)\dot{t}^2 + (1 + 4\mu)a^2\dot{\rho}^2 + (1 + 4\mu)a^2\rho^2\dot{\phi}^2 \,, \tag{A1}$$

where the dot denotes derivation with respect to the affine parameter λ. Since $\xi_{(\phi)} = \delta^\mu_\phi \partial_\mu$ is a Killing vector, there exists the following conserved quantity:

$$L = g_{\mu\nu}P^\mu \xi^\nu_{(\phi)} = (1 + 4\mu)a^2\rho^2\dot{\phi} \,, \tag{A2}$$

where P^μ is the photon four-momentum. The geodesic equation for t, cf. Equation (24), can be cast as follows:

$$(1 - 4\mu)\ddot{t} - 4\frac{\partial\mu}{\partial\rho}\dot{t}\dot{\rho} + \frac{1}{a}\frac{da}{dt}(1 - 4\mu)\dot{t}^2 = 0 \,, \tag{A3}$$

and written in the following compact form:

$$\frac{d}{d\lambda}\left[(1 - 4\mu)a\dot{t}\right] + 4a\dot{\mu}\dot{t}^2 = 0 \,. \tag{A4}$$

Recalling the definition of μ in Equation (10), i.e., $\mu = M/(2a\rho)$, one can easily determine that $\dot{\mu} = -H\mu$. We neglect this contribution since indeed the ratio between the gravitational radius of the lens and the Hubble radius must be very small, as we discussed after Equation (40).

Therefore, neglecting HM, Equation (A4) can be exactly integrated, giving the following result:

$$\left(1 - \frac{2M}{a\rho}\right)\dot{t} = \frac{E}{a} + \mathcal{O}(HM) \,, \tag{A5}$$

where E is an integration constant. We found a mixture of the known results for the Schwarzschild metric and for the FLRW one. Indeed, if $H = 0$, then a is an unimportant constant that we can incorporate into the definitions of ρ and E, and we recover the result for the Schwarzschild metric. On the other hand, with $M = 0$, we recover the usual cosmological decay of the energy of a photon, which is inversely proportional to the scale factor.

Combining Equation (A1) with Equation (A5), we can write the following equation for ρ:

$$\frac{a^4}{E^2}\dot{\rho}^2 = 1 - \frac{D^2}{\rho^2}(1 - 8\mu) \,, \tag{A6}$$

where $D \equiv L/E$ is a parameter associated to the closest approach distance ρ_L, defined as the one for which $\dot{\rho}_L = 0$, i.e.,

$$\rho_L = D(1 - 4\mu_L) \,, \tag{A7}$$

where μ_L is μ evaluated at the closest approach distance, i.e., $\mu_L = M/(2a_L\rho_L)$.

We use now the definition of the bending angle proposed by Rindler and Ishak in Ref. [1], based on the following formula:

$$\tan\psi = \frac{\sqrt{g_{\phi\phi}}}{\sqrt{g_{\rho\rho}}}\left|\frac{d\phi}{d\rho}\right| \,, \tag{A8}$$

which represents the angle between the radial and the tangential directions of the photon trajectory (see Figure A2). Using Equations (A2) and (A6) we find:

$$\tan \psi = \frac{D}{\rho} \frac{1 - 4\mu}{\sqrt{1 - \frac{D^2}{\rho^2}(1 - 8\mu)}} .$$ (A9)

This expression can be rewritten in terms of the closest approach radius ρ_L as follows:

$$\tan \psi = \frac{\rho_L/\rho}{\sqrt{1 - \rho_L^2/\rho^2}} \left(1 - \frac{2M}{a\rho} + \frac{2M}{a_L\rho_L}\right) .$$ (A10)

For $M = 0$, we obtain from Equation (A10) that

$$\tan \psi = \frac{\rho_L/\rho}{\sqrt{1 - \rho_L^2/\rho^2}} = \tan \phi ,$$ (A11)

i.e., we recover the straight trajectory. Therefore, at any given position along the trajectory, $\psi - \phi$ gives the local bending angle, i.e., the deviation from the straight-line trajectory. See Figure A2.

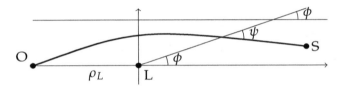

Figure A2. Schematic definition of the angle ψ, defined in Equation (A8). See also Figure 2 of Ref. [1].

The total bending angle is given by:

$$\delta = \psi_S + \psi_O - \phi_S - \phi_O .$$ (A12)

If we assume ρ_S and ρ_O are much larger than ρ_L, then the contributions from ψ_S and ψ_O are very small and practically negligible. Therefore, the dominant contribution to δ comes from $\phi_S + \phi_O$. In order to determine this sum, we must analyse the equation for the trajectory, i.e.,

$$\frac{d\rho}{d\phi} = \pm\rho \left(1 + \frac{4\mu - 4\mu_L}{1 - \rho_L^2/\rho^2}\right) \sqrt{\frac{\rho^2}{\rho_L^2} - 1} .$$ (A13)

For $\rho \gg \rho_L$, one can simplify this equation as follows:

$$\frac{d\rho}{d\phi} = \pm\frac{\rho^2}{\rho_L} (1 - 4\mu_L) ,$$ (A14)

where we have considered only the leading-order correction to the equation for the straight line. The above equation tells us that the trajectory still is a straight line, far away from the lens, but tilted of an angle $4\mu_L$ from each side with respect to the horizontal. Therefore, the bending angle is

$$\delta = 8\mu_L = \frac{4M}{a_L\rho_L} ,$$ (A15)

which is identical to the result of Equation (47) and also valid for a time-dependent H.

Appendix B. Bending Angle in a Matter-Dominated Universe

We check here Formula (47) in the case of a matter-dominated universe, described by the Friedmann equation $H^2 = H_0^2/a^3$. The scale factor as a function of the comoving distance x can be calculated as follows:

$$x = \int_a^1 \frac{da'}{H(a')a^2} = \frac{2}{H_0}\left(1 - \sqrt{a}\right), \tag{B1}$$

which implies $a(x) = (1 - H_0 x/2)^2$. With this $a(x)$, Equation (43) can be solved exactly and the bending angle is the following:

$$\delta = \frac{2\alpha}{Y_S} \frac{4}{(H_0 x_L - 2)^2} + \mathcal{O}(Y_S) = \frac{2\alpha}{a_L Y_S} + \mathcal{O}(Y_S), \tag{B2}$$

i.e., the same result found in Equation (47).

References

1. Rindler, W.; Ishak, M. Contribution of the cosmological constant to the relativistic bending of light revisited. *Phys. Rev. D* **2007**, *76*, 043006.
2. Kottler, F. Über die physikalischen grundlagen der Einsteinschen gravitationstheorie. *Ann. Phys.* **1918**, *361*, 401–462.
3. Islam, J. The cosmological constant and classical tests of general relativity. *Phys. Lett. A* **1983**, *97*, 239–241.
4. Schucker, T. Cosmological constant and lensing. *Gen. Relativ. Gravit.* **2009**, *41*, 67–75.
5. Ishak, M. Light Deflection, Lensing, and Time Delays from Gravitational Potentials and Fermat's Principle in the Presence of a Cosmological Constant. *Phys. Rev. D* **2008**, *78*, 103006.
6. Ishak, M.; Rindler, W.; Dossett, J. More on Lensing by a Cosmological Constant. *Mon. Not. Roy. Astron. Soc.* **2010**, *403*, 2152–2156.
7. Sereno, M. The role of Lambda in the cosmological lens equation. *Phys. Rev. Lett.* **2009**, *102*, 021301.
8. Kantowski, R.; Chen, B.; Dai, X. Gravitational Lensing Corrections in Flat ΛCDM Cosmology. *Astrophys. J.* **2010**, *718*, 913–919.
9. Ishak, M.; Rindler, W. The Relevance of the Cosmological Constant for Lensing. *Gen. Relativ. Gravit.* **2010**, *42*, 2247–2268.
10. Biressa, T.; de Freitas Pacheco, J. The Cosmological Constant and the Gravitational Light Bending. *Gen. Relativ. Gravit.* **2011**, *43*, 2649–2659.
11. Hammad, F. A note on the effect of the cosmological constant on the bending of light. *Mod. Phys. Lett.* **2013**, *A28*, 1350181.
12. Arakida, H. Effect of the Cosmological Constant on Light Deflection: Time Transfer Function Approach. *Universe* **2016**, *2*, 5.
13. Park, M. Rigorous Approach to the Gravitational Lensing. *Phys. Rev. D* **2008**, *78*, 023014.
14. Khriplovich, I.B.; Pomeransky, A.A. Does Cosmological Term Influence Gravitational Lensing? *Int. J. Mod. Phys. D* **2008**, *17*, 2255–2259.
15. Simpson, F.; Peacock, J.A.; Heavens, A.F. On lensing by a cosmological constant. *Mon. Not. Roy. Astron. Soc.* **2010**, *402*, 2009.
16. Butcher, L.M. Lambda does not Lens: Deflection of Light in the Schwarzschild-de Sitter Spacetime. 2016, arXiv:gr-qc/1602.02751.
17. Weinberg, S. *Cosmology*; Oxford University Press: Oxford, UK, 2008.
18. Piattella, O.F. Lensing in the McVittie metric. *Phys. Rev. D* **2016**, *93*, 024020.
19. McVittie, G. The mass-particle in an expanding universe. *Mon. Not. Roy. Astron. Soc.* **1933**, *93*, 325–339.
20. Nolan, B.C. A Point mass in an isotropic universe: Existence, uniqueness and basic properties. *Phys. Rev. D* **1998**, *58*, 064006.
21. Nolan, B. A Point mass in an isotropic universe. 2. Global properties. *Class. Quantum Gravity* **1999**, *16*, 1227–1254.
22. Nolan, B.C. A Point mass in an isotropic universe. 3. The region R less than or = to 2m. *Class. Quantum Gravity* **1999**, *16*, 3183–3191.

23. Kaloper, N.; Kleban, M.; Martin, D. McVittie's Legacy: Black Holes in an Expanding Universe. *Phys. Rev. D* **2010**, *81*, 104044.

24. Lake, K.; Abdelqader, M. More on McVittie's Legacy: A Schwarzschild—De Sitter black and white hole embedded in an asymptotically ΛCDM cosmology. *Phys. Rev. D* **2011**, *84*, 044045.

25. Nandra, R.; Lasenby, A.N.; Hobson, M.P. The effect of a massive object on an expanding universe. *Mon. Not. Roy. Astron. Soc.* **2012**, *422*, 2931–2944.

26. Nandra, R.; Lasenby, A.N.; Hobson, M.P. The effect of an expanding universe on massive objects. *Mon. Not. Roy. Astron. Soc.* **2012**, *422*, 2945–2959.

27. Nolan, B.C. Particle and photon orbits in McVittie spacetimes. *Class. Quantum Gravity* **2014**, *31*, 235008.

28. Faraoni, V. *Cosmological and Black Hole Apparent Horizons*; Springer: Cham, Switzerland, 2015.

29. Aghili, M.E.; Bolen, B.; Bombelli, L. Effect of Accelerated Global Expansion on Bending of Light. 2014, arXiv:gr-qc/1408.0786.

30. Faraoni, V.; Lapierre-Leonard, M. Beyond Lensing by the Cosmological Constant. 2016, arXiv:gr-qc/1608.03164.

31. Misner, C.W.; Sharp, D.H. Relativistic equations for adiabatic, spherically symmetric gravitational collapse. *Phys. Rev.* **1964**, *136*, B571–B576.

32. Hernandez, W.C.; Misner, C.W. Observer Time as a Coordinate in Relativistic Spherical Hydrodynamics. *Astrophys. J.* **1966**, *143*, 452.

33. Dodelson, S. *Modern Cosmology*; Academic Press: Cambridge, MA, USA, 2003.

34. Kastor, D. Komar Integrals in Higher (and Lower) Derivative Gravity. *Class. Quantum Gravity* **2008**, *25*, 175007.

35. Bartelmann, M.; Schneider, P. Weak gravitational lensing. *Phys. Rep.* **2001**, *340*, 291–472.

10

Symplectic Structure of Intrinsic Time Gravity

Eyo Eyo Ita III [1,*] and Amos S. Kubeka [2]

[1]   Physics Department, U.S. Naval Academy, Annapolis, MD 21401, USA
[2]   Mathematical Sciences Department, University of South Africa, Pretoria 0002, South Africa;
      kubekas@unisa.ac.za
*   Correspondence: ita@usna.edu

Academic Editor: Lorenzo Iorio

Abstract: The Poisson structure of intrinsic time gravity is analysed. With the starting point comprising a unimodular three-metric with traceless momentum, a trace-induced anomaly results upon quantization. This leads to a revision of the choice of momentum variable to the (mixed index) traceless momentric. This latter choice unitarily implements the fundamental commutation relations, which now take on the form of an affine algebra with SU(3) Lie algebra amongst the momentric variables. The resulting relations unitarily implement tracelessness upon quantization. The associated Poisson brackets and Hamiltonian dynamics are studied.

Keywords: intrinsic time; quantum gravity; canonical; quantization; symmetry

1. Introduction

A crucially important question in the quantization of gravity in 3+1 dimensions, as for any theory, is the choice of the fundamental dynamical variables of the classical theory, which upon quantization become promoted to quantum operators. In Loop Quantum Gravity (LQG) [1] the starting point for the classical theory are the Ashtekar variables, where a SU(2) gauge connection and a densitized triad form a canonically conjugate pair. This choice of variables turns the initial value constraints of GR from intractable non polynomial phase space functions, as they appear in the Arnowitt Deser Misner (ADM) theory [2], into polynomial form at the expense of an additional set of constraints related to the SU(2) gauge symmetry inherent in the theory. It is hoped that the polynomial form of the constraints in LQG make the constraints more tractable for quantization and the construction of a physical Hilbert space. The actual configuration variable in LQG which is subject to the quantization procedure is not the connection itself, but rather the holonomy of the connection, since the latter is well-defined in the quantum theory whereas the connection fails to exist [3]. Furthermore, the transformation properties of the holonomy more aptly are representative, at the kinematical level, of the symmetry properties of the theory [4]. Consequently, upon quantization in LQG all constraints and quantities must be rewritten in terms of the holonomies and the densitized triad, which themselves no longer form a canonical pair.

In LQG there exists only a manifold structure with no metric, and the metric is no longer fundamental, but becomes a derived quantity in terms of more fundamental variables. A main difficulty in LQG is the construction of a physical Hilbert space from solution of the Hamiltonian constraint. Whether one utilizes the self-dual version of the connection or its real counterparts as in the Barbero variables [5], the solution to the Hamiltonian constraint and its subsequent delineation of the physical Hilbert space, is a long and standing unresolved problem [2]. Consequently, the quantization of LQG remains complete only at the kinematical level (which is more suitably adapted to the fundamental variables), and the physical dynamics of gravity remain to be completely encoded within this procedure [4]. LQG can be contrasted with the standard ADM approach [4], wherein the fundamental variables are the spatial three metric and its conjugate momentum, constructed from the extrinsic curvature of the spatial slice of four-dimensional spacetime upon which the quantization must be performed. The corresponding initial value constraints are intractable due to various technical issues particularly related to ultraviolet divergences associated with operator products, which in LQG are absent. The choice of fundamental variables in the ADM approach poses the problem that as canonically conjugate variables, the momentum generates translations of the spatial three metric. Since the spectrum of both variables is the real line, then the positivity of the metric in the quantum theory cannot be guaranteed while having self-adjoint variables. Positivity of the spatial three metric is a crucially important condition that any quantum theory of gravity must satisfy, since spatial distances as measured by the theory must always be positive.

The theory of Intrinsic time quantum gravity (ITQG) [6] presented in this paper is driven by the motivation to solve all of the above difficulties. The choice of the configuration space variable in ITQG will be a unimodular spatial three metric metric and a momentum variable (ultimately known as a momentric) which generates dilations (more precisely SU(3) and SL(3,R) transformations of the metric). The importance of this choice of fundamental variables is that they will be self-adjoint in the quantum theory, while preserving the positivity and the unimodularity of the spatial three metric forming the configuration space variable. A common misconception of the price for such a result is that the variables cannot be canonically related, resulting in complications in their quantization. However, in the case of ITQG we will see that it is precisely their non canonical nature that makes them perfectly suited for quantization, and admits a group-theoretical interpretation as such, which resolves all of the aforementioned difficulties in the LQG and ADM approaches in one stroke.

In [6], a new formulation for quantization of the gravitational field in ITQG, is presented. The basic idea, as introduced in [7] and [8], is the concept of a new phase space for gravity which breaks the paradigm of four-dimensional spacetime covariance, shifting the emphasis to three dimensional spatial diffeomorphism invariance combined with a physical Hamiltonian which generates evolution with respect to intrinsic time. Through the constructive interference of wavefronts, classical spacetime emerges from the formalism, with direct correlation between intrinsic time intervals and proper time intervals of spacetime. In the present paper we will take a step back to analyse the motivations and canonical structure of ITQG, and then construct the fundamental variables and their commutation relations of the theory. These relations are noncanonical, which lead to the uncovering of an inherent $SU(3)$ structure for gravity. This presents certain advantages from the standpoint of quantization. The paper is thus structures as follows: Section 2 discusses the Poisson structure of the barred classical variables, Section 3 highlights the prelude to the quantum theory, Section 4 discusses the momentric operators and the SU(3) Lie algebra, Section 5 revisits the classical theory, and then lastly, Section 6 concludes the paper with some recommendations for similar future work in this direction.

2. Poisson Structure of the Barred Classical Variables

Let q_{ij}, $\widetilde{\pi}^{ij}$ denote the spatial 3-metric and its conjugate momentum defined on a spatial slice Σ of a four dimensional spacetime of topology $M = \Sigma \times R$. In the ADM metric theory, the basic variables provide a canonical one form

$$\Theta_{ADM} = \int_{\Sigma} d^3x \; \widetilde{\pi}^{ij}(x)\delta q_{ij}(x). \tag{1}$$

Starting from this canonically conjugate pair, let us define as fundamental classical variables the following barred quantities $\bar{q}_{ij}$, a unimodular metric with $\det \bar{q}_{ij} = 1$, and a traceless momentum variable $\overline{\pi}^{ij}$ via the relations [7,8]

$$\bar{q}_{ij} = q^{-1/3}q_{ij}; \quad \overline{\pi}^{ij} = q^{1/3}\big(\widetilde{\pi}^{ij} - \frac{1}{3}q^{ij}\widetilde{\pi}\big), \tag{2}$$

where $\widetilde{\pi} = q_{ij}\widetilde{\pi}^{ij}$ with $\bar{q}_{ij}\overline{\pi}^{ij} = 0$. From Equation (2) we get the following cotangent space decomposition

$$\delta q_{ij} = q^{1/3}\big(\bar{q}_{ij}\delta \ln q^{1/3} + \delta \bar{q}_{ij}\big) \longrightarrow \delta \bar{q}_{ij} = \overline{P}^{kl}_{ij}\delta q_{kl} \tag{3}$$

where we have defined the traceless projector $\overline{P}_{kl}^{ij} = \frac{1}{2}\left(\delta_k^i\delta_l^j + \delta_k^j\delta_l^i\right) - \frac{1}{3}\overline{q}^{ij}\overline{q}_{kl}$, with $\overline{P}_{kl}^{ij}\overline{q}^{kl} = \overline{q}_{ij}\overline{P}_{kl}^{ij} = 0$. So we have $\overline{q}^{ij}\delta\overline{q}_{ij} = 0$, namely that the cotangent space elements $\delta\overline{q}^{ij}$ are traceless. The inverse relations

$$q_{ij} = q^{1/3}\overline{q}_{ij}; \quad \widetilde{\pi}^{ij} = q^{-1/3}\left(\overline{\pi}^{ij} + \frac{1}{3}\overline{q}^{ij}\widetilde{\pi}\right) \tag{4}$$

take us from the barred back to the unbarred variables. Substitution of the left side of the arrow of Equation (3) into Equation (1) provides a clean separation of the barred gravitational degrees of freedom with canonical one-form [7]

$$\Theta = \int_\Sigma d^3x\,\widetilde{\pi}^{ij}\delta q_{ij} = \int_\Sigma d^3x\left(\widetilde{\pi}\delta\ln q^{1/3} + \overline{\pi}^{ij}\delta\overline{q}_{ij}\right), \tag{5}$$

where we have used $\overline{\pi}^{ij}\overline{q}_{ij} = \overline{q}^{ij}\delta\overline{q}_{ij} = 0$. Equation (5) yields a corresponding symplectic two-form

$$\Omega = \delta\Theta = \int_\Sigma d^3x\left(\delta\widetilde{\pi}\wedge\delta\ln q^{1/3} + \delta\overline{\pi}^{ij}\wedge\delta\overline{q}_{ij}\right). \tag{6}$$

While this may be the case, as we will see, the Poisson brackets which can arise from (6) are not unique, on account of subtleties due to the implementation of tracelessness of $\overline{\pi}^{ij}$.

A necessary condition for a consistent canonical quantization of the theory is that the correct Poisson brackets comprise the starting point at the classical level. So let us directly calculate via Equation (2) barred Poisson brackets with respect to the unbarred canonical structure, which is clearly known to be unambiguous. For the metric components we have $\{\overline{q}_{ij}(x),\overline{q}_{kl}(y)\} = 0$ which is encouraging, as the unbarred metric clearly is devoid of any momentum dependence. However, using the following relations

$$\frac{\delta q^{1/3}}{\delta q_{ij}} = \frac{1}{3}q^{1/3}q^{ij}; \quad \frac{\delta q^{ij}}{\delta q_{mn}} = -q^{(im}q^{j)n}; \quad \frac{\delta\overline{q}_{kl}}{\delta q_{ij}} = q^{-1/3}\overline{P}_{kl}^{ij}; \quad \frac{\delta\overline{\pi}^{ij}}{\delta\widetilde{\pi}^{kl}} = q^{1/3}\overline{P}_{kl}^{ij}, \tag{7}$$

in conjunction with

$$\frac{\delta\overline{\pi}^{ij}}{\delta q_{kl}} = \frac{1}{3}\left(q^{kl}\overline{\pi}^{ij} + q^{1/3}\left(q^{(ik}q^{j)l}q_{rs}\widetilde{\pi}^{rs} - q^{ij}\widetilde{\pi}^{kl}\right)\right) = \frac{1}{3}q^{-1/3}\left(\overline{q}^{kl}\overline{\pi}^{ij} - \overline{q}^{ij}\overline{\pi}^{kl}\right) + \frac{1}{3}q^{1/3}q^{(ik}q^{j)l}\widetilde{\pi}, \tag{8}$$

we obtain the following Poisson bracket relations between barred metric and momentum

$$\{\overline{q}_{ij}(x),\overline{\pi}^{kl}(y)\} = \int_\Sigma d^3z\left(\frac{\delta\overline{q}_{ij}(x)}{\delta q_{mn}(z)}\frac{\delta\overline{\pi}^{kl}(y)}{\delta\widetilde{\pi}^{mn}(z)} - \frac{\delta\overline{\pi}^{kl}(y)}{\delta q_{mn}(z)}\frac{\delta\overline{q}_{ij}(x)}{\delta\widetilde{\pi}^{mn}(z)}\right) = \overline{P}_{ij}^{kl}\delta^{(3)}(x,y). \tag{9}$$

Finally, we obtain the following relation amongst the barred momentum components

$$\{\overline{\pi}^{ij}(x),\overline{\pi}^{kl}(y)\} = \int_\Sigma d^3z\left(\frac{\delta\overline{\pi}^{ij}(x)}{\delta q_{mn}(z)}\frac{\delta\overline{\pi}^{kl}(y)}{\delta\widetilde{\pi}^{mn}(z)} - \frac{\delta\overline{\pi}^{kl}(y)}{\delta q_{mn}(z)}\frac{\delta\overline{\pi}^{ij}(x)}{\delta\widetilde{\pi}^{mn}(z)}\right) = \frac{1}{3}\left(\overline{q}^{kl}\overline{\pi}^{ij} - \overline{q}^{ij}\overline{\pi}^{kl}\right)\delta^{(3)}(x,y). \tag{10}$$

The Poisson brackets between barred variables are noncanonical. But we will show that they yield the same barred contribution as the symplectic two form (6) which can be seen as follows. From the calculated Poisson brackets the following Poisson matrix can be constructed

$$P^{IJ} = \begin{pmatrix} \{\overline{q}_{ij}(x),\overline{q}_{kl}(y)\} & \{\overline{q}_{ij}(x),\overline{\pi}^{kl}(y)\} \\ \{\overline{\pi}^{kl}(y),\overline{q}_{ij}(x)\} & \{\overline{\pi}^{ij}(x),\overline{\pi}^{kl}(y)\} \end{pmatrix} = \begin{pmatrix} 0 & \overline{P}_{ij}^{kl} \\ -\overline{P}_{ij}^{kl} & \frac{1}{3}\left(\overline{q}^{kl}\overline{\pi}^{ij} - \overline{q}^{ij}\overline{\pi}^{kl}\right) \end{pmatrix}\delta^{(3)}(x,y). \tag{11}$$

In Poisson geometry, a two form $\Omega = \frac{1}{2}\Omega_{IJ}\delta q^I\wedge\delta q^J$ on the phase space $q^I \equiv \overline{q}_{ij}, \overline{\pi}^{ij}$ can be constructed whose components are the inverse of the Poisson matrix. If Ω is closed ($\delta\Omega = 0$) and nondegenerate, then it is said to be a symplectic two form. Making the identifications $\{\bar{q}, \bar{\pi}\} \sim \beta$ and $\{\bar{\pi}, \bar{\pi}\} \sim \alpha$, then the inverse of the Poisson matrix for the barred variables is of the form

$$P^{-1} = \begin{pmatrix} 0 & \beta \\ -\beta & \alpha \end{pmatrix}^{-1} = \begin{pmatrix} \beta^{-1}\alpha\beta^{-1} & -\beta^{-1} \\ \beta^{-1} & 0 \end{pmatrix}, \tag{12}$$

which does not exist since the projector $\bar{P}_{ij}^{kl}$ is uninvertible. This suggests, naively, that the symplectic structure associated with the above Poisson brackets does not exist.

One method of quantization of a theory is to promote Poisson brackets directly into quantum commutators. The Poisson brackets for a generic theory can be read off directly from its symplectic two form, and which in turn is defined from the Poisson matrix by constructing the inverse of the latter. We would like to construct the symplectic two form for ITQG by inverting the Poisson matrix P^{IJ} constructed in equation Equation (11). The Poisson matrix in its present form is uninvertible since it consists of projectors P_{ij}^{kl} in its block off-diagonal positions denoted by the symbol β. In the process of inversion of P^{IJ}, as shown above with P^{-1}, it is necessary to have β^{-1}. But β^{-1} does not exist on account of the fact that projectors are not invertible, which suggests, naively, that ITQG does not have a well-defined symplectic structure.

To get around this technical difficulty we will add a trace part to the Poisson matrix, parametrized by a parameter γ which we will ultimately remove after all calculations have been performed. While this distorts the theory of ITQG to a new theory parametrized by γ, it renders the resulting Poisson matrix invertible to allow progress to the corresponding symplectic two form, parametrized by γ, since the previously offending terms β now become β_γ, which as in Equation (12) are now invertible. Thus we have

$$\beta_\gamma \equiv (P_\gamma)_{kl}^{ij} = P_{kl}^{ij} + \gamma \bar{q}^{ij}\bar{q}_{kl} \longrightarrow \beta_\gamma^{-1} = P_{mn}^{kl} + \frac{1}{9\gamma}\bar{q}_{kl}\bar{q}_{mn}. \tag{13}$$

So now, we can invert the resulting object, and we have that

$$\beta_\gamma^{-1}\alpha\beta_\gamma^{-1} = \frac{1}{3}\left(\bar{P}_{kl}^{mn} + \frac{1}{9\gamma}\bar{q}^{mn}\bar{q}_{kl}\right)\left(\bar{q}^{kl}\bar{\pi}^{ij} - \bar{q}^{ij}\bar{\pi}^{kl}\right)\left(\bar{P}_{ij}^{rs} + \frac{1}{9\gamma}\bar{q}^{rs}\bar{q}_{ij}\right) = -\frac{1}{9\gamma}\left(\bar{\pi}^{mn}\bar{q}^{rs} - \bar{q}^{mn}\bar{\pi}^{rs}\right), \tag{14}$$

where we have used $\bar{P}_{kl}^{ij}\bar{q}_{ij} = \bar{q}^{kl}\bar{\pi}_{kl} = 0$ and $\bar{P}_{kl}^{ij}\bar{\pi}^{kl} = \bar{\pi}^{ij}$, which assumes that $\bar{\pi}^{ij}$ is traceless. So the inverse of the Poisson matrix parametrized by γ is given by

$$P^{-1} = \begin{pmatrix} -\frac{1}{9\gamma}\left(\bar{q}^{kl}\bar{\pi}^{ij} - \bar{q}^{ij}\bar{\pi}^{kl}\right) & -\left(\bar{P}_{mn}^{kl} + \frac{1}{9\gamma}\bar{q}^{kl}\bar{q}_{mn}\right) \\ \bar{P}_{rs}^{ij} + \frac{1}{9\gamma}\bar{q}^{ij}\bar{q}_{rs} & 0 \end{pmatrix}\delta^{(3)}(x,y),$$

and the associated two form Ω inherits the γ dependence

$$\Omega^\gamma = \frac{1}{2}\Omega_{IJ}^\gamma \delta q^I \delta q^J$$
$$= \int_\Sigma d^3x \left[-\frac{1}{18\gamma}\left(\bar{q}^{kl}\bar{\pi}^{ij} - \bar{q}^{ij}\bar{\pi}^{kl}\right)\delta\bar{q}_{ij} \wedge \delta\bar{q}_{kl} + \left(\bar{P}_{kl}^{ij} + \frac{1}{9\gamma}\bar{q}^{ij}\bar{q}_{kl}\right)\delta\bar{q}_{ij} \wedge \delta\bar{\pi}_{kl} + \frac{1}{2}(0)_{ijkl}\delta\bar{\pi}^{ij} \wedge \delta\bar{\pi}^{kl} \right]. \tag{15}$$

But $\bar{q}^{ij}\delta\bar{q}_{ij} = 0$, causing the $\delta\bar{q} \wedge \delta\bar{q}$ term and the γ contribution to the $\delta\bar{q} \wedge \delta\bar{\pi}$ term of (15) vanish. The quantity $(0)_{ijkl}$ in Equation (15) is basically to highlight the fact that that term, while zero is nontrivially so. Rather than omit this term, we wanted to highlight the fact that it is a tensorial quantity forming the coefficient of the $\delta\bar{\pi} \wedge \delta\bar{\pi}$ two form. This facilitates the keeping track for the reader of each individual term, of which there should be of the type including $\delta\bar{q} \wedge \delta\bar{q}$ and $\delta\bar{q} \wedge \delta\bar{\pi}$. There is no $\delta\bar{\pi} \wedge \delta\bar{\pi}$ term since $\{\bar{q}_{ij}, \bar{q}_{kl}\} = 0$. All explicit γ dependence in the symplectic form has disappeared, so the $\gamma \to 0$ limit can be safely taken, yielding

$$\lim_{\gamma \to 0} \Omega^\gamma = \int_\Sigma d^3 x P^{ij}_{kl} \delta \bar{q}_{ij} \wedge \delta \bar{\pi}^{kl}$$
$$= \int_\Sigma d^3 x \delta \bar{q}_{ij} \wedge \delta \bar{\pi}^{ij} - \frac{1}{3} \int_\Sigma d^3 x (\bar{q}^{ij} \delta \bar{q}_{ij}) \wedge (\bar{q}_{kl} \delta \bar{\pi}^{kl}) = \int_\Sigma d^3 x \delta \bar{q}_{ij} \wedge \delta \bar{\pi}^{ij}. \tag{16}$$

The vanishing of the $\frac{1}{3}$ term is due to $\bar{q}^{ij} \delta \bar{q}_{ij} = 0$ or alternatively by the Leibniz rule for the momentum term

$$(\bar{q}^{ij} \delta \bar{q}_{ij}) \wedge (\bar{q}_{kl} \delta \bar{\pi}^{kl}) = \delta \bar{q}_{ij} \wedge \delta (\bar{q}^{ij} \bar{q}^{kl} \bar{\pi}_{kl}) - \delta q_{ij} \wedge \delta q^{ij} (\bar{q}_{kl} \bar{\pi}^{kl}) - (\bar{q}^{ij} \delta \bar{q}_{ij}) \wedge (\bar{\pi}^{kl} \delta \bar{q}_{kl}) = 0 \tag{17}$$

due additionally to $\bar{q}_{ij} \bar{\pi}^{ij} = 0$. This implies that the tracelessness of $\bar{\pi}^{ij}$ must be conjugate to the fact that infinitesimal variations in $\bar{q}_{ij}$ are traceless. Hence (16) is the same as the barred contribution to (6), with the difference that the tracelessness of $\bar{\pi}^{ij}$ has been implicitly enforced due to a unimodular metric. This calculation demonstrates that extreme care must be exercised when extracting Poisson brackets from a symplectic two form, particular when the index structure of the fundamental variables has implicit symmetries. The requirement to implement the noncanonical Poisson brackets at the quantum level will pose nontrivial issues, which we will address in the next few sections. Let us display, for completeness, the fundamental Poisson brackets for the barred phase space

$$\{\bar{q}_{ij}(x), \bar{q}_{kl}(y)\} = 0; \quad \{\bar{q}_{ij}(x), \bar{\pi}^{kl}(y)\} = \bar{P}^{ij}_{kl}(x, y); \quad \{\bar{\pi}^{ij}(x), \bar{\pi}^{kl}(y)\} = \frac{1}{3}(\bar{q}^{kl} \bar{\pi}^{ij} - \bar{q}^{ij} \bar{\pi}^{kl}) \delta^{(3)}(x, y). \tag{18}$$

The basic Poisson brackets are noncanonical, which can be seen as the price to be paid for choosing $\bar{\pi}^{ij}$ to be traceless at the classical level, or alternatively, the price for choosing unimodular metric variables.

The original motivation was to obtain a symplectic form parametrized by γ and then to take the limit as γ approaches zero. But as one can see from the above that the wedge products in the resulting symplectic two form have coefficients proportional to γ^{-1}, which in the limit as γ approaches zero would be ill-defined. However, note form the arguments provided from Equation (15) through to Equation (18), that the individual wedge products of the fundamental variables all vanish on account of the unimodularity of the configuration space variable $\bar{q}_{ij}$ and the tracelessness of the momentum $\bar{pi}^{ij}$. Hence the proper procedure is to leave γ arbitrary in the symplectic two form, which is immaterial since all terms which depend on γ automatically vanish. The result is that the symplectic two form reduces to $\delta \bar{q} \wedge \delta \bar{\pi}$ form as in Equation (17), whence γ is conspicuously absent. So the justification that the parametrization of the Poisson matrix by the parameter does not affect the results of the symplectic two form is that for all nonzero γ, we can transition from the Poisson matrix to the symplectic two form by inversion as per the standard procedure, yielding a symplectic two form which is independent of the parameter γ. It is the unique choice of unimodular and traceless variables, which makes this the case, which admits a complete quantization of these variables.

3. A Prelude into the Quantum Theory

Having determined the Poisson brackets for the barred phase space, the next step is to implement them at the quantum level. In proceeding to the quantum theory according to the Heisenberg–Dirac prescription, we must promote all classical variables A, B to operators $\widehat{A}, \widehat{B}$ and all Poisson brackets to commutators $\{A, B\} \to \frac{1}{(i\hbar)}[\widehat{A}, \widehat{B}]$. So the fundamental Poisson brackets (18) yield the following equal-time commutation relations

$$[\bar{q}_{ij}(x, t), \bar{q}_{kl}(y, t)] = 0; [\bar{q}_{ij}(x, t), \widehat{\bar{\pi}}^{kl}(y, t)] = i\hbar \bar{P}^{ij}_{kl}(x, y); \quad [\widehat{\bar{\pi}}^{ij}(x, t), \widehat{\bar{\pi}}^{kl}(y, t)] = \frac{i\hbar}{3}(\bar{q}^{kl} \widehat{\bar{\pi}}^{ij} - \bar{q}^{ij} \widehat{\bar{\pi}}^{kl}) \delta^{(3)}(x, y), \tag{19}$$

where we have chosen an operator ordering with the momenta to the right. Since the momentum components fail to commute, then we are restricted to wavefunctionals $\psi[\bar{q}]$ in the metric representation. A representation of the classically traceless momentum as a vector field

$$\widehat{\overline{\pi}}^{ij}(x)\psi[\overline{q}] \longrightarrow \frac{\hbar}{i}\left[\overline{P}^{ij}_{kl}\frac{\delta}{\delta\overline{q}_{kl}} + \frac{1}{3}(\overline{q}^{ij}\overline{\pi}^{kl} - \overline{q}^{kl}\overline{\pi}^{ij})\frac{\delta}{\delta\overline{\pi}^{kl}}\right]\psi[\overline{q}] = \frac{\hbar}{i}\overline{P}^{ij}_{kl}(x)\frac{\delta\psi[\overline{q}]}{\delta\overline{q}_{kl}(x)} \qquad (20)$$

correctly reproduces the commutation relations (19) (The term of (20) from the $\overline{\pi}^{ij}, \overline{\pi}^{kl}$ commutation relation does not contribute for wavefunctionals $\psi[\overline{q}]$ polarized in the metric representation.). However, Equation (20) does not constitute a self-adjoint operator since

$$\frac{\hbar}{i}\frac{\delta}{\delta\overline{q}_{kl}(x)}\overline{P}^{ij}_{kl}(x) = \frac{\hbar}{i}\overline{P}^{ij}_{kl}(x)\frac{\delta}{\delta\overline{q}_{kl}(x)} - \frac{2\hbar}{3i}\overline{q}^{ij}\delta^{(3)}(0). \qquad (21)$$

So $\overline{q}_{ij}\widehat{\overline{\pi}}^{ij} = 0 \neq \widehat{\overline{\pi}}^{ij}\overline{q}_{ij}$, namely that the momentum in (20) is left-traceless, but is not right-traceless. A self-adjoint operator can be constructed by averaging the left-traceless and right-traceless versions $\frac{1}{2}(\frac{\delta}{\delta\overline{q}_{ij}}\overline{P}^{kl}_{ij} + \overline{P}^{kl}_{ij}\frac{\delta}{\delta\overline{q}_{ij}})$. However, the resulting operator, while self-adjoint, is neither traceless from the left nor from the right. So it appears that tracelessness is a property which is nontrivial to enforce at the quantum level in the $\overline{q}_{ij}, \overline{\pi}^{ij}$ variables.

The quantity $\delta^{(3)}(0)$ in Equation (21) is an ultraviolet singularity in field theory, which results from evaluating the commutation relations at the same spatial point. It is a formal expression more rigorously defined by a limiting procedure in the coincidence limit of the arguments x and y. It is necessary to perform the commutation relations at the same spatial point in order to reorder the fundamental operators in Equations (20) and (21), which are defined at the same spatial point, which is necessary in order to evaluate self-adjointness. This operator ordering induced ambiguity, parametrized by $\delta^{(3)}(0)$, highlights that the variables in their present form, while solving the aforementioned problem of the symplectic structure, are still not ideally suited for quantization. This will ultimately lead us to the choice of the momentric $\overline{\pi}^i_j$, in lieu of the momentum variable $\overline{\pi}^{ij}$, which being self adjoint as we will demonstrate in the remainder of this paper, will eliminate the presence of any such $\delta^{(3)}(0)$ divergences in the quantum theory.

4. Momentric Operators and the SU(3) Lie Algebra

Let us define a mixed-index version of the momentum, namely the momentric variables $\overline{P}^i_j = \overline{q}_{jm}\overline{\pi}^{im}$. We first compute the commutator of $\overline{P}^i_j$ with the barred metric. This is given by

$$[\overline{P}^i_j(x), \overline{q}_{kl}(y)] = [\overline{q}_{jm}(x)\overline{\pi}^{mi}(x), \overline{q}_{kl}(y)] = \overline{q}_{jm}(x)[\overline{\pi}^{mi}(x), \overline{q}_{kl}(y)] = -i\hbar\overline{q}_{jm}\overline{P}^{mi}_{kl}\delta^{(3)}(x,y) \equiv \frac{\hbar}{i}\overline{E}^i_{j(kl)}\delta^{(3)}(x,y) \qquad (22)$$

where we have used (19), with the "superspace vielbein" defined as $\overline{E}^i_{j(kl)} = \frac{1}{2}(\delta^i_k\overline{q}_{jl} + \delta^i_l\overline{q}_{jk}) - \frac{1}{3}\delta^i_j\overline{q}_{kl}$. So we will rather adopt the pair $\overline{q}_{ij}, \overline{P}^i_j$ as the fundamental variables, and recompute the fundamental relations (19) with respect to them.

For the commutators amongst the momentric components themselves the following identity involving commutation relations regarding generic operators $\widehat{A}, \widehat{B}, \widehat{C}, \widehat{D}$ will be useful

$$[\widehat{A}\widehat{B}, \widehat{C}\widehat{D}] = \widehat{A}[\widehat{B}, \widehat{C}]\widehat{D} + \widehat{C}[\widehat{A}, \widehat{D}]\widehat{B} + [\widehat{A}, \widehat{C}]\widehat{B}\widehat{D} + \widehat{C}\widehat{A}[\widehat{B}, \widehat{D}]. \qquad (23)$$

Note that the proper operator ordering has been preserved in (23). So we have the following, suppressing the x-y dependence in the intermediate steps and suppressing the hats to avoid cluttering up the notation,

$$[\overline{P}^i_j(x), \overline{P}^k_l(y)] = [\overline{q}_{jm}(x)\overline{\pi}^{im}(x), \overline{q}_{ln}(y)\overline{\pi}^{kn}(y)]$$

$$= \overline{q}_{jm}[\overline{\pi}^{im}, \overline{q}_{ln}]\overline{\pi}^{kn} + \overline{q}_{ln}[\overline{q}_{jm}, \overline{\pi}^{kn}]\overline{\pi}^{im} + [\overline{q}_{jm}, \overline{q}_{ln}]\overline{\pi}^{im}\overline{\pi}^{kn} + \overline{q}_{ln}\overline{q}_{jm}[\overline{\pi}^{im}, \overline{\pi}^{kn}] \qquad (24)$$

$$= \frac{\hbar}{i}\Big[\overline{q}_{jm}\overline{P}^{im}_{ln}\overline{\pi}^{kn} - \overline{q}_{ln}\overline{P}^{kn}_{jm}\overline{\pi}^{im} + 0 + \frac{1}{3}\overline{q}_{ln}\overline{q}_{jm}(\overline{q}^{kn}\overline{\pi}^{im} - \overline{q}^{im}\overline{\pi}^{kn})\Big]\delta^{(3)}(x, y).$$

In the third line of Equation (24) we have used the fundamental equal time commutation relations (19). For completeness, let us display some of the intermediate steps from Equation (24). For the first term on the right hand side we have

$$\overline{q}_{jm}\overline{P}^{im}_{ln}\overline{\pi}^{kn} = \overline{q}_{jm}\Big(\frac{1}{2}(\delta^i_l\delta^m_n + \delta^i_n\delta^m_l) - \frac{1}{3}\overline{q}^{im}\overline{q}_{ln}\Big)\overline{\pi}^{kn} = \frac{1}{2}(\delta^i_l\overline{P}^k_j + \overline{q}_{jl}\overline{\pi}^{ki}) - \frac{1}{3}\delta^i_j\overline{P}^k_l. \qquad (25)$$

For the middle term we have

$$\overline{q}_{ln}\overline{P}^{kn}_{jm}\overline{\pi}^{im} = \overline{q}_{ln}\Big(\frac{1}{2}(\delta^k_j\delta^n_m + \delta^k_m\delta^n_j) - \frac{1}{3}\overline{q}^{kn}\overline{q}_{jm}\Big)\overline{\pi}^{im} = \frac{1}{2}(\delta^k_j\overline{P}^i_l + \overline{q}_{lj}\overline{\pi}^{ik}) - \frac{1}{3}\delta^k_l\overline{P}^i_j. \qquad (26)$$

For the last term on the right hand side of (24) we have

$$\frac{1}{3}\overline{q}_{ln}\overline{q}_{jm}(\overline{q}^{kn}\overline{\pi}^{im} - \overline{q}^{im}\overline{\pi}^{kn}) = \frac{1}{3}(\delta^k_l\overline{P}^i_j - \delta^i_j\overline{P}^k_l). \qquad (27)$$

Substitution of Equations (25)–(27) into Equation (24) yields the result that

$$[\overline{P}^i_j(x), \overline{P}^k_l(y)] = \frac{\hbar}{i}\Big[\frac{1}{2}(\delta^i_l\overline{P}^k_j - \delta^k_j\overline{P}^i_l) + \frac{2}{3}(\delta^k_l\overline{P}^i_j - \delta^i_j\overline{P}^k_l)\Big]\delta^{(3)}(x, y). \qquad (28)$$

Note that the algebra closes (if not for the precise cancellation of terms of the form $\overline{q}_{jl}\overline{\pi}^{ki}$, this would not be the case). While the algebra (28) closes on the momentric variables $\overline{P}^i_j$, it does not enforce the vanishing of the trace $\overline{P} = \delta^j_i\overline{P}^i_j$. This can be seen by contraction of (28) with δ^j_i, wherein

$$[\overline{P}(x), \overline{P}^i_j(y)] = -\frac{2\hbar}{i}\Big(\overline{P}^i_j - \frac{1}{3}\delta^i_j\overline{P}\Big)\delta^{(3)}(x, y) \equiv 2i\hbar\overline{\pi}^i_j\delta^{(3)}(x, y), \qquad (29)$$

where $\overline{\pi}^i_j$ denotes the traceless part of the momentric. Note that $\overline{P} = 0$ in Equation (29) leads to a contradiction, whereas the relation (22) implies $[\overline{P}, \overline{q}_{ij}] = 0$ due to tracelessless of $\overline{E}^i_{j(kl)}$.

Still, it is interesting in Equation (29) that the commutator of $\overline{P}^i_j$ with its trace yields it traceless part $\overline{\pi}^i_j$. So let us evaluate the commutation relations involving the traceless part (suppressing the coordinate dependence for simplicity)

$$[\overline{\pi}^i_j(x), \overline{\pi}^k_l(y)] = [\overline{P}^i_j - \frac{1}{3}\delta^i_j\overline{P}, \overline{P}^k_l - \frac{1}{3}\delta^k_l\overline{P}] = [\overline{P}^i_j, \overline{P}^k_l] - \frac{1}{3}\delta^k_l[\overline{P}^i_j, \overline{P}] - \frac{1}{3}\delta^i_j[\overline{P}, \overline{P}^k_l] + \frac{1}{9}\delta^i_j\delta^k_l[\overline{P}, \overline{P}]$$

$$= \frac{i\hbar}{2}(\delta^i_l\overline{P}^k_j - \delta^k_j\overline{P}^i_l)\delta^{(3)}(x, y) \qquad (30)$$

where we have used (28) and (29). We can now make the substitution $\overline{P}^i_j = \overline{\pi}^i_j + \frac{1}{3}\delta^i_j\overline{P}$, and the trace part cancels out to yield $[\overline{\pi}^i_j, \overline{\pi}^k_l] = \frac{i\hbar}{2}(\delta^i_l\overline{\pi}^k_j - \delta^k_j\overline{\pi}^i_l)$. The final result of our commutation relations (19), in terms of the traceless momentric variables $\overline{\pi}^i_j$ is given by

$$[\overline{q}_{ij}(x), \overline{q}_{kl}(y)] = 0; \quad [\widehat{\overline{\pi}}^i_j(x), \overline{q}_{kl}(y)] = \frac{\hbar}{i}\overline{E}^i_{j(kl)}\delta^{(3)}(x, y); \quad [\widehat{\overline{\pi}}^i_j(x), \widehat{\overline{\pi}}^k_l(y)] = \frac{i\hbar}{2}(\delta^i_l\overline{\pi}^k_j - \delta^k_j\overline{\pi}^i_l)\delta^{(3)}(x, y). \qquad (31)$$

Note that Equation (31) implies a representation of the momentric as a vector field

$$\widehat{\overline{P}}_j^i = \frac{\hbar}{i}\frac{\delta}{\delta\overline{q}_{kl}}\overline{E}_{j(kl)}^i = \frac{\hbar}{i}\overline{E}_{j(kl)}^i\frac{\delta}{\delta\overline{q}_{kl}} + \frac{\hbar}{i}\left[\frac{\delta}{\delta\overline{q}_{kl}},\overline{E}_{j(kl)}^i\right] = \frac{\hbar}{i}\overline{E}_{j(kl)}^i\frac{\delta}{\delta\overline{q}_{kl}},\tag{32}$$

which is both self-adjoint and left-right traceless, implements the commutation relations, and is traceless in the sense that $\delta_j^i\widehat{\overline{\pi}}_j^i = \widehat{\overline{\pi}}_j^i\delta_i^j = 0$. There are a few things to note regarding (31). First, upon contraction with δ_i^j, yields consistently that the trace $\delta_i^j\overline{\pi}_j^i = 0$ vanishes as well as its comutator with all quantities. Secondly, the traceless momentric variables by themselves form a $SU(3)$ current algebra, and also generate an affine algebra with the metric, which unlike (19) preserves the positivity of the metric $\overline{q}_{ij}$. Thus, the fundamental variables $\overline{q}_{ij}, \overline{\pi}_j^i$ will be the prime choice for the quantum theory which, at the kinematical level, will involve constructing unitary, irreducible representations of the $SU(3)$ Lie algebra. Also of note is that the the object $\Delta = \overline{\pi}_i^j\overline{\pi}_j^i$ encodes to the quadratic Casimir of $SU(3)$, which by definition must commute with all traceless momentric components $[\Delta,\overline{\pi}_j^i] = 0$.

The Gell–Mann matrices satisfy the relations

$$[\lambda_A,\lambda_B]_j^i = if_{AB}^C(\lambda_C)_j^i; \quad \{\lambda_A,\lambda_B\}_j^i = d_{ABC}(\lambda_C)_j^i\tag{33}$$

with totally antisymmetric structure constants f_{ABC}, and totally symmetric d_{ABC}. We will exploit the aforementioned index structure by projection of the momentric onto the Gell–Mann matrices

$$T^A = (\lambda^A)_i^j\overline{\pi}_j^i \longrightarrow \overline{\pi}_j^i = 2T^A(\lambda_A)_j^i,\tag{34}$$

where we have used the $SU(3)$ completeness relation $(\lambda^A)_j^i(\lambda^A)_l^k = \frac{1}{2}(\delta_j^k\delta_l^i - \frac{1}{3}\delta_j^i\delta_l^k)$. The $SU(3)$ Lie algebra is of rank 2, and therefore has two Casimir operators, $C^{(2)}$ and $C^{(3)}$ given by

$$C^{(2)} = (\lambda_A)_i^j(\lambda_A)_j^i = T^AT^A; \quad C^{(3)} = d_{ABC}(\lambda_A)_j^i(\lambda_B)_k^j(\lambda_C)_i^k = \epsilon^{ijk}\epsilon_{mnl}\overline{\pi}_i^m\overline{\pi}_j^n\overline{\pi}_k^l \propto 6\det\overline{\pi}_j^i.\tag{35}$$

Note for $C^{(3)}$ that the pair of epsilon symbols is totally symmetric under interchange of any index pair (i,m), (j,n), (k,l), which is consistent with the total symmetry of d_{ABC}.

5. The Classical Theory, Revisited

Having determined the ideal variables for quantization as the unimodular- traceless momentric pair $\overline{q}_{ij}, \overline{\pi}_j^i$, we will now re-evaluate the Poisson brackets of the theory. This provides a basis for correlation of quantum predictions to the classical dynamics. First, the fundamental Poisson brackets are given by

$$\{\overline{q}_{ij}(x),\overline{q}_{kl}(y)\} = 0, \{\overline{q}_{ij}(x),\overline{\pi}_l^k(y)\} = \overline{E}_{j(kl)}^i\delta^{(3)}(x,y); \quad \{\overline{\pi}_j^i(x),\overline{\pi}_l^k(y)\} = \frac{1}{2}(\delta_l^i\overline{\pi}_j^k - \delta_j^k\overline{\pi}_l^i)\delta^{(3)}(x,y).\tag{36}$$

So the Poisson brackets between phase space functions A and B is given by

$$\{A,B\} = \int_\Sigma d^3x\int_\Sigma d^3y\left[\frac{\delta A}{\delta\overline{q}_{ij}(x)}\{\overline{q}_{ij}(x),\overline{q}_{kl}(y)\}\frac{\delta B}{\delta\overline{q}_{kl}(y)} + \frac{\delta A}{\delta\overline{q}_{ij}(x)}\{\overline{q}_{ij}(x),\overline{\pi}_l^k(y)\}\frac{\delta B}{\delta\overline{\pi}_l^k(y)}\right.$$
$$\left. + \frac{\delta A}{\delta\overline{\pi}_j^i(x)}\{\overline{\pi}_j^i(x),\overline{q}_{kl}(y)\}\frac{\delta B}{\delta\overline{q}_{kl}(y)} + \frac{\delta A}{\delta\overline{\pi}_j^i(x)}\{\overline{\pi}_j^i(x),\overline{\pi}_l^k(y)\}\frac{\delta B}{\delta\overline{\pi}_l^k(y)}\right]\tag{37}$$
$$= \int_\Sigma d^3z\left[\overline{E}_{j(ij)}^k\left(\frac{\delta A}{\delta\overline{q}_{ij}}\frac{\delta B}{\delta\overline{\pi}_l^k} - \frac{\delta B}{\delta\overline{q}_{ij}}\frac{\delta A}{\delta\overline{\pi}_l^k}\right) + \frac{\delta A}{\delta\overline{\pi}_j^i}\overline{\pi}_l^i\frac{\delta B}{\delta\overline{\pi}_l^j} - \frac{\delta A}{\delta\overline{\pi}_j^i}\overline{\pi}_j^k\frac{\delta B}{\delta\overline{\pi}_i^k}\right].$$

In General relativity, we will be interested in the evolution of the basic variables with respect to T, gauge-invariant part of intrinsic time $\ln q^{1/3}$, under the action of a physical Hamiltonian

$$H_{Phys} = \int_\Sigma d^3x \bar{H}(x) = \int_\Sigma d^3x \sqrt{\bar{\pi}^j_i \bar{\pi}^i_j + \mathcal{V}[q_{ij}]}, \tag{38}$$

where $\mathcal{V}$ is a potential term which depends on the metric. The Hamilton's equations for the basic variables with respect to the Poisson brackets (37) are given by

$$\frac{\delta \bar{q}_{ij}(x)}{\delta T} = \{\bar{q}_{ij}(x), H_{Phys}\} = \frac{1}{\bar{H}} \bar{E}^k_{l(ij)} \bar{\pi}^l_k;$$

$$\frac{\delta \bar{\pi}^i_j(x)}{\delta T} = \{\bar{\pi}^i_j(x), H_{Phys}\} = \frac{1}{\bar{H}}\left[\frac{1}{2}\bar{E}^i_{j(kl)}\frac{\delta \mathcal{V}}{\delta \bar{q}_{kl}} + \bar{\pi}^i_l\bar{\pi}^l_j - \bar{\pi}^k_j\bar{\pi}^i_k\right] = \frac{1}{2\bar{H}}\bar{E}^i_{j(kl)}\frac{\delta \mathcal{V}}{\delta \bar{q}_{kl}}. \tag{39}$$

As a quick consistency check, contraction of the first equation of (39) with $\bar{q}^{ij}$ and contraction of the second equation with δ^j_i shows that if $\bar{q}_{ij}$ is unimodular and $\bar{\pi}^i_j$ is traceless at time T_0, then these properties will be preserved under evolution in intrinsic time by the Hamilton's equations.

6. Conclusions

The consistent quantization of 3+1 gravity is one of the biggest unsolved problems in theoretical physics spanning the past 100 years of approaches which, while leading to insights into certain often complementary aspects of the problem, have so far not provided a complete solution due to various technical and conceptual difficulties and issues. The novelty of the author's approach is the claim that with ITQG, one has a complete and consistent quantization of gravity which provides a possible resolution to the long-standing problem, while solving the difficulties inherent in all of the approaches so far, in one stroke.

For future work, we aim to follow the work of this paper with a similar work by focusing on some 2+1 aspect of ITQG, with the aim of studying the thermodynamic aspects of the BTZ black hole. Also, looking at the initial wave function, one difference from the case of 3+1 gravity seems to be the observation that there is no Cotton-York tensor in two spatial dimensions. So we should expect just a Ricci curvature-squared higher derivative rendition of the theory. This then will help us to be able to exploit the SU(2) structure of the theory, which will go a long way towards learning about the physical Hilbert space.

Acknowledgments: Eyo Eyo Ita III and Amos S. Kubeka would like to thank the U.S. Naval Academy and the University of South Africa for the financial support.

Author Contributions: Both authors contributed equally to this work.

References

1. Ashtekar, A. New Variables for Classical and Quantum Gravity. *Phys. Rev. Lett.* **1986**, *57*, 2244–2247.
2. Arnowitt, R.; Deser, S.; Misner, C. Dynamical Structure and Definition of Energy in General Relativity. *Phys. Rev.* **1959**, *5*, 1322–1330.
3. Gambini, R.; Pullin, J. *A First Course in Loop Quantum Gravity*; Oxford University Press: Oxford, UK, 2011.
4. Rovelli C. *Quantum Gravity*; Cambridge University Press: Cambridge, UK, 2004.
5. Fatibene, L.; Francaviglia, M.; Rovelli, C. On a covariant formulation of the Barbero–Immirzi connection. *Class. Quantum Grav.* **2007**, *24*, 11.
6. Ita, E.; Soo, C.; Yu, H.-L. Intrinsic time quantum geometrodynamics. *Prog. Theor. Exp. Phys.* **2015**, *2015*, 083E01.
7. Soo, C.; Yu, H.L. General Relativity without paradigm of space-time covariance, and resolution of the problem of time. *Prog. Theor. Phys.* **2014**, *2014*, 013E01.
8. O'Murchada, N.; Soo, C.; Yu, H.L. Intrinsic time gravity and the Lichnerowicz-York equation. *Class. Quantum Grav.* **2013**, *30*, 095016.

11

Virial Theorem in Nonlocal Newtonian Gravity

Bahram Mashhoon

Department of Physics and Astronomy, University of Missouri, Columbia, MO 65211, USA;
mashhoonb@missouri.edu

Academic Editors: Lorenzo Iorio and Elias C. Vagenas

Abstract: Nonlocal gravity is the recent classical nonlocal generalization of Einstein's theory of gravitation in which the past history of the gravitational field is taken into account. In this theory, nonlocality appears to simulate dark matter. The virial theorem for the Newtonian regime of nonlocal gravity theory is derived and its consequences for "isolated" astronomical systems in virial equilibrium at the present epoch are investigated. In particular, for a sufficiently isolated nearby *galaxy* in virial equilibrium, the galaxy's baryonic diameter $\mathcal{D}_0$—namely, the diameter of the smallest sphere that completely surrounds the baryonic system at the present time—is predicted to be larger than the effective dark matter fraction f_{DM} times a universal length that is the basic nonlocality length scale $\lambda_0 \approx 3 \pm 2$ kpc.

Keywords: nonlocal gravity; celestial mechanics; dark matter

1. Introduction

In the standard theory of relativity, physics is local in the sense that a postulate of locality permeates through the special and general theories of relativity. First, Lorentz invariance is extended in a pointwise manner to actual, namely, accelerated, observers in Minkowski spacetime. This *hypothesis of locality* is then employed crucially in Einstein's local principle of equivalence to render observers pointwise inertial in a gravitational field [1]. Field measurements are intrinsically nonlocal, however. To go beyond the locality postulate in Minkowski spacetime, the past history of the accelerated observer must be taken into account. The observer in general carries the memory of its past acceleration. The deep connection between inertia and gravitation suggests that gravity could be nonlocal as well, and, in nonlocal gravity, the gravitational memory of past events must then be taken into account. Along this line of thought, a classical nonlocal generalization of Einstein's theory of gravitation has recently been developed [2–13]. In this theory, the gravitational field is local but satisfies partial integro-differential field equations. Moreover, a significant observational consequence of this theory is that the nonlocal aspect of gravity appears to simulate dark matter. The physical foundations of this classical theory, from nonlocal special relativity theory to nonlocal general relativity, sets it completely apart from purely phenomenological and *ad hoc* approaches to the problem of dark matter.

Dark matter is currently required in astrophysics for explaining the gravitational dynamics of galaxies as well as clusters of galaxies [9], gravitational lensing observations [10] and structure formation in cosmology [13]. We emphasize that only some of the implications of nonlocal gravity theory have thus far been confronted with observation [9,12]. It is also important to mention here that many other approaches to nonlocal gravitation theory exist that are, however, inspired by developments in quantum field theory. The consideration of such theories is well beyond the scope of this purely classical work.

In this paper, we are concerned with the Newtonian regime of nonlocal gravity, where Poisson's equation of Newtonian gravity is modified by the addition of a certain average over the gravitational field. This nonlocal term involves a kernel function q whose functional form can perhaps be derived from a future more complete theory, but, at the present stage of the development of nonlocal gravity, must be determined using observational data. It is necessary that a unique kernel be eventually chosen in this way, but kernel q at the present time could be either q_1 or q_2 [6]. Each of these kernels is spherically symmetric in space and contains three length scales a_0, λ_0, and μ_0^{-1} such that $a_0 < \lambda_0 < \mu_0^{-1}$. The basic scale of nonlocality is a galactic length λ_0 of order 1 kpc, while a_0 is a short-range parameter that controls the behavior of $q(r)$ as $r \to 0$. At the other extreme, $r \to \infty$, $q(r)$ decays exponentially as $\exp(-\mu_0 r)$, indicating the fading of spatial memory with distance. The short-range parameter a_0 is necessary in dealing with the gravitational physics of the Solar System, globular clusters and isolated dwarf galaxies; however, it may be safely neglected in dealing with larger systems such as clusters of galaxies. When $a_0 = 0$, q_1 and q_2 reduce to a single kernel q_0, $q_1 = q_2 = q_0$, and the remaining parameters (λ_0 and μ_0) have been determined from a comparison of the theory with the astronomical data regarding a sample of 12 spiral galaxies from the THINGS catalog—see reference [9] for a detailed treatment. The results can be expressed, for the sake of convenience, as $\lambda_0 \approx 3$ kpc and $\mu_0^{-1} \approx 17$ kpc. Moreover, lower limits have been placed on a_0 from the study of the precession of perihelia of planetary orbits in the Solar System [12,14,15].

It is interesting to explore the implications of the virial theorem for nonlocal gravity. In general, the virial theorem of Newtonian physics establishes a simple linear relation between the time averages of the kinetic and potential energies of an isolated material system for which the potential energy is a homogeneous function of spatial coordinates. For an isolated *gravitational* N-body system, the significance of the virial theorem has to do with the circumstance that the kinetic energy is a sum of terms each proportional to the mass of a body in the system, while the potential energy is a sum of terms each proportional to the product of two masses in the system. Thus, under favorable conditions, the virial theorem can be used to connect the total dynamic mass of an isolated relaxed gravitational system with its average internal motion.

The main purpose of the present paper is to discuss, within the Newtonian regime of nonlocal gravity, the consequences of the extension of the virial theorem to nonlocal gravity. Though such an extension is technically straightforward, it is nevertheless physically quite significant as it allows the possibility of making *predictions* regarding the effective dark-matter content of cosmologically nearby isolated N-body gravitational systems in virial equilibrium.

2. Modification of the Inverse Square Force Law

It can be shown [12] that, in the Newtonian regime of nonlocal gravity, the force of gravity on point mass m due to point mass m' is given by:

$$\mathbf{F}(\mathbf{r}) = -Gmm' \frac{\hat{\mathbf{r}}}{r^2} \left\{ [1 - \mathcal{E}(r) + \alpha_0] - \alpha_0 \left(1 + \frac{1}{2} \mu_0 r\right) e^{-\mu_0 r} \right\} \tag{1}$$

where $\mathbf{r} = \mathbf{x}_m - \mathbf{x}_{m'}$, $r = |\mathbf{r}|$ and $\hat{\mathbf{r}} = \mathbf{r}/r$. The quantity in curly brackets is henceforth denoted by $1 + \mathbb{N}$, where $\mathbb{N}$ is the contribution of nonlocality to the force law and depends upon three parameters, namely, α_0, μ_0 and a short-range parameter a_0 that is contained in $\mathcal{E}$; in fact, $\mathcal{E} = 0$ when $a_0 = 0$. We will show in the next section that $\mathbb{N}$ starts out from zero at $r = 0$ with vanishing slope and monotonically increases toward an asymptotic value of about 10 as $r \to \infty$. Thus, the gravitational force in Equation (1) is *always attractive*; moreover, this force is central, conservative and satisfies Newton's third law of motion.

Nonlocal gravity is in the early stages of development and, depending on whether we choose kernel q_1 or kernel q_2, $\mathcal{E}(r)$ at the present time can be either

$$\mathcal{E}_1(r) = \frac{a_0}{\lambda_0} e^p \left[E_1(p) - E_1(p + \mu_0 r) \right] \tag{2}$$

or

$$\mathcal{E}_2(r) = \frac{a_0}{\lambda_0} \left\{ -\frac{r}{r+a_0} e^{-\mu_0 r} + 2e^p \left[E_1(p) - E_1(p+\mu_0 r) \right] \right\}$$ (3)

respectively, where $p = \mu_0 a_0$, $\lambda_0 = 2/(\alpha_0 \mu_0)$ and $E_1(u)$ is the *exponential integral function* [16]:

$$E_1(u) = \int_u^\infty \frac{e^{-t}}{t} dt$$ (4)

For $u : 0 \to \infty$, $E_1(u) > 0$ monotonically decreases from infinity to zero. In fact, near $u = 0$, $E_1(u)$ behaves like $-\ln u$ and as $u \to \infty$, $E_1(u)$ vanishes exponentially. Furthermore,

$$E_1(x) = -C - \ln x - \sum_{n=1}^\infty \frac{(-x)^n}{n\, n!}$$ (5)

where $C = 0.577 \ldots$ is Euler's constant. It is useful to note that

$$\frac{e^{-u}}{u+1} < E_1(u) \leq \frac{e^{-u}}{u}$$ (6)

(see Equation 5.1.19 in reference [16]).

It is clear from Equation (1) that α_0 is dimensionless, while μ_0^{-1}, λ_0 and a_0 have dimensions of length. In fact, we expect that $a_0 < \lambda_0 < \mu_0^{-1}$; moreover, the short-range parameter a_0 and $\mathcal{E}$ may be neglected in Equation (1) when dealing with the rotation curves of spiral galaxies and the internal gravitational physics of clusters of galaxies. In this way, α_0 and μ_0 have been *tentatively* determined from a detailed comparison of nonlocal gravity with observational data [9]:

$$\alpha_0 = 10.94 \pm 2.56, \qquad \mu_0 = 0.059 \pm 0.028 \text{ kpc}^{-1}$$ (7)

Hence, we find $\lambda_0 = 2/(\alpha_0 \mu_0) \approx 3 \pm 2$ kpc. It is important to mention here that λ_0 *is the fundamental length scale of nonlocal gravity at the present epoch*; indeed, for $\lambda_0 \to \infty$, $\mathbb{N} \to 0$ and Equation (1) reduces to Newton's inverse square force law. In what follows, we usually assume $\alpha_0 \approx 11$ and $\mu_0^{-1} \approx 17$ kpc for the sake of convenience. Furthermore, we expect that $p = \mu_0 a_0$ is such that $0 < p < \frac{1}{5}$. In reference [12], preliminary lower limits have been placed on a_0 on the basis of current data regarding planetary orbits in the Solar System. For instance, using the data for the orbit of Saturn, a preliminary lower limit of $a_0 \gtrsim 2 \times 10^{15}$ cm can be established if we use $\mathcal{E}_1$, while $a_0 \gtrsim 5.5 \times 10^{14}$ cm if we use $\mathcal{E}_2$.

Let us note that

$$\frac{d\mathcal{E}_1}{dr} = \frac{a_0}{\lambda_0} \frac{1}{a_0 + r} e^{-\mu_0 r}$$ (8)

and

$$\frac{d\mathcal{E}_2}{dr} = \frac{a_0}{\lambda_0} \frac{a_0 + 2r + \mu_0 r(a_0 + r)}{(a_0 + r)^2} e^{-\mu_0 r}$$ (9)

Therefore, $\mathcal{E}_1(r)$ and $\mathcal{E}_2(r)$ start from zero at $r = 0$ and monotonically increase as $r \to \infty$; furthermore, they asymptotically approach $\mathcal{E}_1(\infty) = \mathcal{E}_\infty$ and $\mathcal{E}_2(\infty) = 2\mathcal{E}_\infty$, respectively, where

$$\mathcal{E}_\infty = \frac{1}{2} \alpha_0 p\, e^p E_1(p)$$ (10)

It is a consequence of (6) that $\mathcal{E}_\infty < \alpha_0/2$, so that, in the gravitational force in Equation (1),

$$\alpha_0 - \mathcal{E}(r) > 0$$ (11)

In the Newtonian regime, where we formally let the speed of light $c \to \infty$, retardation effects vanish and gravitational memory is purely spatial. The resulting gravitational force in Equation (1) thus consists of two parts: an enhanced attractive "Newtonian" part and a repulsive fading spatial memory ("Yukawa") part with an exponential decay length of $\mu_0^{-1} \approx 17$ kpc. Equation (1) is such that it reduces to Newton's inverse square force law for $r \to 0$, as it should [17–21], and on galactic scales, it is a generalization of the phenomenological Tohline-Kuhn modified gravity approach to the flat rotation curves of spiral galaxies [22–25]. An excellent review of the Tohline-Kuhn work is contained in the paper of Bekenstein [26].

For $r \gg \mu_0^{-1}$, the exponentially decaying ("fading memory") part of Equation (1) can be neglected and

$$\mathbf{F}(\mathbf{r}) \approx -\frac{Gmm' \left[1 + \alpha_0 - \mathcal{E}(\infty) \right]}{r^2} \, \hat{\mathbf{r}} \tag{12}$$

so that $m' \left[\alpha_0 - \mathcal{E}(\infty) \right]$ has the interpretation of the *total effective dark mass* associated with m'. For $a_0 = 0$, the net effective dark mass associated with point mass m' is simply $\alpha_0 \, m'$, where $\alpha_0 \approx 11$ [9]. On the other hand, for $a_0 \neq 0$, the corresponding result is $\alpha_0 \, \epsilon(p) \, m'$, where

$$\epsilon_1(p) = 1 - \frac{1}{2} \, p \, e^p \, E_1(p) \,, \qquad \epsilon_2(p) = 1 - p \, e^p \, E_1(p) \tag{13}$$

depending on whether we use $\mathcal{E}_1$ or $\mathcal{E}_2$, respectively. The functions in Equation (13) start from unity at $p = 0$ and decrease monotonically to $\epsilon_1(0.2) \approx 0.85$ and $\epsilon_2(0.2) \approx 0.70$ at $p = 0.2$; they are plotted in Figure 1 of reference [12] for $p : 0 \to 0.2$. If a_0 turns out to be just a few parsecs or smaller, for instance, then $\epsilon_1 \approx \epsilon_2 \approx 1$.

A detailed investigation reveals that it is possible to approximate the exterior gravitational force due to a star or a planet by assuming that its mass is concentrated at its center [12]. In this connection, we note that the radius of a star or a planet is generally much smaller than the length scales a_0, λ_0 and μ_0^{-1} that appear in the nonlocal contribution to the gravitational force. Therefore, one can employ Equation (1) in the approximate treatment of the two-body problem in astronomical systems such as binary pulsars and the Solar System, where possible deviations from general relativity may become measurable in the future.

Consider, for instance, the deviation from the Newtonian inverse square force law, namely,

$$\delta \, \mathbf{F}(\mathbf{r}) = -\frac{Gmm' \, \hat{\mathbf{r}}}{r^2} \, \mathbb{N}(r) \tag{14}$$

For $r < a_0$, it is possible to show via an expansion in powers of r/a_0 that [12]

$$\delta \, \mathbf{F}_1(\mathbf{r}) = -\frac{1}{2} \frac{Gmm'}{\lambda_0 \, a_0} \, (1+p) \, \hat{\mathbf{r}} + \frac{1}{3} \frac{Gmm'}{\lambda_0 \, a_0} \, (1+p+p^2) \, \frac{r}{a_0} \, \hat{\mathbf{r}} + \cdots \tag{15}$$

if $\mathcal{E}_1$ is employed, or

$$\delta \, \mathbf{F}_2(\mathbf{r}) = -\frac{1}{3} \frac{Gmm'}{\lambda_0 \, a_0} \, (1+p) \, \frac{r}{a_0} \, \hat{\mathbf{r}} + \cdots \tag{16}$$

if $\mathcal{E}_2$ is employed. Perhaps dedicated missions, such as ESA's Gaia mission that was launched in 2013, can measure the imprint of nonlocal gravity in the Solar System [27,28]. In this connection, we note that

$$\frac{1}{2} \frac{G \, M_\odot}{\lambda_0 \, a_0} \, (1+p) \approx \left(\frac{10^{18} \, \text{cm}}{a_0} \right) 10^{-14} \, \text{cm s}^{-2} \tag{17}$$

which, combined with lower limits on a_0 established in reference [12], is at least three orders of magnitude smaller than the acceleration involved in the Pioneer anomaly ($\sim 10^{-7}$ cm s^{-2}). It follows from these results that nonlocal gravity is consistent with the gravitational physics of the Solar System.

3. Virial Theorem

Consider an idealized isolated system of N Newtonian point particles with fixed masses m_i, $i = 1, 2, \ldots, N$. We assume that the particles occupy a finite region of space and interact with each other only gravitationally such that the center of mass of the isolated system is at rest in a global inertial frame and the isolated system permanently occupies a compact region of space. The equation of motion of the particle with mass m_i and state $(\mathbf{x}_i, \mathbf{v}_i)$ is then

$$m_i \frac{d\,\mathbf{v}_i}{dt} = -\sum_j{}' \frac{G\,m_i\,m_j\,(\mathbf{x}_i - \mathbf{x}_j)}{|\mathbf{x}_i - \mathbf{x}_j|^3} \left[1 + \mathbb{N}(|\mathbf{x}_i - \mathbf{x}_j|)\right] \tag{18}$$

for $j = 1, 2, \ldots, N$, but the case $j = i$ is excluded in the sum by convention. In fact, a prime over the summation sign indicates that in the sum $j \neq i$. Here, $1 + \mathbb{N}(r)$ is a *universal* function that is inside the curly brackets in Equation (1) and the contribution of nonlocality, $\mathbb{N}(r)$, is given by

$$\mathbb{N}(r) = \alpha_0 \left[1 - \left(1 + \frac{1}{2}\mu_0 r\right) e^{-\mu_0 r}\right] - \mathcal{E}(r) \tag{19}$$

Consider next the quantities

$$\mathbb{I} = \frac{1}{2}\sum_i m_i\,x_i^2\,, \qquad \frac{d\,\mathbb{I}}{dt} = \sum_i m_i\,\mathbf{x}_i \cdot \mathbf{v}_i \tag{20}$$

where $x_i = |\mathbf{x}_i|$ and

$$\frac{d^2\,\mathbb{I}}{dt^2} = \sum_i m_i\,v_i^2 + \sum_i m_i\,\mathbf{x}_i \cdot \frac{d\,\mathbf{v}_i}{dt} \tag{21}$$

It follows from Equation (18) that

$$\sum_i m_i\,\mathbf{x}_i \cdot \frac{d\,\mathbf{v}_i}{dt} = -\sum_{i,j}{}' \frac{G\,m_i\,m_j\,(\mathbf{x}_i - \mathbf{x}_j) \cdot \mathbf{x}_i}{|\mathbf{x}_i - \mathbf{x}_j|^3}\left[1 + \mathbb{N}(|\mathbf{x}_i - \mathbf{x}_j|)\right] \tag{22}$$

Exchanging i and j in the expression on the right-hand side of Equation (22), we get

$$\sum_i m_i\,\mathbf{x}_i \cdot \frac{d\,\mathbf{v}_i}{dt} = \sum_{i,j}{}' \frac{G\,m_i\,m_j\,(\mathbf{x}_i - \mathbf{x}_j) \cdot \mathbf{x}_j}{|\mathbf{x}_i - \mathbf{x}_j|^3}\left[1 + \mathbb{N}(|\mathbf{x}_i - \mathbf{x}_j|)\right] \tag{23}$$

Adding Equations (22) and (23) results in

$$\sum_i m_i\,\mathbf{x}_i \cdot \frac{d\,\mathbf{v}_i}{dt} = -\frac{1}{2}\sum_{i,j}{}' \frac{G\,m_i\,m_j}{|\mathbf{x}_i - \mathbf{x}_j|}\left[1 + \mathbb{N}(|\mathbf{x}_i - \mathbf{x}_j|)\right] \tag{24}$$

Using this result, Equation (21) takes the form

$$\frac{d^2\,\mathbb{I}}{dt^2} = \sum_i m_i\,v_i^2 - \frac{1}{2}\sum_{i,j}{}' \frac{G\,m_i\,m_j}{|\mathbf{x}_i - \mathbf{x}_j|}\left[1 + \mathbb{N}(|\mathbf{x}_i - \mathbf{x}_j|)\right] \tag{25}$$

Let us recall that the net kinetic energy and the Newtonian gravitational potential energy of the system are given by

$$\mathbb{T} = \frac{1}{2}\sum_i m_i\,v_i^2\,, \qquad \mathbb{W}_N = -\frac{1}{2}\sum_{i,j}{}' \frac{G\,m_i\,m_j}{|\mathbf{x}_i - \mathbf{x}_j|} \tag{26}$$

Hence,

$$\frac{d^2\,\mathbb{I}}{dt^2} = 2\,\mathbb{T} + \mathbb{W}_N + \mathbb{D} \tag{27}$$

where

$$\mathbb{D} = -\frac{1}{2} \sum_{i,j}' \frac{G \, m_i \, m_j}{|\mathbf{x}_i - \mathbf{x}_j|} \, \mathbb{N}(|\mathbf{x}_i - \mathbf{x}_j|) \tag{28}$$

and $\mathbb{N}$ is given by Equation (19).

Finally, we are interested in the average of Equation (27) over time. Let $< f >$ denote the time average of f, where

$$< f > \ = \ \lim_{\tau \to \infty} \frac{1}{\tau} \int_0^\tau f(t) \, dt \tag{29}$$

Then, it follows from averaging Equation (27) over time that

$$2 < \mathbb{T} > \ = - < \mathbb{W}_N > - < \mathbb{D} > \tag{30}$$

since $d\mathbb{I}/dt$, which is the sum of $m\,\mathbf{x} \cdot \mathbf{v}$ over all particles in the system, is a bounded function of time and hence the time average of $d^2\mathbb{I}/dt^2$ vanishes. This is clearly based on the assumption that the spatial coordinates and velocities of all particles indeed remain finite for all time. Equation (30) expresses the *virial theorem* in nonlocal Newtonian gravity.

It is important to digress here and re-examine some of the assumptions involved in our derivation of the virial theorem. In general, any consequence of the gravitational interaction involves the whole mass-energy content of the universe due to the universality of the gravitational interaction; therefore, an astronomical system may be considered isolated only to the extent that the tidal influence of the rest of the universe on the internal dynamics of the system can be neglected. Moreover, the parameters of the force law in Equation (1) refer to the present epoch and hence the virial theorem in Equation (30) ignores cosmological evolution. Thus, the temporal average over an infinite period of time in Equation (30) must be reinterpreted here to mean that the relatively isolated system under consideration has evolved under its own gravity such that it is at the present epoch in a steady equilibrium state. That is, the system is currently in virial equilibrium. Finally, we recall that a point particle of mass m in Equation (30) could reasonably represent a star of mass m as well, where the mass of the star is assumed to be concentrated at its center.

The deviation of the virial theorem in Equation (30) from the Newtonian result is contained in $< \mathbb{D} >$, where $\mathbb{D}$ is given by Equation (28). More explicitly, we have

$$\mathbb{D} = -\frac{1}{2} \sum_{i,j}' \frac{G \, m_i \, m_j}{|\mathbf{x}_i - \mathbf{x}_j|} \left[\alpha_0 - \alpha_0 \left(1 + \frac{1}{2} \mu_0 |\mathbf{x}_i - \mathbf{x}_j|\right) e^{-\mu_0 |\mathbf{x}_i - \mathbf{x}_j|} - \mathcal{E}(|\mathbf{x}_i - \mathbf{x}_j|) \right] \tag{31}$$

It proves useful at this point to study some of the properties of the function $\mathbb{N}$, which is the contribution of nonlocality that is inside the square brackets in Equation (31). The argument of this function is $|\mathbf{x}_i - \mathbf{x}_j| > 0$ for $i \neq j$; therefore, $|\mathbf{x}_i - \mathbf{x}_j|$ varies over the interval $(0, \mathcal{D}_0]$, where $\mathcal{D}_0$ is the largest possible distance between any two baryonic point masses in the system. Thus, $\mathbb{N}(r)$, in the context of the virial theorem, is defined for the interval $0 < r \leq \mathcal{D}_0$, where $\mathcal{D}_0$ is the diameter of the smallest sphere that completely encloses the *baryonic* system for all time. In general, however, $\mathbb{N}(0) = 0$ and $\mathbb{N}(\infty) = \alpha_0 - \mathcal{E}(\infty) > 0$, where $\mathcal{E}(\infty) = \mathcal{E}_\infty$ or $2\mathcal{E}_\infty$, depending on whether we use $\mathcal{E}_1$ or $\mathcal{E}_2$, respectively. Moreover, $d\mathbb{N}(r)/dr$ is given by

$$\frac{d}{dr} \mathbb{N}_1(r) = \frac{1}{2} \alpha_0 \, \mu_0 \, \frac{r \, [1 + \mu_0 \, (a_0 + r)]}{a_0 + r} \, e^{-\mu_0 r} \tag{32}$$

if we use $\mathcal{E}_1$ or

$$\frac{d}{dr} \mathbb{N}_2(r) = \frac{1}{2} \alpha_0 \, \mu_0 \, \frac{r^2 \, [1 + \mu_0 \, (a_0 + r)]}{(a_0 + r)^2} \, e^{-\mu_0 r} \tag{33}$$

if we use $\mathcal{E}_2$. Writing $\exp(\mu_0 r) = 1 + \mu_0 r + \mathcal{R}$, where $\mathcal{R} > 0$ represents the remainder of the power series, it is straightforward to see that for $r \geq 0$ and $n = 1, 2, \ldots,$

$$e^{\mu_0 r} (a_0 + r)^n > r^n \left[1 + \mu_0 (a_0 + r)\right] \tag{34}$$

This result, for $n = 1$ and $n = 2$, implies that the right-hand sides of Equations (32) and (33), respectively, are less than $\alpha_0 \mu_0 / 2$. Therefore, it follows that, in general,

$$\frac{d}{dr} \mathbb{N}(r) < \frac{1}{2} \alpha_0 \mu_0 \tag{35}$$

Moreover, for $r > 0$, (35) implies:

$$\mathbb{N}(r) = \int_0^r \frac{d\,\mathbb{N}(x)}{dx} \, dx < \frac{1}{2} \alpha_0 \mu_0 \, r \tag{36}$$

We conclude that $\mathbb{N}$ is a monotonically increasing function of r that is zero at $r = 0$ with a slope that vanishes at $r = 0$. For $r \gg \mu_0^{-1}$, $\mathbb{N}(r)$ asymptotically approaches a constant $\alpha_0 \epsilon := \alpha_0 - \mathcal{E}(\infty)$. Here, $\epsilon(p)$ is either $\epsilon_1(p)$ or $\epsilon_2(p)$ depending on whether we use $\mathcal{E}_1$ or $\mathcal{E}_2$, respectively. The functions $\epsilon_1(p)$ and $\epsilon_2(p)$ are defined in Equation (13).

4. Dark Matter

Most of the matter in the universe is currently thought to be in the form of certain elusive particles that have not been directly detected [29–32]. The existence and properties of this *dark matter* have thus far been deduced only through its gravity. We are interested here in dark matter only as it pertains to stellar systems such as galaxies and clusters of galaxies [33–39]. We mention that dark matter is also essential in the explanation of gravitational lensing observations [40,41] and in the solution of the problem of structure formation in cosmology [13,42]; however, these topics are beyond the scope of this work.

Actual (mainly baryonic) mass is observationally estimated for astronomical systems using the mass-to-light ratio M/L. However, it turns out that the dynamic mass of the system is usually larger and this observational fact is normally attributed to the possible existence of nonbaryonic dark matter. Let M be the baryonic mass and M_{DM} be the mass of the nonbaryonic dark matter needed to explain the gravitational dynamics of the system. Then,

$$f_{DM} = \frac{M_{DM}}{M} \tag{37}$$

is the dark matter fraction and $M + M_{DM} = M\,(1 + f_{DM})$ is the dynamic mass of the system.

In observational astrophysics, the virial theorem of Newtonian gravity is interpreted to be a relationship between the dynamic (virial) mass of the entire system and its average internal motion deduced from the rotation curve or velocity dispersion of the bound collection of masses in virial equilibrium. Therefore, regardless of how the net amount of dark matter in galaxies and clusters of galaxies is operationally estimated and the corresponding f_{DM} is thereby determined, for sufficiently isolated self-gravitating astronomical systems in virial equilibrium, we must have

$$2 < \mathbb{T} > \ = -(1 + f_{DM}) < \mathbb{W}_N > \tag{38}$$

That is, virial theorem Equation (38) is employed in astronomy to infer in some way the total dynamic mass of the system. Indeed, Zwicky first noted the need for dark matter in his application of the standard virial theorem of Newtonian gravity to the Coma cluster of galaxies [33,34].

5. Effective Dark Matter

A significant physical consequence of nonlocal gravity theory is that it appears to simulate dark matter [9]. In particular, in the Newtonian regime of nonlocal gravity, the Poisson equation is modified such that the density of ordinary matter ρ is accompanied by a term ρ_D that is obtained from the folding (convolution) of ρ with the reciprocal kernel of nonlocal gravity. Thus, ρ_D has the interpretation of the density of *effective dark matter* and $\rho + \rho_D$ is the density of the *effective dynamic mass*.

The virial theorem makes it possible to elucidate in a simple way the manner in which nonlocality can simulate dark matter. It follows from a comparison of Equations (30) and (38) that nonlocal gravity can account for this "excess mass" if

$$< \mathbb{D} > \ = f_{DM} < \mathbb{W}_N > \tag{39}$$

where $\mathbb{W}_N$ and $\mathbb{D}$ are given in Equations (26) and (28), respectively.

It is interesting to apply the virial theorem of nonlocal gravity to sufficiently isolated astronomical N-body systems. The configurations that we briefly consider below consist of clusters of galaxies with diameters $\mathcal{D}_0 \gg \mu_0^{-1} \approx 17$ kpc, galaxies with $\mathcal{D}_0 \sim \mu_0^{-1}$ and globular star clusters with $\mathcal{D}_0 \ll \mu_0^{-1}$. The results presented in this section follow from certain general properties of the function $\mathbb{N}(r)$ and are completely independent of how the baryonic matter is distributed within the astronomical system under consideration.

We emphasize that, after setting the short-range parameter $a_0 = 0$, the parameters α_0 and μ_0, and hence λ_0, were originally determined from the combined observational data for the rotation curves of a sample of 12 nearby spiral galaxies from the THINGS catalog [9]. These tentative values are given in Equation (7). These parameter values were then found to be in reasonable agreement with the internal dynamics of a sample of 10 rich nearby clusters of galaxies from the Chandra X-ray catalog [9]. In the present paper, we use these parameter values to make predictions about *all* nearby isolated N-body gravitational systems that are in virial equilibrium.

5.1. Clusters of Galaxies: $f_{DM} \approx \alpha_0 \, \epsilon(p)$

Consider, for example, a cluster of galaxies, where nearly all of the relevant distances are much larger than $\mu_0^{-1} \approx 17$ kpc. In this case, $\mu_0 r \gg 1$ and hence $\mathbb{N}$ approaches its asymptotic value, namely,

$$\mathbb{N} \approx \alpha_0 \, \epsilon(p) \tag{40}$$

where $\epsilon = \epsilon_1$ or ϵ_2, defined in Equation (13), depending on whether we use $\mathcal{E}_1$ or $\mathcal{E}_2$, respectively. Hence, Equation (28) can be written as:

$$< \mathbb{D} > \approx \alpha_0 \, \epsilon(p) < \mathbb{W}_N > \tag{41}$$

It then follows from Equation (39) that, for galaxy clusters,

$$f_{DM} \approx \alpha_0 \, \epsilon(p) \tag{42}$$

in nonlocal gravity. We recall that ϵ is only weakly sensitive to the magnitude of a_0. It follows from $\alpha_0 \approx 11$ that f_{DM} for galaxy clusters is about 10, in general agreement with observational data [9]. This theoretical result is essentially equivalent to the work on galaxy clusters contained in reference [9], except that Equation (42) takes into account the existence of the short-range parameter a_0.

Nonlocal gravity thus predicts that the effective dark matter fraction f_{DM} has approximately the same constant value of about 10 for all isolated nearby clusters of galaxies that are in equilibrium.

5.2. Galaxies: $f_{DM} < \mathcal{D}_0/\lambda_0$

Consider next a sufficiently isolated galaxy of diameter $\mathcal{D}_0$ in virial equilibrium. In this case, we recall that $\mathbb{N}(r)$ is a monotonically increasing function of r, so that for $0 < r \leq \mathcal{D}_0$, Equation (36) implies

$$\mathbb{N}(r) \leq \mathbb{N}(\mathcal{D}_0) < \frac{1}{2} \alpha_0 \mu_0 \mathcal{D}_0 \tag{43}$$

Therefore, it follows from Equation (28) that, in this case,

$$\mathbb{D} > (\frac{1}{2} \alpha_0 \mu_0 \mathcal{D}_0) \mathbb{W}_N \tag{44}$$

The virial theorem for nonlocal gravity in the case of an isolated galaxy is then

$$2 < \mathbb{T} > + < \mathbb{W}_N > \; < \; -(\frac{1}{2} \alpha_0 \mu_0 \mathcal{D}_0) < \mathbb{W}_N > \tag{45}$$

which means, when compared with Equation (38), that

$$f_{DM} < \frac{1}{2} \alpha_0 \mu_0 \mathcal{D}_0 \tag{46}$$

Let us note that

$$\frac{1}{2} \alpha_0 \mu_0 = \frac{1}{\lambda_0} \tag{47}$$

where λ_0 is the basic nonlocality length scale. Its exact value is not known; however, from the results of reference [9], we have $\lambda_0 \approx 3 \pm 2$ kpc. If we formally let $\lambda_0 \to \infty$, then (46), namely, $f_{DM} < \mathcal{D}_0 / \lambda_0$, implies that in this case nonlocality and the effective dark matter both disappear, as expected. Therefore, for a sufficiently isolated galaxy in virial equilibrium, the ratio of its baryonic diameter to dark matter fraction f_{DM} must always be above a fixed length λ_0 of about 3 ± 2 kpc; that is,

$$\frac{\mathcal{D}_0}{f_{DM}} > \lambda_0 \tag{48}$$

To illustrate (48), consider, for instance, the Andromeda Galaxy (M31) with a diameter $\mathcal{D}_0$ of about 67 kpc. In this case, we have $f_{DM} \approx 12.7$ [43,44], so that for this spiral galaxy

$$\frac{\mathcal{D}_0}{f_{DM}} \text{ (Andromeda Galaxy)} \approx 5.3 \text{ kpc} \tag{49}$$

More recently, the distribution of dark matter in M31 has been further studied in reference [45]. Similarly, for the Triangulum Galaxy (M33), we have $\mathcal{D}_0 \approx 34$ kpc and $f_{DM} \approx 5$ [46], so that

$$\frac{\mathcal{D}_0}{f_{DM}} \text{ (Triangulum Galaxy)} \approx 6.8 \text{ kpc} \tag{50}$$

Turning next to an elliptical galaxy, namely, the massive E0 galaxy NGC 1407, we have $\mathcal{D}_0 \approx 160$ kpc and $f_{DM} \approx 31$ [47], so that

$$\frac{\mathcal{D}_0}{f_{DM}} \text{ (NGC 1407)} \approx 5.2 \text{ kpc} \tag{51}$$

Moreover, for the intermediate-luminosity elliptical galaxy NGC 4494, which has a half-light radius of $R_e \approx 3.77$ kpc, the dark matter fraction has been found to be $f_{DM} = 0.6 \pm 0.1$ [48]. Assuming that the baryonic system has a radius of $2 R_e$, we have $\mathcal{D}_0 = 4 R_e \approx 15$ kpc and $f_{DM} \approx 0.6$; hence,

$$\frac{\mathcal{D}_0}{f_{DM}} \text{ (NGC 4494)} \approx 25 \text{ kpc} \tag{52}$$

Let us note that the results presented here are essentially for the present epoch in the expansion of the universe. Observations indicate, however, that the diameters of massive galaxies can increase with decreasing redshift z. For a discussion of such *massive compact galaxies*, see reference [49].

Finally, it is interesting to consider f_{DM} at the other extreme, namely, for the case of globular star clusters and isolated dwarf galaxies. The diameter of a globular star cluster is about 40 pc. We can therefore conclude from (48) with $\lambda_0 \approx 3$ kpc that for globular star clusters:

$$f_{DM} \text{ (globular star cluster)} \lesssim 10^{-2} \tag{53}$$

Thus, according to the virial theorem of nonlocal gravity, less than about one percent of the mass of a globular star cluster must appear as effective dark matter if the system is sufficiently isolated and is in virial equilibrium. It is not clear to what extent such systems can be considered isolated. It is usually assumed that observational data are consistent with the existence of almost no dark matter in globular star clusters. However, a recent investigation of six galactic globular clusters has led to the conclusion that $f_{DM} \approx 0.4$ [50]. The resolution of this discrepancy is beyond the scope of the present work.

Isolated dwarf galaxies with diameters $\mathcal{D}_0 \ll \mu_0^{-1}$ would similarly be expected to contain a relatively small percentage of effective dark matter. There is a significant discrepancy here as well, see reference [51]; again, the resolution of this difficulty is beyond the scope of this paper. In dwarf systems that are not isolated, the tidal influence of a much larger neighboring galaxy on the dynamics of the dwarf spheroidal galaxy cannot be ignored [52–54].

6. Discussion

Nonlocal gravity theory predicts that the amount of effective dark matter in a sufficiently isolated nearby galaxy in virial equilibrium is such that f_{DM} has an upper bound, $\mathcal{D}_0/\lambda_0$, that is completely independent of the distribution of baryonic matter in the galaxy. However, it is possible to derive an *improved* upper bound for f_{DM}, which does depend on how baryons are distributed within the galaxy. To this end, we note that Equation (28) for $\mathbb{D}$ and $\mathbb{N}(r) < r/\lambda_0$ imply:

$$\mathbb{D} > -\frac{1}{2} \sum_{i,j}' \frac{G \, m_i \, m_j}{\lambda_0} \tag{54}$$

If follows from this result together with Equation (39) that

$$< \mathbb{W}_N > f_{DM} > -\frac{1}{2} \sum_{i,j}' \frac{G \, m_i \, m_j}{\lambda_0} \tag{55}$$

Let us define a characteristic length, R_{av}, for the average extent of the distribution of baryons in the galaxy via

$$R_{av} < \mathbb{W}_N >= -\frac{1}{2} \sum_{i,j}' G \, m_i \, m_j \tag{56}$$

Then, it follows from (55) and Equation (56), that

$$f_{DM} < \frac{R_{av}}{\lambda_0} \tag{57}$$

Clearly, R_{av} depends upon the density of baryons in the galaxy. In the Newtonian gravitational potential energy in Equation (56), $0 < |\mathbf{x}_i - \mathbf{x}_j| \leq \mathcal{D}_0$; therefore, in general, $R_{av} \leq \mathcal{D}_0$; hence, we recover from the new inequality, namely, $f_{DM} < R_{av}/\lambda_0$, our previous less tight but more general result $f_{DM} < \mathcal{D}_0/\lambda_0$.

Acknowledgments: I am grateful to Jeffrey Kuhn, Sohrab Rahvar and Haojing Yan for valuable discussions.

References

1.   Einstein, A. *The Meaning of Relativity*; Princeton University Press: Princeton, NJ, USA, 1955.
2.   Hehl, F.W.; Mashhoon, B. Nonlocal gravity simulates dark matter. *Phys. Lett. B* **2009**, *673*, 279–282.

3.   Hehl, F.W.; Mashhoon, B. Formal framework for a nonlocal generalization of Einstein's theory of gravitation. *Phys. Rev. D* **2009**, *79*, 064028.

4.   Blome, H.-J.; Chicone, C.; Hehl, F.W.; Mashhoon, B. Nonlocal modification of Newtonian gravity. *Phys. Rev. D* **2010**, *81*, 065020.

5.   Mashhoon, B. Nonlocal gravity. In *Cosmology and Gravitation*; Novello, M., Begliaffa, S.E.P., Eds.; Cambridge Scientific Publishers: Cambridge, UK, 2011; pp. 1–9.

6.   Chicone, C.; Mashhoon, B. Nonlocal gravity: Modified Poisson's equation. *J. Math. Phys.* **2012**, *53*, 042501.

7.   Chicone, C.; Mashhoon, B. Linearized gravitational waves in nonlocal general relativity. *Phys. Rev. D* **2013**, *87*, 064015.

8.   Mashhoon, B. Nonlocal gravity: Damping of linearized gravitational waves. *Class. Quantum Gravity* **2013**, *30*, 155008.

9.   Rahvar, S.; Mashhoon, B. Observational tests of nonlocal gravity: Galaxy rotation curves and clusters of galaxies. *Phys. Rev. D* **2014**, *89*, 104011.

10. Mashhoon, B. Nonlocal gravity: The general linear approximation. *Phys. Rev. D* **2014**, *90*, 124031.

11. Mashhoon, B. Nonlocal general relativity. *Galaxies* **2015**, *3*, 1–17.

12. Chicone, C.; Mashhoon, B. Nonlocal gravity in the Solar System. *Class. Quantum Gravity* **2016**, *33*, 075005.

13. Chicone, C.; Mashhoon, B. Nonlocal Newtonian cosmology. 2015, arXiv:1510.07316 [gr-qc].

14. Iorio, L. Gravitational Anomalies in the Solar System? *Int. J. Mod. Phys. D* **2015**, *24*, 1530015.

15. Deng, X.-M.; Xie, Y. Solar System test of the nonlocal gravity and the necessity for a screening mechanism. *Ann. Phys.* **2015**, *361*, 62–71.

16. Abramowitz, M.; Stegun, I.A. *Handbook of Mathematical Functions*; National Bureau of Standards: Washington, DC, USA, 1964.

17. Adelberger, E.G.; Heckel, B.R.; Nelson, A.E. Tests of the Gravitational Inverse-Square Law. *Ann. Rev. Nucl. Part. Sci.* **2003**, *53*, 77–121.

18. Hoyle, C.D.; Kapner, D.J.; Heckel, B.R.; Adelberger, E.G.; Gundlach, J.H.; Schmidt, U.; Swanson, H.E. Sub-millimeter tests of the gravitational inverse-square law. *Phys. Rev. D* **2004**, *70*, 042004.

19. Adelberger, E.G.; Heckel, B.R.; Hoedl, S.A.; Hoyle, C.D.; Kapner, D.J.; Upadhye, A. Particle-Physics Implications of a Recent Test of the Gravitational Inverse-Square Law. *Phys. Rev. Lett.* **2007**, *98*, 131104.

20. Kapner, D.J.; Cook, T.S.; Adelberger, E.G.; Gundlach, J.H.; Heckel, B.R.; Hoyle, C.D.; Swanson, H.E. Tests of the Gravitational Inverse-Square Law below the Dark-Energy Length Scale. *Phys. Rev. Lett.* **2007**, *98*, 021101.

21. Little, S.; Little, M. Laboratory test of Newton's law of gravity for small accelerations. *Class. Quantum Gravity* **2014**, *31*, 195008.

22. Tohline, J.E. Stabilizing a Cold Disk with a 1/r Force Law. In *IAU Symposium 100, Internal Kinematics and Dynamics of Galaxies*; Athanassoula, E., Ed.; Reidel: Dordrecht, The Netherlands, 1983; pp. 205–206.

23. Tohline, J.E. Does Gravity Exhibit a 1/r Force on the Scale of Galaxies? *Ann. N. Y. Acad. Sci.* **1984**, *422*, 390–390.

24. Kuhn, J.R.; Burns, C.A.; Schorr, A.J. Numerical Coincidences, Fictional Forces, and the Galactic Dark Matter Distribution. 1986, unpublished work.

25. Kuhn, J.R.; Kruglyak, L. Non-Newtonian forces and the invisible mass problem. *Astrophys. J.* **1987**, *313*, 1–12.

26. Bekenstein, J.D. *Second Canadian Conference on General Relativity and Relativistic Astrophysics*; Coley, A., Dyer, C., Tupper, T., Eds.; World Scientific: Singapore, 1988; p. 68.

27. Hees, A.; Hestroffer, D.; Le Poncin-Lafitte, C.; David, P. Tests of gravitation with Gaia observations of Solar System Objects. 2015, arXiv: 1509.06868.

28. Buscaino, B.; DeBra, D.; Graham, P.W.; Gratta, G.; Wiser, T.D. Testing long-distance modifications of gravity to 100 astronomical units. *Phys. Rev. D* **2015**, *92*, 104048.

29. Aprile, E.; Alfonsi, M.; Arisaka, K.; Arneodo, F.; Balan, C.; Baudis, L.; Bauermeister, B.; Behrens, A.; Beltrame, P.; Bokeloh, K.; *et al.* Dark Matter Results from 225 Live Days of XENON100 Data. *Phys. Rev. Lett.* **2012**, *109*, 181301.

30. Akerib, D.S.; Araújo, H.M.; Bai, X.; Bailey, A.J.; Balajthy, J.; Bedikian, S.; Bernard, E.; Bernstein, A.; Bolozdynya, A.; Bradley, A.; *et al.* First Results from the LUX Dark Matter Experiment at the Sanford Underground Research Facility. *Phys. Rev. Lett.* **2014**, *112*, 091303.

31. Agnese, R.; Anderson, A.J.; Asai, M.; Balakishiyeva, D.; Basu Thakur, R.; Bauer, D.A.; Beaty, J.; Billard, J.; Borgland, A.; Bowles, M.A.; *et al.* Search for Low-Mass Weakly Interacting Massive Particles with SuperCDMS. *Phys. Rev. Lett.* **2014**, *112*, 241302.

32. Baudis, L. Dark matter searches. *Ann. Phys.* **2016**, *528*, 74–83.

33. Zwicky, F. Die Rotverschiebung von extragalaktischen Nebeln. *Helv. Phys. Acta* **1933**, *6*, 110–127.

34. Zwicky, F. On the Masses of Nebulae and of Clusters of Nebulae. *Astrophys. J.* **1937**, *86*, 217–246.

35. Rubin, V.C.; Ford, W.K. Rotation of the Andromeda Nebula from a Spectroscopic Survey of Emission Regions. *Astrophys. J.* **1970**, *159*, 379–403.

36. Roberts, M.S.; Whitehurst, R.N. The rotation curve and geometry of M31 at large galactocentric distances. *Astrophys. J.* **1975**, *201*, 327–346.

37. Sofue, Y.; Rubin, V. Rotation Curves of Spiral Galaxies. *Annu. Rev. Astron. Astrophys.* **2001**, *39*, 137–174.

38. Seigar, M.S. *Dark Matter in the Universe*; Morgan and Claypool: San Rafael, CA, USA, 2015.

39. Harvey, D.; Massey, R.; Kitching, T.; Taylor, A.; Tittley, E. The nongravitational interactions of dark matter in colliding galaxy clusters. *Science* **2015**, *347*, 1462–1465.

40. Clowe, D.; Bradač, M.; Gonzalez, A.H.; Markevitch, M.; Randall, S.W.; Jones, C.; Zaritsky, D. A direct empirical proof of the existence of dark matter. *Astrophys. J. Lett.* **2006**, *648*, L109–L113.

41. Clowe, D.; Randall, S.W.; Markevitch, M. Catching a bullet: direct evidence for the existence of dark matter. *Nucl. Phys. B Proc. Suppl.* **2007**, *173*, 28–31.

42. Bini, D.; Mashhoon, B. Nonlocal gravity: Conformally flat spacetimes. *Int. J. Geom. Methods Mod. Phys.* **2016**, *13*, 1650081.

43. Barmby, P.; Ashby, M.L.N.; Bianchi, L.; Engelbracht, C.W.; Gehrz, R.D.; Gordon, K.D.; Hinz, J.L.; Huchra, J.P.; Humphreys, R.M.; Pahre, M.A.; *et al.* Dusty waves on a starry sea: The mid-infrared view of M31 *Astrophys. J.* **2006**, *650*, L45–L49.

44. Barmby, P.; Ashby, M.L.N.; Bianchi, L.; Engelbracht, C.W.; Gehrz, R.D.; Gordon, K.D.; Hinz, J.L.; Huchra, J.P.; Humphreys, R.M.; Pahre, M.A.; *et al.* Erratum: ''Dusty Waves on a Starry Sea: The Mid-Infrared View of M31''. *Astrophys. J.* **2007**, *655*, L61–L61.

45. Tamm, A.; Tempel, E.; Tenjes, P.; Tihhonova, O.; Tuvikene, T. Stellar mass map and dark matter distribution in M31. *Astron. Astrophys.* **2012**, *546*, A4.

46. Corbelli, E. Dark matter and visible baryons in M33. *Mon. Not. R. Astron. Soc.* **2003**, *342*, 199–207.

47. Pota, V.; Romanowsky, A.J.; Brodie, J.P.; Peñarrubia, J.; Forbes, D.A.; Napolitano, N.R.; Foster, C.; Walker, M.G.; Strader, J.; Roediger, J.C. The SLUGGS survey: Multipopulation dynamical modelling of the elliptical galaxy NGC 1407 from stars and globular clusters. *Mon. Not. R. Astron. Soc.* **2015**, *450*, 3345–3358.

48. Morganti, L.; Gerhard, O.; Coccato, L.; Martinez-Valpuesta, I.; Arnaboldi, M. Elliptical galaxies with rapidly decreasing velocity dispersion profiles: NMAGIC models and dark halo parameter estimates for NGC 4494 *Mon. Not. R. Astron. Soc.* **2013**, *431*, 3570–3588.

49. De Arriba, L.P.; Balcells, M.; Falcón-Barroso, J.; Trujillo, I. The discrepancy between dynamical and stellar masses in massive compact galaxies traces non-homology. *Mon. Not. R. Astron. Soc.* **2014**, *440*, 1634–1648.

50. Sollima, A.; Bellazzini, M.; Lee, J.-W. A comparison between the stellar and dynamical masses of six globular clusters. *Astrophys. J.* **2012**, *755*, 156.

51. Oh, S.-H.; Hunter, D.A.; Brinks, E.; Elmegreen, B.G.; Schruba, A.; Walter, F.; Rupen, M.P.; Young, L.M.; Simpson, C.E.; Johnson, M.C. High-resolution mass models of dwarf galaxies from LITTLE THINGS. *Astron. J.* **2015**, *149*, 180.

52. Kuhn, J.R.; Miller, R.H. Dwarf spheroidal galaxies and resonant orbital coupling. *Astrophys. J. Lett.* **1989**, *341*, L41–L45.

53. Fleck, J.-J.; Kuhn, J.R. Parametric dwarf spheroidal tidal interaction. *Astrophys. J.* **2003**, *592*, 147–160.

54. Muñoz, R.R.; Frinchaboy, P.M.; Majewski, S.R.; Kuhn, J.R.; Chou, M-Y.; Palma, C.; Sohn, S.T.; Patterson, R.J.; Siegel, M.H. Exploring Halo Substructure with Giant Stars: The Velocity Dispersion Profiles of the Ursa Minor and Draco Dwarf Spheroidal Galaxies at Large Angular Separations. *Astrophys. J. Lett.* **2005**, *631*, L137–L141.

12

Strategies to Ascertain the Sign of the Spatial Curvature

Pedro C. Ferreira [1] **and Diego Pavón** [2,*]

[1] Escola de Ciências e Tecnologia, Universidade Federal do Rio Grande do Norte, Natal 59072-970, Rio Grande do Norte, Brazil; pedro.ferreira@ect.ufrn.br

[2] Departamento de Física, Universidad Autónoma de Barcelona, Bellaterra, Barcelona 08193, Spain

* Correspondence: diego.pavon@uab.es

Academic Editors: Lorenzo Iorio and Elias C. Vagenas

Abstract: The second law of thermodynamics, in the presence of gravity, is known to hold at small scales, as in the case of black holes and self-gravitating radiation spheres. Using the Friedmann–Lemaître–Robertson–Walker metric and the history of the Hubble factor, we argue that this law also holds at cosmological scales. Based on this, we study the connection between the deceleration parameter and the spatial curvature of the metric, Ω_k, and set limits on the latter, valid for any homogeneous and isotropic cosmological model. Likewise, we devise strategies to determine the sign of the spatial curvature index k. Finally, assuming the lambda cold dark matter model is correct, we find that the acceleration of the cosmic expansion is increasing today.

Keywords: mathematical cosmology; spatial curvatur; thermodynamics

1. Introduction

The validity of the second law of thermodynamics for systems dominated by gravity should not be taken for granted. Gravity is a long-ranged interaction while the formulation of the second is based on the observation of ordinary systems, i.e., those dominated by short-ranged interactions. In actual fact, its validity for the former systems was studied only recently, notably in the case of black holes and self-gravitating radiation spheres. In the former case, Bekenstein demonstrated that the black-hole entropy, in addition to the entropy of the black-hole exterior, never decreases [1,2]. In the latter, it was shown that the static stable configurations of a sphere of self-gravitating radiation are those that maximize the radiation entropy [3,4]. Both instances correspond to small scale systems. Although different authors assumed it to be in order to constrain the evolution of cosmological models (see, e.g., [5] and references therein), as far as we know, the validity of the said law at large (i.e., cosmic) scales has not been explored as yet. The main purpose of this work is to fill this gap. Our study analysis rests on the simplest realistic large-scale space-time metric, namely, the Friedmann–Lemaître–Robertson–Walker (FLRW) one alongside a selected set of observational data about the history of cosmic expansion.

Homogeneous and isotropic universe models are usually described by the FLRW metric

$$ds^2 = -c^2 dt^2 + a^2(t) \left\{ \frac{dr^2}{1 - kr^2} + r^2 \left(d\theta^2 + \sin^2\theta \, d\phi^2 \right) \right\}, \tag{1}$$

coupled to the sources of the gravitational field. This metric relies on the cosmological principle [6–8] whose validity, at large scales, has not been contradicted thus far [9] and it looks rather robust [10–12]. The curvature index, k, is either 0, $+1$, or -1 depending on whether the spatial part of the metric is flat, positively curved (closed), or negatively curved (hyperbolic), respectively.

This constant index, like the scale factor $a(t)$, is not a directly observable quantity. In principle, however, it can be determined through the knowledge of the dimensionless, fractional curvature density, $\Omega_k \equiv -k/(a^2 H^2)$, which is accessible to observation, albeit indirectly. As usual, $H = c\, d\ln a/dt$ denotes the Hubble factor. Current measurements of Ω_k only indicate that its present absolute value is small ($| \Omega_{k0} | \lesssim 10^{-3}$ [13,14]). Note that this constraint was obtained under the assumption that the universe is accurately described by the ΛCDM model. Thus the sign of k remains unknown.

The aim of this research is fourfold: (i) To determine whether the second law of thermodynamics is fulfilled at cosmological scales and; if so, (ii) constrain Ω_k as much as possible and (iii) determine the sign of k; finally, (iv) to derive a thermodynamic constraint relating the present value of the deceleration and jerk parameters. For the first three objectives, neither a cosmological model nor theory of gravity will be assumed. We shall just use the FLRW metric, the history $H(z)$ of the Hubble factor and the second law of thermodynamics. For the fourth objective, we will assume Einstein gravity and the ΛCDM model. As is customary, a subindex zero attached to any quantity means that the latter should be evaluated at present time.

2. Cosmological Consequences of the Second Law

Given the strong connection between gravity and thermodynamics [1,2,15–17], it is natural to expect that the universe behaves as a normal thermodynamic system; it therefore must tend to a state of maximum entropy in the long run [18,19].

For comoving observers, FLRW models entail "normal", "trapped" and "anti-trapped" regions. In the first one, the expansion of outgoing null geodesic congruences, normal to the spatial two-sphere of radius $\tilde{r}(= ra(t))$ centered at the origin (i.e., at the position of the observer), is positive, and negative for ingoing null geodesic congruences. In the trapped region, both kind of geodesic congruences have negative expansion. By contrast, in the anti-trapped region the expansion of both congruences is positive. The boundary hyper-surface of the space-time anti-trapped region is called the apparent horizon; its radius is $\tilde{r}_A = [(H/c)^2 + ka^{-2}]^{-1/2}$. Since the observer has no information about what might be going on beyond the horizon, the latter has an entropy, namely: $S_A = k_B \pi \tilde{r}_A^2 / \ell_{pl}^2$, where ℓ_{pl} is Planck's length. For details, see [20]. (Bear in mind that $\tilde{r}$ and H have dimensions of length and length^{-1}, respectively, k of length^{-2}, and a is dimensionless.)

A rather reasonable assumption concerning the entropy of the observable universe is that it is dominated by the entropy of the cosmic horizon. In the current universe, the entropy of the horizon exceeds that of supermassive black holes, stellar black holes, relic neutrinos and CMB photons by 18, 25, 33 and 33 orders of magnitude, respectively [21]. There are several possible choices for the cosmic horizon: the particle horizon, the event horizon, the apparent horizon and the Hubble horizon. Given that the first one does not exist for accelerating universes and the second only exists if the universe accelerates forever in the future, we take the apparent horizon, which, on the one hand, always exists, both for ever-expanding and ever-contracting universes, and, on the other hand, by contrast to the other mentioned possibilities, the laws of thermodynamics are fulfilled on it [22]. The Hubble horizon is a particular case of the apparent horizon when $k = 0$.

To support the above claim that the entropy of the horizon dominates over the entropy of any form of energy inside the horizon, especially at late times, we shall consider the entropy of pressureless matter. The latter is given by $S_m = k_B n V_k$ [23], with $n = n_0 a^{-3}$, being n_0 the present number density of matter particles, and

$$V_k = 2\pi a^2 \left[\sqrt{|k|}\, a \sin^{-1}(\sqrt{|k|}\, a^{-1} \tilde{r}_A) - k \tilde{r}_A^2 H \right] \tag{2}$$

the volume enclosed by the apparent horizon for $k = +1$ and -1 (for the flat case, $V_{k=0} = (4\pi/3)\,\tilde{r}_{\mathcal{A}}^3$). For $k = -1$ one follows $S_m(a \gg 1) \to 2k_B\,n_0\,\pi a^{-1}\tilde{r}_{\mathcal{A}}^2 H$. Hence, when $a \gg 1$ the ratio $S_{\mathcal{A}}/S_m$ results proportional to a/H. For $k = +1$ one has $S_m(a \gg 1) \to 2k_B\,n_0\,a^{-1}\left(1 - \sqrt{1 - \tilde{r}_{\mathcal{A}}^2 a^{-1}}\right)$, hence

$$\frac{S_{\mathcal{A}}}{S_m} \propto \frac{a\,\tilde{r}_{\mathcal{A}}^2}{1 - \sqrt{1 - \frac{\tilde{r}_{\mathcal{A}}^2}{a}}}.$$

Accordingly, in all three cases ($k = 0, +1, -1$) the entropy of the horizon overwhelms that of the matter inside it, especially at late times.

Recalling that $S_{\mathcal{A}} \propto \mathcal{A}$ with $\mathcal{A} = 4\pi(H^2 + k a^{-2})^{-1}$ the area of the horizon (henceforward we set $c = 1$), the second law of thermodynamics $S'_{\mathcal{A}} \geq 0$ leads to

$$\mathcal{A}' = -\frac{\mathcal{A}^2}{2\pi}\left(HH' - \frac{k}{a^3}\right) \geq 0 \quad \Rightarrow \quad HH' \leq \frac{k}{a^3}, \tag{3}$$

where the prime means d/da.

The second inequality tells us that if H' is or has been positive at any stage of cosmic expansion (excluding, possibly, the pre-Planckian era), then $k = +1$ and that, in principle, any sign of k is compatible with $H' < 0$. Multiplying the said inequality by $-aH^{-2}$ produces $-aH'/H \geq \Omega_k$, which can be recast in terms of the redshift as

$$(1 + z)\frac{d \ln H}{dz} \geq \Omega_k. \tag{4}$$

Thus, if $dH/dz > 0$ for all $z \geq 0$, then both $k = +1$ and $k = 0$ are consistent with the second law of thermodynamics at large scales. However, given the present ample uncertainties in the observational data regarding the Hubble history, if k were -1, then the said law could break down at cosmic scales. To explore this, we set $k = -1$ in Equation (4) and integrate the resulting expression in the interval $z_1 \leq z \leq z_2$ to get

$$H_2^2 - H_1^2 \geq 2(z_2 - z_1) + (z_2^2 - z_1^2). \tag{5}$$

Therefore, if this relationship failed for whatever pair of points (z_i, H_i), with $i = 1, 2$, it should mean that the choice $k = -1$ would not be consistent with the second law at the said scales.

We use Equation (5) alongside the 28 experimental data H vs. z, in the interval $0.1 \leq z \leq 2.36$, with their 1σ error bars, compiled by Farook et al. [24] and listed in Table 1 (see also Figure 1) for the reader convenience, to draw Figure 2. The latter suggests that, given the experimental uncertainties, the possibility $k = -1$ also appears compatible with the inequality $S'_{\mathcal{A}} \geq 0$. While wider compilations of $H(z)$ are available, we believe this one is preferable because it does not include any obviously correlated data, nor does it contain older, less reliable data, some with much weight from anomalously small error bars.

Equation (4) can alternatively be written as

$$1 + q \geq \Omega_k, \tag{6}$$

where $q = -\ddot{a}/(aH^2)$ is the dimensionless deceleration parameter. The last equation, like (4), imposes an upper bound (that depends on redshift) on Ω_k. In the radiation dominated era q was close to 1; a result that, in spite of having been derived for spatially flat universes described by general relativity, should hold irrespective of the sign of the curvature and the gravity theory employed. Notice that even a mild deviation of $q \simeq 1$ at that time would conflict with the observational results about the primordial nucleosynthesis of light elements [25]. This suggests an easily verifiable test on modified gravity theories, namely, that they should be consistent with the bound $\Omega_k \leq 2$ at the radiation era. However, if general relativity is the right theory of gravity, the first Friedmann equation implies the stronger bound $\Omega_k < 1$ at all epochs. Nevertheless, even if one uses general relativity, Equation (6) might provide a useful bound when $q < 0$.

Table 1. Hubble Parameter vs. Redshift Data.

z	$H(z)$ (km·s^{-1}·Mpc^{-1})	Reference
0.100	69 ± 12	[26]
0.170	83 ± 8	[26]
0.179	75 ± 4	[27]
0.199	75 ± 5	[27]
0.270	77 ± 14	[26]
0.320	79.2 ± 5.6	[28]
0.352	83 ± 14	[27]
0.400	95 ± 17	[26]
0.440	82.6 ± 7.8	[29]
0.480	97 ± 62	[30]
0.570	100.3 ± 3.7	[28]
0.593	104 ± 13	[27]
0.600	87.9 ± 6.1	[29]
0.680	92 ± 8	[27]
0.730	97.3 ± 7	[29]
0.781	105 ± 12	[27]
0.875	125 ± 17	[27]
0.880	90 ± 40	[30]
0.900	117 ± 23	[26]
1.037	154 ± 20	[27]
1.300	168 ± 17	[26]
1.363	160 ± 33.6	[31]
1.430	177 ± 18	[26]
1.530	140 ± 14	[26]
1.750	202 ± 40	[26]
1.965	186.5 ± 50.4	[31]
2.340	222 ± 7	[32]
2.360	226 ± 8	[33]

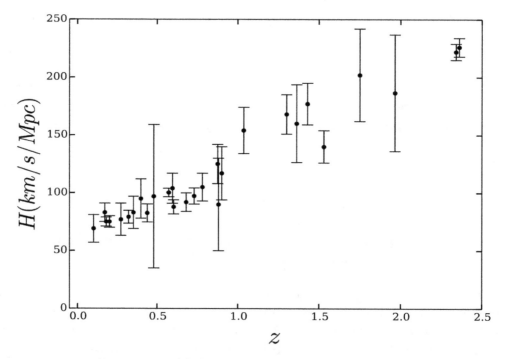

Figure 1. 28 $H(z)$ data points with their 1σ uncertainty.

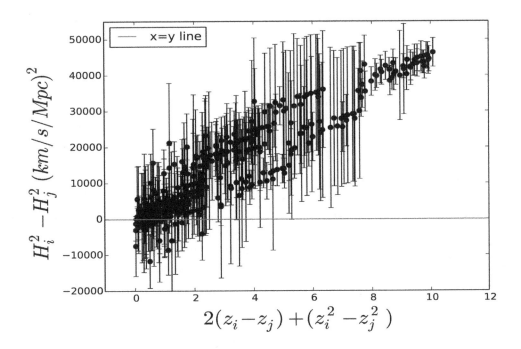

Figure 2. Left-hand side vs. right-hand side of Equation (5) for all possible $i > j$ combinations of the data shown in Table 1. The error bars denote 1σ confidence level.

We can draw further consequences from the thermodynamic bound (6). To this end, we first apply the model independent Gaussian process (GP) introduced by Seikel et al. [34] to smooth the 28 observational $H(z)$ data depicted in Figure 1. Figure 3 shows the outcome.

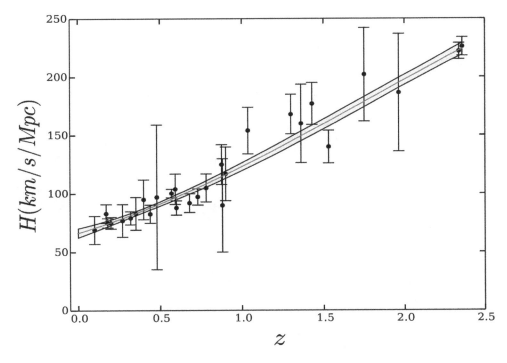

Figure 3. Gaussian process reconstruction of the history of the Hubble factor from the raw $H(z)$ data depicted in Figure 1, as well as here for convenience of the reader. The blue shaded region shows the 1σ uncertainty.

Inspection of the latter suggests that $dH/dz \geq 0$ in the redshift range be considered. If this gets confirmed by future $H(z)$ data of much higher quality, any sign of the curvature scalar index k will be consistent with the second law of thermodynamics. The following analysis, based on the smoothed data shown in Figure 3, allows the quantification of the gap between $1 + q$ and Ω_k.

The quantity $1 + q$ alongside its 1σ, uncertainty is obtained by computing the quantity in the left-hand side of (4) using the smoothed $H(z)$ data, and similarly Ω_k by computing $-k(1+z)^2/H^2(z)$ using the same data. Figures 4 and 5 summarize the results for $k = +1$ and -1, respectively. It is apparent that, whatever the sign of k, the second law is fulfilled by a generous margin. Likewise, inspection of the left panels of the aforesaid figures indicates that $\Omega_{k0} \leq 0.64$. Obviously, this upper bound is much more loose than the one obtained in [14] ($6.5 \times 10^{-3} \leq \Omega_{k0} \leq -6.6 \times 10^{-3}$), but the latter is based on a particular (though so far successful) cosmological model—the ΛCDM—that rests on a number of assumptions, some of which can be justified only a posteriori. By contrast, this other rests just on the FLRW metric and the second law of thermodynamics. Combining the readings on the vertical axes of the right panels of the same figures yields the constraint $2 \times 10^{-4} \leq \Omega_{k0} \leq -2.6 \times 10^{-4}$.

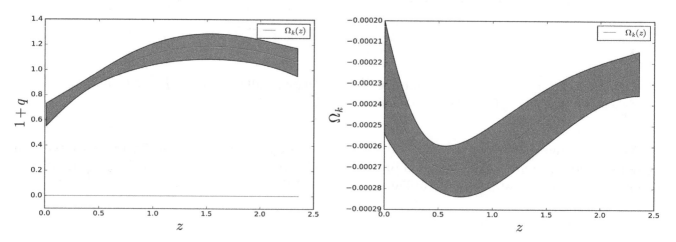

Figure 4. Left panel: $1 + q$ vs. redshift after smoothing the 28 $H(z)$ data as depicted in Figure 3. Also shown is Ω_k for $k = +1$. Clearly, the latter is practically zero. Right panel: Zoom of Ω_k and its 1σ uncertainty interval.

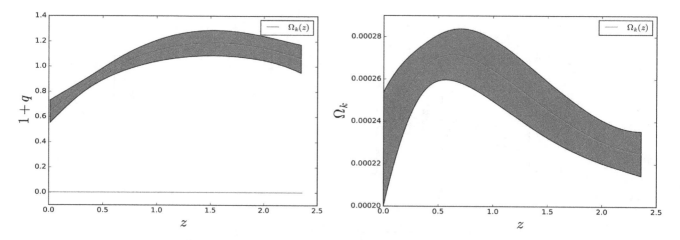

Figure 5. Same as Figure 4 but for $k = -1$.

Regrettably, as hinted above, the quality of the available sets of $H(z)$ data is not good enough to directly constrain Ω_k into a small range, much less to discriminate the sign of k. One has to apply some smoothing procedure to the data of the Hubble history (the GP process in our case) to downsize the error bars and thus obtain a tighter constraint. However, one should not be fully confident about the outcome since the said procedure, though efficient, is not exempt of potential shortcomings.

Nevertheless, the situation is expected to improve greatly in the not so distant future thanks to the Sandage–Loeb (SL) test [35,36] based on the Mc-Vittie formula [37]

$$H(z_s) = H_0[1 + z_s(t_0)] - \frac{\Delta z_s}{\Delta t_0} \tag{7}$$

that governs the drift of the redshift. Here, z_s stands for the redshift of the source (e.g., quasar, globular cluster, HI region, ...). With the use of high precision spectrographs, such as CODEX [38], and extremely large telescopes, as the ELT [39], the SL test will provide us accurate $H(z)$ data sets at different redshift intervals. These data will be free of any assumption whatsoever about the spatial curvature, gravity theory or cosmological model.

Observational data in the $0 < z < 1.0$ interval will be provided by the square kilometer array (SKA) radio-telescope [40], likewise the wide radio-sky survey PARKES will scan 21-cm radio-sources [41] as well as the experiment CHIME in the $0.8 < z < 2.5$ interval [42]. To collect a useful sample of $H(z)$ data will take between one and four decades, approximately. Details can be found in References [43,44].

If the data revealed that, in some redshift, interval H decreased with increasing z, it would immediately imply $k = +1$. On the contrary, if H always increased in every z interval, the application of (4) would require more effort, but in any case it will (hopefully) permit one to discern the sign of k.

If the above strategy would fail, for instance if the data would indicate different signs for Ω_k in separate intervals, it would mean either that the second law of thermodynamics does fail at large scales or that the FLRW metric should not be trusted after all.

3. The Jerk Parameter

By expanding the scale factor in terms of its successive derivatives we can write

$$a(t) = a_0 \left\{ 1 + H_0\,(t - t_0) - \tfrac{1}{2}q_0 H_0^2(t - t_0)^2 + \tfrac{1}{6}j_0 H_0^3(t - t_0)^3 + \tfrac{1}{24}s_0 H_0^4(t - t_0)^4 + \mathcal{O}([t - t_0]^5) \right\}, \tag{8}$$

where $j = \dddot{a}/(aH^3)$ and $s = (aH^4)^{-1}d^4a/dt^4$ are the dimensionless jerk and snap parameters, respectively.

Here, we shall focus on the current value of the jerk parameter of a universe dominated by pressureless matter and the cosmological constant (subindexes m and Λ, respectively). Thus far, we did not specialize to any cosmological model nor theory of gravity. In what follows, to constrain the theoretical value of j_0, we adopt general relativity and the ΛCDM model because they are the simplest theory and model, respectively, that comply, at least at the background level, with the observational data [14]. In this model, the Hubble factor, as well as the deceleration and jerk parameters, read in terms of the redshift

$$H(z) = H_0 \sqrt{\Omega_{m0}(1+z)^3 + \Omega_{\Lambda 0} + \Omega_{k0}(1+z)^2}, \tag{9}$$

$$q(z) = \frac{1}{2} \frac{\Omega_{m0}(1+z)^3 - 2\Omega_{\Lambda 0}}{\Omega_{m0}(1+z)^3 + \Omega_{\Lambda 0} + \Omega_{k0}(1+z)^2}, \tag{10}$$

$$j(z) = 1 - \frac{\Omega_{k0}(1+z)^2}{\Omega_{m0}(1+z)^3 + \Omega_{\Lambda 0} + \Omega_{k0}(1+z)^2}, \tag{11}$$

where the various Ω_{i0}, with $i = m, \Lambda$, and k, stand for the current values of the fractional energy densities. Bearing in mind the Friedmann constraint $\Omega_m + \Omega_\Lambda + \Omega_k = 1$ we readily get

$$j_0 = 1 - \Omega_{k0} \tag{12}$$

from Equation (11). Thereby if future accurate measurements show that j_0 deviates from unity, we will know that our universe (modulo the FLRW metric and the ΛCDM are correct) is not spatially flat, and the deviation will coincide with minus the present value of the spatial curvature. Unfortunately, current measurements of j_0 come along only with great latitude, $-7.6 \leq j_0 \leq 8.5$ [45]. However, this wide observational uncertainty gets substantially reduced after combining (12) with Equation (6), specialized to the ΛCDM model. It readily yields $q_0 + j_0 \geq 0$. For instance, using the experimental constraint on q_0 of Daly et al. [46], $q_0 = -0.48 \pm 0.11$, we find (within 1σ)

$$j_0 \geq 0.37. \tag{13}$$

The simple fact that, observationally, q_0 is negative [24,46–48], renders j_0 positive in the said model; i.e., cosmic acceleration should be increasing nowadays.

4. Concluding Remarks

The validity of the second law, in the presence of gravity, is well supported at small scales by the thermodynamics of astrophysical-sized collapsed objects, in particular of black holes [1,2], and of self-gravitating radiation spheres [3,4] but, to the best of our knowledge, this law had not been tested at cosmological scales thus far. Here, assuming the correctness of the FLRW metric at large scales and using the history of the Hubble factor—see Equations (4) and (5) and Figure 2—we found that the second law likely holds at these scales as well. However, due to the sizable error bars of the $H(z)$ data, the thermodynamic constraint on $| \Omega_{k0} |$ is rather loose. As we have shown, the situation greatly improves by applying the GP procedure of Reference [34] to these data. Then, $| \Omega_{k0} | \sim 10^{-4}$—see the right-hand panel of Figures 4 and 5. However, although the procedure rests on very reasonable assumptions, these are hard to test. On the other hand, we could not determine the sign of k. Nevertheless, we suggested that by means of Mc Vittie formula, Equation (7) of the drift of the redshift [37] and the use of advanced telescopes and spectrographs that will be in service soon, it will be possible to obtain accurate $H(z)$ data capable of discerning it. Further, in the context of the ΛCDM model, we demonstrated a very simple relationship, Equation (12), between the present value of the jerk parameter and Ω_{k0}. Finally, we showed that the second law drastically reduces the ample uncertainty about the current value of the jerk and using current constraints on q_0 sets a lower bound on it.

Author Contributions: The authors contributed equally to this paper.

References

1.   Bekenstein, J.D. Generalized second law of thermodynamics in black-hole physics. *Phys. Rev. D* **1974**, *9*, 3292–3300.
2.   Bekenstein, J.D. Statistical Black Hole Thermodynamics. *Phys. Rev. D* **1975**, *12*, 3077–3085.
3.   Sorkin, R.D.; Wald, R.M.; Jiu, Z.Z. Entropy of Self-Gravitating Radiation. *Gen. Relativ. Gravit.* **1981**, *13*, 1127–1146.
4.   Pavón, D.; Landsberg, P.T. Heat capacity of a self-gravitating radiation sphere. *Gen. Relativ. Gravit.* **1988**, *20*, 457–461.
5.   Ferreira, P.C.; Pavón, D. Thermodynamics of nonsingular bouncing universes. *Eur. Phys. J. C* **2016**, *76*, 37.
6.   Robertson, H.P. Kinematics and World-Structure. *Astrophys. J.* **1935**, *82*, 284–301.
7.   Robertson, H.P. Kinematics and World-Structure III. *Astrophys. J.* **1936**, *83*, 257–271.
8.   Walker, A.G. On Milne's theory of world-structure. *Proc. Lond. Math. Soc.* **1936**, *s2-42*, 90–127.
9.   Clarkson, C.; Basset, B.; Lu, T.H.-C. A general test of the Copernican principle *Phys. Rev. Lett.* **2008**, *101*, 011301.
10.   Zhang, P.; Stebbins, A. Confirmation of the Copernican principle through the anisotropic kinetic Sunyaev Zel'dovich effect. *Phil. Trans. R. Soc. A* **2011**, *369*, 5138–5145.
11.   Bentivegna, E.; Bruni, M. Effects of Nonlinear Inhomogeneity on the Cosmic Expansion with Numerical Relativity. *Phys. Rev. Lett.* **2016**, *116*, 251302.

12. Saadeh, D.; Feeney, S.M.; Pontzen, A.; Peiris, H.V.; McEwen, J.D. How Isotropic is the Universe? *Phys. Rev. Lett.* **2016**, *117*, 131302.

13. Komatsu, E.; Smith, K.M.; Dunkley, J.; Bennett, C.L.; Gold, B.; Hinshaw, G.; Jarosik, N.; Larson, D.; Nolta, M.R.; Page, L.; et al. Seven-year Wilkinson Microwave Anisotropy Probe (WMAP) Observations: Cosmological Interpretation. *Astrophys. J. Suppl. Ser.* **2011**, *192*, 18.

14. Ade, P.R.; Aghanim, N.; Armitage-Caplan, C.; Arnaud, M.; Ashdown, M.; Atrio-Barandela, F.; Aumont, J.; Baccigalupi, C.; Banday, A.J.; Barreiro, R.B.; et al. Planck 2013 results. XVI. Cosmological parameters. *Astron. Astrophys.* **2014**, *571*, A16.

15. Hawking, S.W. Black hole explosions? *Nature* **1974**, *248*, 30–31.

16. Jacobson, T. Thermodynamics of Spacetime: The Einstein Equation of State. *Phys. Rev. Lett.* **1995**, *75*, 1260–1263.

17. Padmanabhan, T. Gravity and the thermodynamics of horizons. *Phys. Rep.* **2005**, *406*, 49–125.

18. Radicella, N.; Pavón, D. A thermodynamic motivation for dark energy. *Gen. Relativ. Grav.* **2012**, *44*, 685–702.

19. Pavón, D.; Radicella, N. Does the entropy of the Universe tend to a maximum? *Gen. Relativ. Grav.* **2013**, *45*, 63–68.

20. Bak, D.; Rey, S.-J. Cosmic holography. *Class. Quantum Grav.* **2000**, *17*, L83.

21. Egan, C.; Lineweaver, C.L. A Larger Estimate of the Entropy of the Universe. *Astrophys. J.* **2010**, *710*, 1825–1834.

22. Wang, B.; Gong, Y.; Abdalla, E. Thermodynamics of an accelerated expanding universe. *Phys. Rev. D* **2006**, *74*, 083520.

23. Frautschi, S. Entropy in an Expanding Universe. *Science* **1982**, *217*, 593–599.

24. Farook, O.; Madiyar, F.R.; Crandall, S.; Ratra, B. Hubble Parameter Measurement Constraints on the Redshift of the Deceleration-Acceleration Transition, Dynamical Dark Energy, and Space Curvature. 2016, arXiv:1607.03537.

25. Cyburt, R.H.; Fields, B.D.; Olive, K.; Yeh, T.H. Big bang nucleosynthesis: Present status. *Rev. Mod. Phys.* **2016**, *88*, 015004.

26. Simon, J.; Verde, L.; Jiménez, R. Constraints on the redshift dependence of the dark energy potential. *Phys. Rev. D* **2005**, *71*, 123001.

27. Moresco, M.; Cimatti, A.; Jimenez, R.; Pozzetti, L.; Zamorani, G.; Bolzonella, M.; Dunlop, J.; Lamareille, F.; Mignoli, M.; Pearce, H.; et al. Improved constraints on the expansion rate of the Universe up to $z \sim 1.1$ from the spectroscopic evolution of cosmic chronometers. *J. Cosmol. Astropart. Phys.* **2012**, *2012*, 006.

28. Cuesta, A.J.; Vargas-Magaña, M.; Beutler, F.; Bolton, A.S.; Brownstein, J.R.; Eisenstein, D.J.; Gil-Marín, H.; Ho, S.; McBride, C.K.; Maraston, C.; et al. The clustering of galaxies in the SDSS-III Baryon Oscillation Spectroscopic Survey: Baryon acoustic oscillations in the correlation function of LOWZ and CMASS galaxies in Data Release 12. *Mont. Not. R. Astron. Soc.* **2016**, *457*, 1770–1785.

29. Blake, C.; Brough, S.; Colless, M.; Contreras, C.; Couch, W.; Croom, S.; Croton, D.; Davis, T.M.; Drinkwater, M.J.; Forster, K.; et al. The WiggleZ Dark Energy Survey: Joint measurements of the expansion and growth history at z < 1. *Mon. Not. R. Astron. Soc.* **2012**, *425*, 405–141.

30. Stern, D.; Jimenez, R.; Verde, L.; Kamionkowski, M.; Adam, S. Cosmic Chronometers: Constraining the Equation of State of Dark Energy. I: H(z) Measurements. *J. Cosmol. Astropart. Phys.* **2010**, *2010*, 008.

31. Moresco, M. Raising the bar: New constraints on the Hubble parameter with cosmic chronometers at z~2. *Mon. Not. R. Astron. Soc.* **2015**, *450*, L16–L20.

32. Delubac, T.; Bautista, J.E.; Busca, N.G.; Rich, J.; Kirkby, D.; Bailey, S.; Font-Ribera, A.; Slosar, A.; Lee, K.-G.; Pieri, M.M.; et al. aryon acoustic oscillations in the Lyα forest of BOSS DR11 quasars. *Astron. Astrophys.* **2015**, *574*, A59.

33. Font-Ribera, A.; Kirkby, D.; Busca, N.; Miralda-Escudé, J.; Ross, N.P.; Slosar, A.; Rich, J.; Aubourg, E.; Bailey, S.; Bhardwaj, V.; et al. Quasar-Lyman α forest cross-correlation from BOSS DR11: Baryon Acoustic Oscillations. *J. Cosmol. Astropart. Phys.* **2014**, *2014*, 027.

34. Seikel, M.; Clarkson, C.; Smith, M. Reconstruction of dark energy and expansion dynamics using Gaussian processes. *J. Cosmol. Astropart. Phys.* **2012**, *2012*, 036.

35. Sandage, A. The Change of Redshift and Apparent Luminosity of Galaxies due to the Deceleration of Selected Expanding Universes. *Astrophys. J.* **1962**, *136*, 319–333.

36. Loeb, A. The Change of Redshift and Apparent Luminosity of Galaxies due to the Deceleration of Selected Expanding Universes. *Astrophys. J.* **1998**, *499*, L111–L114.

37. Vittie, G.C.M. *Cosmological Theory*, 2nd ed.; Wiley: New York, NY, USA, 1949.

38. Spectrograph CODEX. Available online: http://www.iac.es/proyecto/codex/ (accessed on 22 November 2016).

39. The European Extremely Large Telescope. Available online: http://www.eso.org/public/teles-instr/e-elt/ (accessed on 22 November 2016).

40. Klockner, H.R.; Obreschkow, D.; Martins, C.; Raccanelli, A.; Champion, D.; Roy, A.; Lobanov, A.; Wagner, J.; Keller, R. Real time cosmology-A direct measure of the expansion rate of the Universe with the SKA. *Proc. Sci.* **2015**, *AASKA14*, 027.

41. Parkers 21 cm Multibeam Project. Available online:http://www.atnf.csiro.au/research/multibeam/(accessed on 22 November 2016).

42. The Canadian Hydrogen Intensity Mapping Experiment. Available online: chime.phas.ubc.ca/ (accessed on 22 November 2016).

43. Yu, H.R.; Zhang, T.J.; Pen, U.L. Method for Direct Measurement of Cosmic Acceleration by 21-cm Absorption Systems. *Phys. Rev. Lett.* **2014**, *113*, 041303.

44. Liske, J.; Grazian, A.; Vanzella, E.; Dessauges, M.; Viel, M.; Pasquini, L.; Haehnelt, M.; Cristiani, S.; Pepe, F.; Avila, G.; et al. Cosmic dynamics in the era of Extremely Large Telescopes. *Mon. Not. R. Astron. Soc.* **2008**, *386*, 1192–1218.

45. Bochner, B.; Pappas, D.; Dong, M. Testing Lambda and the Limits of. Cosmography with the Union2.1. Supernova Compilation. *Astrophys. J.* **2015**, *814*, 7.

46. Daly, R.; Djorgovski, S.G.; Freeman, K.A.; Mory, M.P.; O'Dea, C.P.; Kharb, P.; Baum, S. Improved Constraints on the Acceleration History of the Universe and the Properties of the Dark Energ. *Astrophys. J.* **2008**, *677*, 1–11.

47. Perlmutter, S.; Aldering, G.; Della Valle, M.; Deustua, S.; Ellis, R.S.; Fabbro, S.; Fruchter, A.; Goldhaber, G; Groom, D.E.; Hook, I.M.; et al. Discovery of a supernova explosion at half the age of the Universe. *Nature* **1998**, *391*, 51–54.

48. Riess, A.G.; Kirshner, R.P.; Schmidt, B.P.; Jha, S.; Challis, P.; Garnavich, P.M.; Esin, A.A.; Carpenter, C.; Grashius, R.; Schild, R.E.; et al. BV RI light curves for 22 type Ia supernovae. *Astron. J.* **1999**, *117*, 707–724.

Permissions

The contributors of this book come from diverse backgrounds, making this book a truly international effort. This book will bring forth new frontiers with its revolutionizing research information and detailed analysis of the nascent developments around the world.

We would like to thank all the contributing authors for lending their expertise to make the book truly unique. They have played a crucial role in the development of this book. Without their invaluable contributions this book wouldn't have been possible. They have made vital efforts to compile up to date information on the varied aspects of this subject to make this book a valuable addition to the collection of many professionals and students.

This book was conceptualized with the vision of imparting up-to-date information and advanced data in this field. To ensure the same, a matchless editorial board was set up. Every individual on the board went through rigorous rounds of assessment to prove their worth. After which they invested a large part of their time researching and compiling the most relevant data for our readers.

The editorial board has been involved in producing this book since its inception. They have spent rigorous hours researching and exploring the diverse topics which have resulted in the successful publishing of this book. They have passed on their knowledge of decades through this book. To expedite this challenging task, the publisher supported the team at every step. A small team of assistant editors was also appointed to further simplify the editing procedure and attain best results for the readers.

Apart from the editorial board, the designing team has also invested a significant amount of their time in understanding the subject and creating the most relevant covers. They scrutinized every image to scout for the most suitable representation of the subject and create an appropriate cover for the book.

The publishing team has been an ardent support to the editorial, designing and production team. Their endless efforts to recruit the best for this project, has resulted in the accomplishment of this book. They are a veteran in the field of academics and their pool of knowledge is as vast as their experience in printing. Their expertise and guidance has proved useful at every step. Their uncompromising quality standards have made this book an exceptional effort. Their encouragement from time to time has been an inspiration for everyone.

The publisher and the editorial board hope that this book will prove to be a valuable piece of knowledge for researchers, students, practitioners and scholars across the globe.

List of Contributors

Ismael Ayuso
Departamento de Física and Instituto de Astrofísica e Ciências do Espaço, Faculdade de Ciências, Universidade de Lisboa, Edifício C8, Campo Grande, 1769-016 Lisboa, Portugal

Diego Sáez-Chillón Gómez
Department of Theoretical Physics, Atomic and Optics, Campus Miguel Delibes, University of Valladolid UVA, Paseo Belén, 7, 47011 Valladolid, Spain

Kirill A. Bronnikov
Center for Gravitation and Fundamental Metrology, VNIIMS, Ozyornaya ul. 46, 119361 Moscow, Russia
Institute for Gravitation and Cosmology, Peoples' Friendship University of Russia (RUDN University), ul. Miklukho-Maklaya 6, 117198 Moscow, Russia
Moscow Engineering Physics Institute, National Research Nuclear University "MEPhI", 105005 Moscow, Russia

Vladimir G. Krechet and Vadim B. Oshurko
Department of Physics, Moscow State Technological University "Stankin", Vadkovsky per. 3A, 127055 Moscow, Russia

Aurélien Hees
Department of Physics and Astronomy, University of California, Los Angeles, CA 90095, USA

Quentin G. Bailey
Physics Department, Embry-Riddle Aeronautical University, Prescott, AZ 86301, USA

Adrien Bourgoin, Hélène Pihan-Le Bars and Christophe Le Poncin-Lafitte
SYRTE, Observatoire de Paris, PSL Research University, CNRS, Sorbonne Universités, UPMC Univ. Paris 06, LNE, 61 avenue de l'Observatoire, 75014 Paris, France

Christine Guerlin
SYRTE, Observatoire de Paris, PSL Research University, CNRS, Sorbonne Universités, UPMC Univ. Paris 06, LNE, 61 avenue de l'Observatoire, 75014 Paris, France
Laboratoire Kastler Brossel, ENS-PSL Research University, CNRS, UPMC-Sorbonne Universités, Collège de France, 75005 Paris, France

Ram Gopal Vishwakarma
Unidad Académica de Matemáticas, Universidad Autónoma de Zacatecas, Zacatecas, ZAC C.P. 98000, Mexico

Giulia Schettino
IFAC-CNR, Via Madonna del Piano 10, 50019 Sesto Fiorentino (FI), Italy

Giacomo Tommei
Department of Mathematics, University of Pisa, Largo Bruno Pontecorvo 5, 56127 Pisa, Italy

Luis Acedo
Instituto Universitario de Matemática Multidisciplinar, Universitat Politècnica de València, Building 8G, 2° Floor, Camino de Vera 46022, Valencia, Spain

Øyvind Grøn
Art and Design, Faculty of Technology, Oslo and Akershus University College of Applied Sciences, Olavs Plass, NO-0130 Oslo, Norway

Jorge L. Cervantes-Cota and Salvador Galindo-Uribarri
Department of Physics, National Institute for Nuclear Research, Km 36.5 Carretera Mexico-Toluca, Ocoyoacac, C.P. 52750 Mexico, Mexico

George F. Smoot
Helmut and Ana Pao Sohmen Professor at Large, Institute for Advanced Study, Hong Kong University of Science and Technology, Clear Water Bay, Kowloon, 999077 Hong Kong, China
Université Sorbonne Paris Cité, Laboratoire APC-PCCP, Université Paris Diderot, 10 rue Alice Domon et Leonie Duquet, 75205 Paris Cedex 13, France
Department of Physics and LBNL, University of California; MS Bldg 50-5505 LBNL, 1 Cyclotron Road Berkeley, 94720 CA, USA

Oliver F. Piattella
Physics Department, Universidade Federal do Espírito Santo, Vitória 29075-910, Brazil

Eyo Eyo Ita III
Physics Department, U.S. Naval Academy, Annapolis, MD 21401, USA

Amos S. Kubeka
Mathematical Sciences Department, University of South Africa, Pretoria 0002, South Africa

Bahram Mashhoon
Department of Physics and Astronomy, University of Missouri, Columbia, MO 65211, USA

Pedro C. Ferreira
Escola de Ciências e Tecnologia, Universidade Federal do Rio Grande do Norte, Natal 59072-970, Rio Grande do Norte, Brazil

Diego Pavón
Departamento de Física, Universidad Autónoma de Barcelona, Bellaterra, Barcelona 08193, Spain

Index

Printed in the USA
CPSIA information can be obtained
at www.ICGtesting.com
JSHW061053121023
49903JS00030B/254